Most galaxies are in clusters, where tidal interactions are not uncommon. Tidal and dynamical interaction in galaxies are of importance in studying evolution. A large amount of data has been collected on dust-lane ellipticals, polar ring galaxies, spirals with extended warps, and galaxies with inclined HI rings or unusual "tails". This book is a record of a meeting which was held at the University of Pittsburgh.

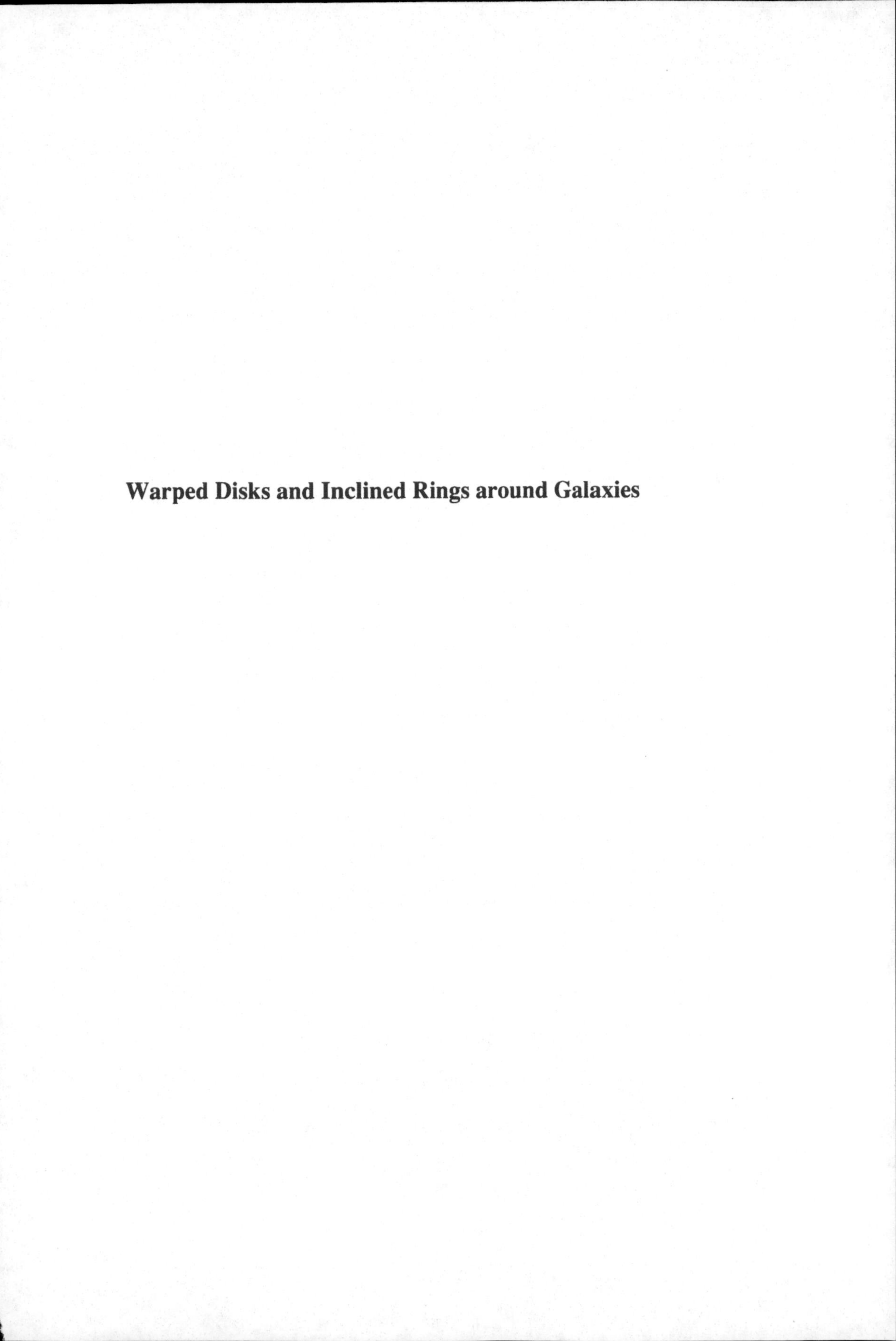

Warped Disks and Inclined Rings around Galaxies

Warped Disks and Inclined Rings around Galaxies

Edited by

STEFANO CASERTANO

PENNY D. SACKETT

FRANKLIN H. BRIGGS

Department of Physics and Astronomy, University of Pittsburgh

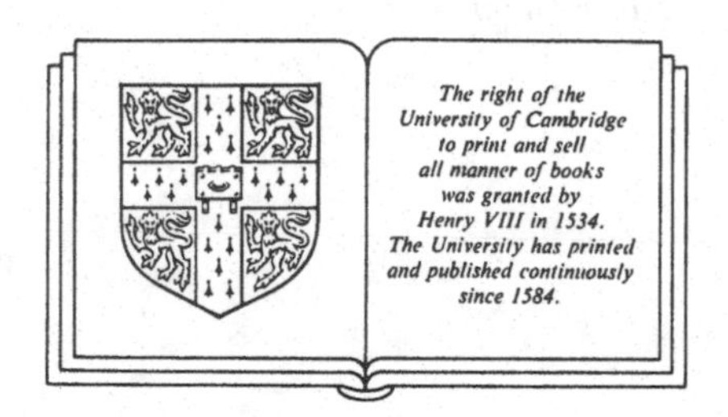

CAMBRIDGE UNIVERSITY PRESS

Cambridge

New York Port Chester

Melbourne Sydney

CAMBRIDGE UNIVERSITY PRESS
Cambridge, New York, Melbourne, Madrid, Cape Town, Singapore, São Paulo

Cambridge University Press
The Edinburgh Building, Cambridge CB2 2RU, UK

Published in the United States of America by Cambridge University Press, New York

www.cambridge.org
Information on this title: www.cambridge.org/9780521401845

© Cambridge University Press 1991

This publication is in copyright. Subject to statutory exception
and to the provisions of relevant collective licensing agreements,
no reproduction of any part may take place without
the written permission of Cambridge University Press.

First published 1991
This digitally printed first paperback version 2006

A catalogue record for this publication is available from the British Library

ISBN-13 978-0-521-40184-5 hardback
ISBN-10 0-521-40184-4 hardback

ISBN-13 978-0-521-03163-9 paperback
ISBN-10 0-521-03163-X paperback

CONTENTS

PREFACE

For decades, galaxies were thought to be isolated islands of stars in a vast sea of space, but research of the past few years has shown that these giant stellar conglomerates are dynamic systems that evolve through a variety of processes including interactions with neighboring galaxies. During the period May 30 to June 1, 1990, astronomers studying the dynamics and morphology of galaxies increased their own interactions at a three-day workshop on the topic "Warped Disks and Inclined Rings around Galaxies" held at the University of Pittsburgh, Pennsylvania, U.S.A. The focus of the workshop was the increasing number of galaxies now being discovered to have some of their stars and gas organized into prominent warps or inclined rings, which destroy the axisymmetric morphology enjoyed by many galaxies.

The Pittsburgh meeting brought together scientists with specialties in a wide range of observational and theoretical techniques but with a common interest in deciphering the information encoded in the structure of these warped and ringed systems. Astronomers studying polar-ring galaxies, dust-lane ellipticals, and warped spirals were present from around the world. Discussion centered on the origins of warps and rings, how these origins relate to galaxy interactions, the frequency of their occurrence, the stability and longevity of such structures, and how their study may provide us with new insights into the evolution of galaxies and the nature of dark matter.

This book presents the proceedings of the workshop organized in the chronological order of the meeting. The first day of the workshop was devoted to discussion of accretion of external gas onto galactic systems as a formation mechanism for warped and polar-ringed systems and on the analysis of observed gas kinematics in non-spherical potentials. On the second day, the meeting convened in the Mellon

Institute at the invitation of the Pittsburgh Supercomputing Center, and the discussion focused on computational studies and numerical simulations of galaxy interactions and the dynamical evolution of material accreted at substantial inclinations to the host galaxies. The theme of the final day of the workshop was warps. Three short informal presentations that do not appear in these proceedings should be mentioned: Neal Katz summarized preliminary numerical simulations of galaxy formation using a "tree + SPH" code which produced flattened dark halos and "primordial warps" as a general consequence; John Hibbard described his ongoing work with Jacqueline van Gorkom studying interacting systems like NGC 262 [=Mkn 348] which display massive and *substantially* extended tidal features in neutral hydrogen; and Ken Freeman talked about his use of the planetary nebulae in Cen A as a tracer of old stellar kinematics, with the result that the kinematic major axis for the PN does not coincide with either the major or the minor axis of the stellar elliptical body. The last day of the workshop ended with a general discussion which has been transcribed for inclusion as the final chapter of this proceedings.

In most cases the presentations at the workshop summarized research still in progress. The workshop participants returned to their home institutions with optimism that observational and computational resources are in hand so that the next few years will bring exciting advances in our understanding of these interesting systems.

Pittsburgh, December 1990

Stefano Casertano
Penny Sackett
Frank Briggs

ACKNOWLEDGEMENTS

The workshop was hosted by the University of Pittsburgh, the Pittsburgh Supercomputing Center and Allegheny Observatory. We are grateful to the three institutions and Gene Engels, Ralph Roskies and George Gatewood for providing facilities and generous financial support. A grant from the National Science Foundation played a crucial role in supplementing participant costs.

This meeting could not have happened without the competent and dedicated support of the staff at the three institutions. Especially warm thanks are due to Wendy Janocha of the PSC for her invaluable help and experience in dealing with the innumerable details attending the organization of this workshop; to Edith Cohen of the University of Pittsburgh, who steered us through the maze of bureaucratic difficulties; and to Nancy Robinson, for arranging an evening reception at Allegheny Observatory. Mike Castelaz took time from his busy research schedule to show workshop participants the facilities at Allegheny Observatory. Dorothy Lain and Mary Beth Kraus were very helpful with the financial arrangements. Angela Mercer helped with the registration and provided general assistance to the participants, and Steve Cunningham set up a workstation for Peter Teuben's live demonstration of the `NEMO` software package during the May 31 session at the PSC.

Thanks are due to all our colleagues who helped with the scientific program of the workshop: to Linda Sparke and Tim de Zeeuw for useful suggestions during the organizing period; to Albert Bosma, Thijs van der Hulst, Joel Tohline, Linda Sparke and Tim de Zeeuw for acting as session chairs; to Ken Freeman for chairing the final session and leading—and occasionally moderating—a lively concluding discussion; and to all participants for coming and sharing their knowledge in a very interesting and productive workshop.

WORKSHOP PARTICIPANTS

E. Battaner	Departamento de Fisica Teorica y del Cosmos, Facultad de Ciencias, Universidad de Granada, Granada, Spain
Albert Bosma	Observatoire de Marseille, Marseille, France
Frank H. Briggs	Department of Physics and Astronomy, University of Pittsburgh, Pittsburgh, PA
Stefano Casertano	Department of Physics and Astronomy, University of Pittsburgh, Pittsburgh, PA
Dimitris M. Christodoulou	Steward Observatory, University of Arizona, Tucson, AZ
Tim de Zeeuw	Theoretical Astrophysics, California Institute of Technology, Pasadena, CA and Sterrewacht Leiden, Leiden, The Netherlands
E. Florido	Departamento de Fisica Teorica y del Cosmos, Facultad de Ciencias, Universidad de Granada, Granada, Spain
Ken C. Freeman	Mt. Stromlo and Siding Springs Observatories, ACT, Australia, and Space Telescope Science Institute, Baltimore, MD
Giuseppe Galletta	Dipartimento di Astronomia, Università di Padova, Padova, Italy
Lars E. Hernquist	Astronomy Department, Princeton University, Princeton, NJ and Board of Studies in Astronomy and Astrophysics, University of California, Santa Cruz, CA

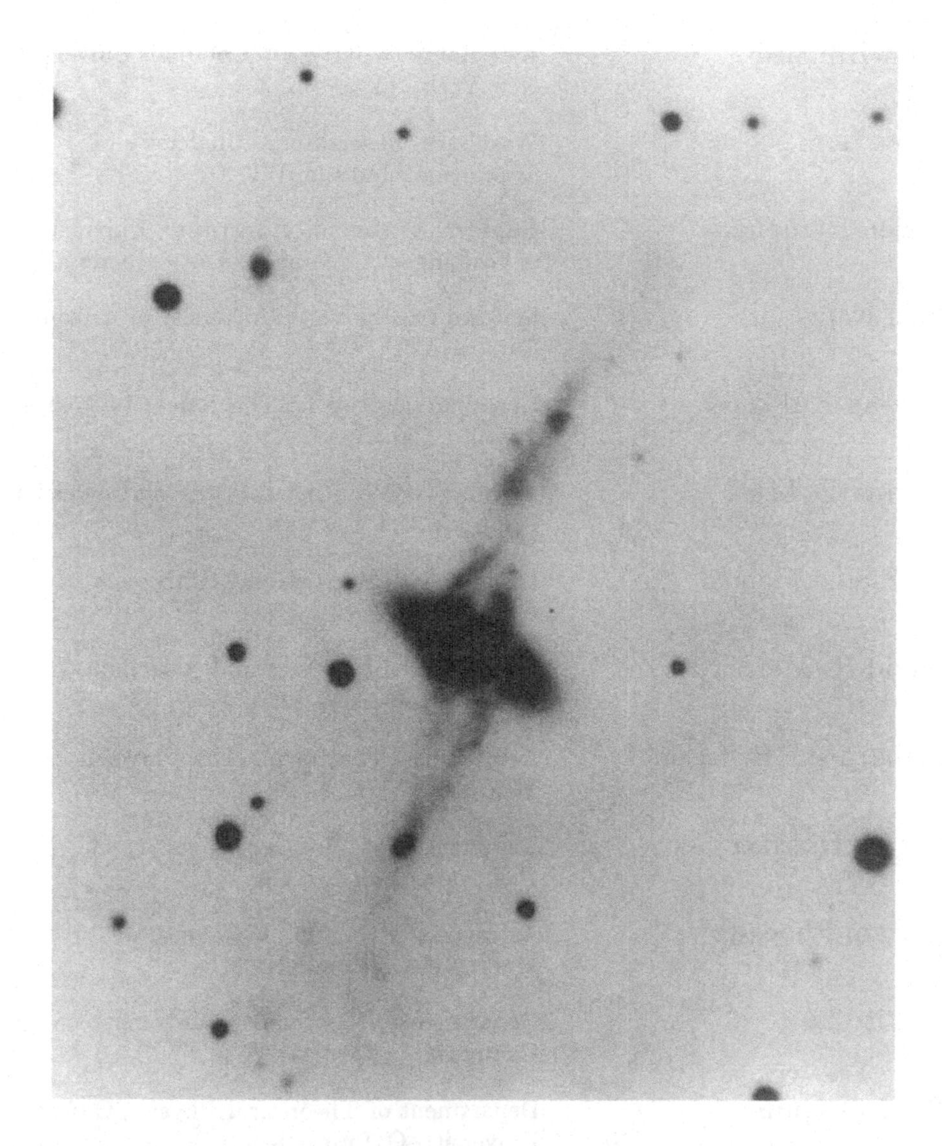

One of the protagonists of the workshop: the "prototypical" polar-ring galaxy NGC 4650A, in a blue CCD image taken at the 4 m telescope at CTIO. Photo courtesy of Brad Whitmore.

On the facing pace: workshop participants on the steps of the Mellon Institute Building, where the Pittsburgh Supercomputing Center is located. *Front row* (left to right): John Hibbard, E. Florido, Albert Bosma, Stefano Casertano, Thijs van der Hulst, Tom Steiman-Cameron, Giuseppe Galletta, Jo Pitesky, Eve Ostriker. *Second row:* Tim de Zeeuw, Bikram Phookun, E. Battaner, Thomas Quinn, Neal Katz, David Merritt, Linda Sparke, Joanna Lees, Steven Schneider. *Third row:* Elaine Mahon, Penny Sackett, Jurjen Kamphuis, Joel Tohline, Hans-Walter Rix, Lars Hernquist, Konrad Kuijken, Frank Briggs, Ken Freeman. *Back row:* Brad Whitmore, Dimitris Christodoulou, Peter Teuben, Richard Nicholson.

John Hibbard	Astronomy Department, Columbia University, New York, NY
Peter Hofner	Washburn Observatory, University of Wisconsin, Madison, WI
Jurjen Kamphuis	Kapteyn Astronomical Institute, University of Groningen, Groningen, The Netherlands
Neal Katz	Steward Observatory, University of Arizona, Tucson, AZ
Konrad Kuijken	Canadian Institute for Theoretical Astrophysics, Toronto, Canada
Joanna F. Lees	Astronomy Department, Princeton University, Princeton, NJ
M. Elaine Mahon	Department of Astronomy, University of Florida, Gainesville, FL
David R. Merritt	Department of Physics and Astronomy, Rutgers University, Piscataway, NJ
Richard A. Nicholson	Astronomy Department, The University, Manchester, UK
Eve C. Ostriker	Department of Physics, University of California, Berkeley, CA
Bikram Phookun	Astronomy Program, University of Maryland, College Park, MD
Jo Pitesky	Department of Astronomy, University of California, Los Angeles, CA
Thomas Quinn	Department of Theoretical Physics, Oxford University, Oxford, UK
Hans-Walter Rix	Steward Observatory, University of Arizona, Tucson, AZ
Penny D. Sackett	Department of Physics and Astronomy, University of Pittsburgh, Pittsburgh, PA
Stephen E. Schneider	Department of Physics and Astronomy, University of Massachusetts, Amherst, MA
Linda S. Sparke	Washburn Observatory, University of Wisconsin, Madison, WI

Thomas Steiman-Cameron	NASA Ames Research Center, Moffett Field, CA
Peter J. Teuben	Astronomy Program, University of Maryland, College Park, MD
Joel E. Tohline	Department of Physics and Astronomy, Louisiana State University, Baton Rouge, LA
J. M. van der Hulst	Kapteyn Astronomical Institute, University of Groningen, Groningen, The Netherlands
Brad C. Whitmore	Space Telescope Science Institute, Baltimore, MD

The intergalactic HI supply

FRANK BRIGGS
University of Pittsburgh

Abstract

An inventory of the HI observed in nearby groups and clusters shows that the bulk of the neutral gas is already bound to large spiral galaxies and is not contained in gas-rich dwarfs. However, small galaxies do significantly outnumber large ones, despite the fact that they do not form a large reservoir of neutral gas. Intergalactic HI clouds with masses in the range $M_{HI} \approx 3\times10^8$ to $10^{10} M_\odot$ must be less common than normal galaxies of similar HI mass by more than a factor of one hundred. HI clouds with low HI masses ($\sim10^7 M_\odot$) could be as abundant as galaxies of comparable mass.

Introduction

Polar ring galaxies and dust-lane ellipticals are systems whose host galaxies could happily be classified along the Hubble Tuning Fork if they did not possess their additional peculiarities – rings or bands of material that appear to encircle the host on stable trajectories. The most impressive of the warped spiral galaxies have disks of neutral gas that can be mapped in the 21cm line far outside the optical image on photographic plates; here again, the host galaxy often fits nicely into the Hubble sequence, and the galaxies with extended gas would not be reliably selected on the basis of optical appearance, although the beginning of a warp can occasionally be recognized in the optical image of a disk galaxy that is viewed edge-on.

In many of these systems, the rings and gas in warped orbits carry very little mass and may be viewed as kinematical tracers of the gravitational potentials. Thus, these special galaxies, which appear to have something *extra* added-on after a normal history of formation for the host, provide fascinating laboratories for studying galactic dynamics; the results may be generalizable to more normal galaxies as well.

The purpose of this paper is to appraise the supply of gas-rich material that is presently available for capture onto host galaxies. Working on the hypothesis that the material is accreted in lumps that are small compared to the host galaxies in order for the hosts to retain their identities, two sources of low mass objects are considered: intergalactic HI clouds and gas-rich dwarf galaxies. The conclusions are based on the study of the local environment of nearby groups and clusters, and the populations of the low mass objects could well have been very different in the distant past.

Another possible source of accretable gas may be the outskirts of large, gas-rich galaxies that find themselves involved in mild encounters with a second system (Schweizer, Whitmore, and Rubin 1983). In this scenario, a portion of the gas-rich galaxy's extended disk might be stripped off and added to an elliptical or S0 host galaxy.

Intergalactic HI clouds

There is no evidence supporting the existence of intergalactic HI clouds at the present epoch. No free-floating, non-luminous, intergalactic HI cloud has been discovered; all candidates have been subsequently detected optically or shown to be associated with galaxies, either bound to the galaxies' gravitational potentials (Schneider 1989) or representing tidal debris from an interacting system (Roberts 1988). The discoveries by Giovanelli and Haynes (1989) and Brinks (1990) of very LSB galaxies with M_{HI} of a few times $10^9 M_\odot$ that are barely visible on the Palomar Sky Survey come tantalizingly close to the detection of isolated HI.

More detailed discussions of the abundance of intergalactic HI center on just how tightly the space density can be constrained. Fisher and Tully (1981) have used the results of a large number of extragalactic surveys that have been conducted in the 21cm line to place limits on the space density of intergalactic clouds. More recently, Briggs (1990) has taken advantage of the extensive survey work performed by a number of investigators during the past decade to update the Fisher and Tully study. He found that intergalactic HI clouds with masses in the range $M_{HI} \approx 3\times10^8$ to $10^{10} M_\odot$ must be less common than normal galaxies of similar HI mass by more than a factor of one hundred. The space density of HI clouds with low HI masses

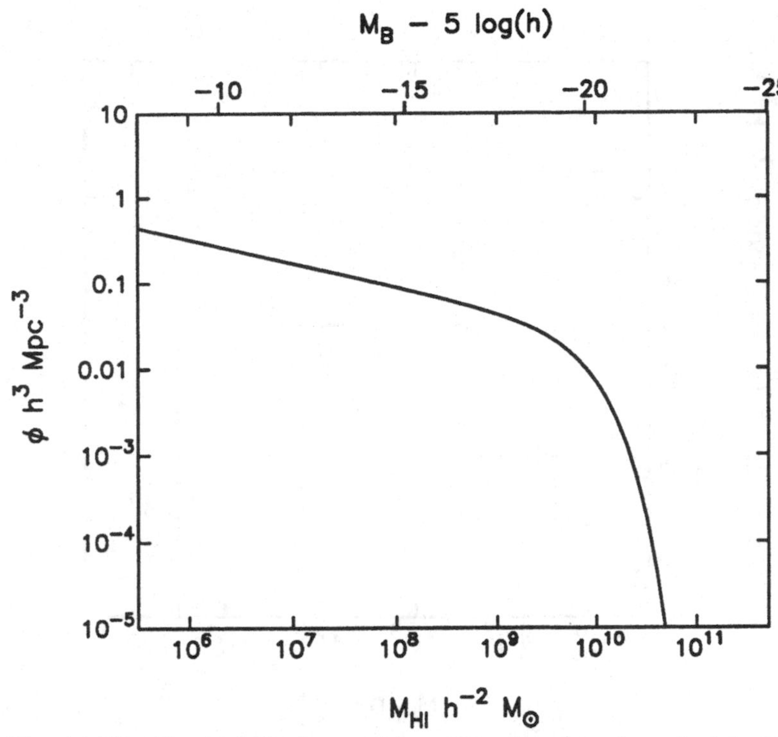

Fig. 1. The Neutral Hydrogen Mass Function (number of objects per Mpc^3 per mass decade) from Briggs (1990). (Here $h = H_o/100$ km s^{-1} Mpc^{-1}, where $H_o =$ the Hubble constant.)

($\sim 10^7 M_\odot$) is not tightly constrained, and they could be as abundant as galaxies of comparable mass.

Gas-rich dwarf galaxies

The population of low surface brightness (LSB) dwarf irregular galaxies is another logical place to look for a reservoir of accretable HI. Briggs (1990) derived an estimate for the *HI mass function* that specifies the space density of galaxies as a function of HI mass. Figure 1 shows the number of objects per Mpc^3 per decade of M_{HI} plotted as a function of M_{HI}. The normalization represents the average over a large volume of space at the present epoch, but the shape is compatible with the Leo Group and Virgo Cluster galaxy populations, once density scaling factors are applied. The resulting HI mass function rises toward low masses, indicating that small gas rich objects are more numerous than big ones.

The cumulative amount of HI mass per volume contained in objects

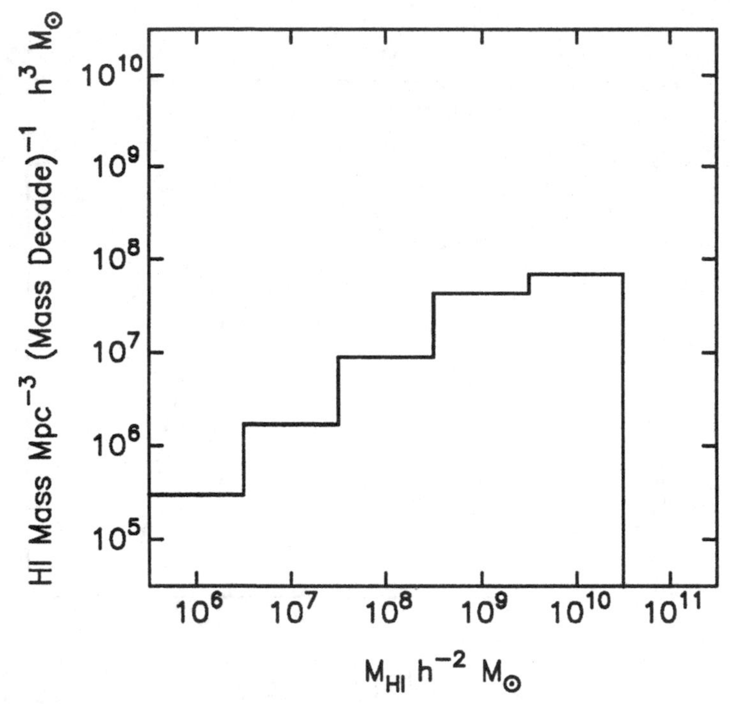

Fig. 2. The space density of neutral hydrogen contained in objects of different HI mass. For each decade of M_{HI}, the histogram indicates the integral of the HI mass per Mpc^3 provided by that range of mass, assuming the HI mass function derived by Briggs (1990).

in decade-sized mass intervals is presented in Figure 2. The distribution of HI mass is strongly skewed toward the large mass galaxies. More than half of the neutral hydrogen observed at the present epoch is tied up in galaxies with HI mass greater than $\sim 3\times 10^9 M_\odot$, and $\sim 90\%$ is contained in galaxies with HI mass greater than $\sim 3\times 10^8 M_\odot$. Only a few percent of the neutral gas at the present epoch is contained in galaxies of less than $3\times 10^7 M_\odot$.

Conclusion

At the present epoch, a large supply of neutral hydrogen for accretion onto host galaxies does not exist in the form of intergalactic clouds or gas-rich dwarf galaxies. Most of the neutral gas is presently bound to large spirals. The smaller galaxies do outnumber the large ones, so small lumps of mass are available for capture, but the net result of merging small dwarfs with larger hosts can not greatly add to the

HI mass of the larger. Of course, the situation may have been very different in the distant past.

Acknowledgements

The author wishes to thank Renzo Sancisi and Morton Roberts for stimulating discussions.

References

Briggs, F.H. (1990). Astron. J. **100**, 999.

Brinks, E. (1990). In *Dynamics and Interactions of Galaxies* (ed. R. Wielen), p. 146. Berlin: Springer.

Fisher, J.R. & Tully, R.B. (1981). Astrophys. J., Lett. **243**, L23.

Giovanelli, R. & Haynes, M.P. (1989). Astrophys. J., Lett. **346**, L5.

Roberts, M.S. (1988). In *New Ideas in Astronomy* (eds. F. Bertola, J.W. Sulentic and B.F. Madore), p. 65. Cambridge: Cambridge University Press.

Schneider, S.E. (1989). Astrophys. J. **343**, 94.

Schweizer, F., Whitmore, B.C. & Rubin, V.C. (1983). Astron. J. **88**, 909.

Neutral gas infall into NGC 628

JURJEN KAMPHUIS
Kapteyn Astronomical Institute, Groningen

FRANK BRIGGS
University of Pittsburgh

Abstract

HI observations of the warped, face-on ScI galaxy NGC 628 with the VLA show the presence of two high velocity gas complexes in the outer parts of the HI distribution and a large extension to the southwest with $\sim 7 \times 10^8$ $M_\odot$ of HI. A possible explanation is that NGC 628 is in the process of accreting a gas-rich, dwarf irregular galaxy.

Introduction

Face-on spirals are excellent candidates for observing HI gas with anomalous velocities out of the plane of the galaxy. A search for this kind of motion led to the discovery of the high velocity complexes in M 101 with HI masses of $2 \times 10^8 M_\odot$ and velocity excesses of about $150\,\mathrm{km\,s^{-1}}$ (van der Hulst and Sancisi 1988). These clouds can be explained as the result of an infalling HI-rich dwarf galaxy punching through the disk of M101. In another face-on galaxy, NGC 6946, several gas complexes with anomalous velocities up to $80\,\mathrm{km\,s^{-1}}$ have been discovered (Kamphuis et al. 1990, Kamphuis, Sancisi and van der Hulst, in preparation). Collective supernova explosions and/or stellar winds could explain both the blue- and redshifted distortions and the large kinetic energies ($\approx 10^{53}$ erg) involved, but an explanation in terms of infalling clouds is also possible.

The large, fairly isolated, ScI galaxy NGC 628 is the third example in which faint gas emission with peculiar velocities out of the plane has been found. Earlier Arecibo HI observations revealed an anomalous behaviour in the kinematics of the gas in the outer parts of NGC 628 (Briggs 1982). The galaxy has a lopsided HI appearance with an extension towards the southwestern side. The velocity field in the outer disk deviates from what is expected for differential rotation in a flat disk. Briggs (1982) ascribed the peculiar velocities in the outer parts to motion of the gas on elliptical orbits in an inclined ring around the galaxy; this explanation does not account, however, for a high velocity cloud complex situated in the southwestern part. The resolution of the Arecibo observations is insufficient for making a detailed analysis of the kinematics. The higher resolution WSRT observations by Shostak and van der Kruit (1984) produced a velocity field that could be well described by a system of inclined rings of gas in circular motion. This ring model must have an onset of warping in the gaseous disk near the Holmberg radius and a gradual increase in the inclination of the warp with increasing radius. Peculiar velocities of the gas in the outer parts were detectable in re-analysis of the WSRT observations (Broeils, private communication), but there was not enough sensitivity for studying the faint, outermost emission.

Results

New HI observations have been carried out with the D array ($\approx 1'$ resolution) of the VLA. Four fields were observed to cover the full extent of the galaxy. A short account of the observations and interpretation will be given in this paper and a more extensive discussion will be presented in a forthcoming paper (Kamphuis and Briggs, in preparation).

Total HI distribution

The HI column density distribution (Figure 1) is very extended (>3 R_{Ho}) and asymmetric. The gas emission in the inner HI disk ($R < R_{Ho}$) is a factor 50 - 100 brighter than in the outer parts. The higher resolution WSRT observations (Shostak and van der Kruit 1984) show a correspondence between the HI structure and the spiral arms seen in optical photographs. Two spiral arms can be traced through the transition region from the bright inner HI to the faint outer emission. These spiral arms are probably continuations of the inner spiral arms and extend clearly beyond the optical image.

At larger radii, the HI is not smoothly distributed, but contains

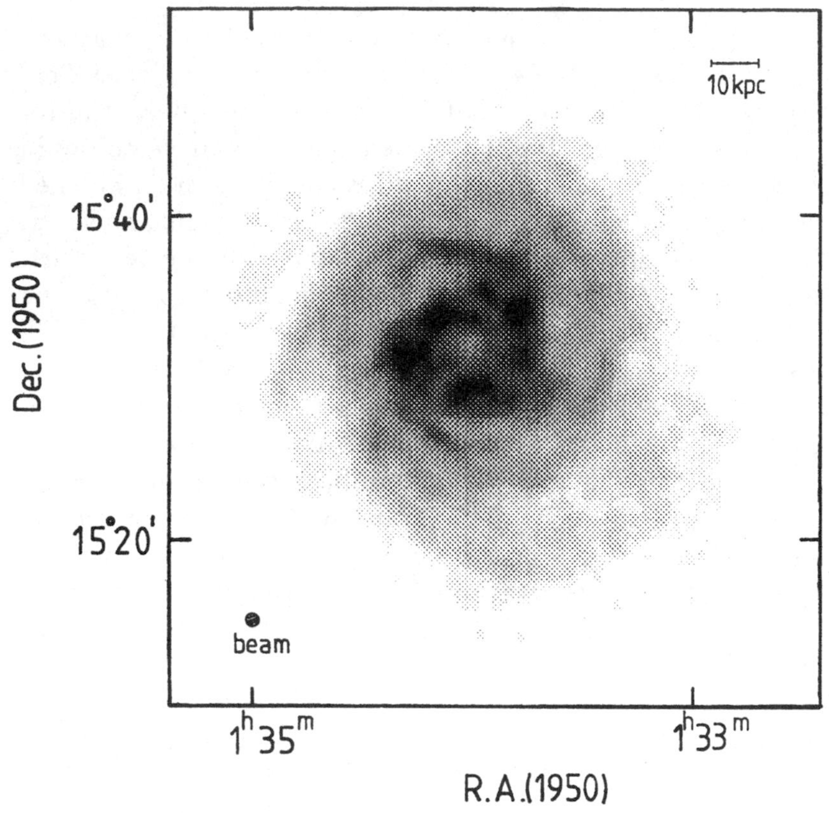

Fig. 1. Grey-scale picture of the total HI brightness distribution of NGC 628 (62″ × 72″ resolution). The highest column densities (dark) are $\approx 10^{21}$ cm^{-2}, the lowest (light shading) $\approx 2 \times 10^{19}$ cm^{-2}. The Holmberg radius is 6′ (= 18 kpc). The HI extension in the southwest is clearly visible.

several structures. Near $R \approx 2\,R_{Ho}$, a large elongated HI filament is visible in the northwest. On the opposite side, spiral arm-like features are visible at the edge of the HI disk. There is no obvious continuation from the inner spiral arms to these structures.

The outer HI is more extended to the southwestern than to the northeastern side ($R = 3.3$ vs. $2.5\,R_{Ho}$). The HI mass in this asymmetric extension is about $7 \times 10^8 M_\odot$.

As pointed out by Briggs (1982), the gas in the outer regions, especially to the southwest, seems to be "added-on" to a more normal HI distribution.

Velocity structure

The intensity-weighted mean velocity field is shown in Figure 2. A

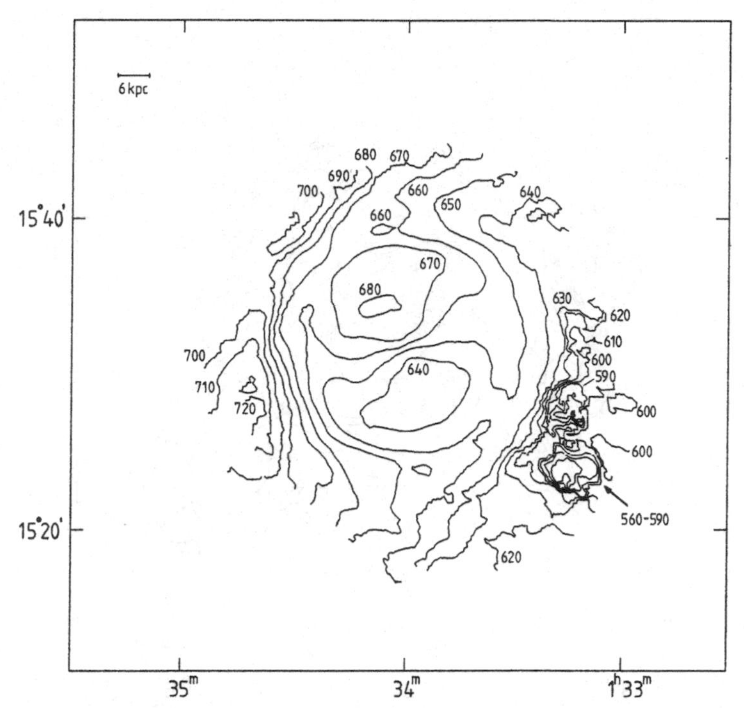

Fig. 2. Contour map (in $\mathrm{km\,s^{-1}}$) of the intensity-weighted mean velocity field. The irregular structure in the southwest is due to the complex structure in the spectra in this region.

tilted ring model, consisting of concentric circular rings in uniform rotation, has been constructed to fit the observed velocity field. The model agrees closely with the one presented by Shostak and van der Kruit (1984) but extends to larger radii and higher inclination angles for the outer, warped orbits. The systematic behavior of the warp is discussed in the contribution by Briggs and Kamphuis later in this volume.

The motion of the gas in the inner parts exhibits the normal pattern of differential rotation in a flat disk. Near $R \approx R_{Ho}$, the velocity field shows a sharp change in the position angle of the major axis, indicating that the HI layer starts to incline away from the flat inner disk. In the outer parts, the HI layer has an inclination of nearly 20° with respect to the flat inner disk.

At $R \approx 1.5$–$2.5 R_{Ho}$, two high velocity gas complexes are visible on the western and eastern sides of the galaxy, symmetric with respect to the centre. These complexes are outlined by dashed lines in the contour map of the total HI distribution (Figure 3). In the position-velocity map (Figure 4), describing a slice through the centre of the

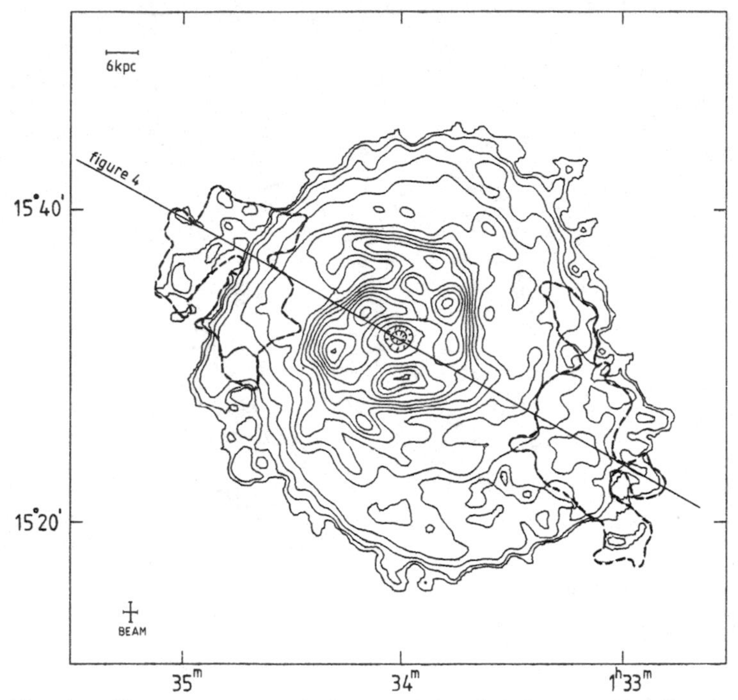

Fig. 3. Contour map of the total HI brightness distribution of NGC 628, with the high velocity complexes outlined. The contour levels are 1.3, 2.5 and $5.1 \times 10^{19}\,\mathrm{cm}^{-2}$, followed by $10.2 \times 10^{19}\,\mathrm{cm}^{-2}$ to $1.22 \times 10^{21}\,\mathrm{cm}^{-2}$ in steps of $10.2 \times 10^{19}\,\mathrm{cm}^{-2}$. The solid line represents the orientation of the position-velocity diagram in Fig. 4.

galaxy at a position angle of 62°, the high velocity gas regions are visible at the extreme velocities. Each of these regions contains a few $10^7 M_\odot$ in HI and has a large angular size. The velocity deviations are about $100\,\mathrm{km\,s}^{-1}$ with respect to the velocities of the HI disk. There is no evidence that the high velocity gas features are part of a system of gas clouds moving on circular or elliptical orbits, since filled orbits of these types cannot be traced in the data completely around the galaxy.

The motions of the bulk HI in the southwestern extension resemble that of the HI at smaller radii, and the tilted ring model shows a smooth rise in inclination from the inner to outer orbits. The high velocity cloud complex (R $\approx$ 12$'$, V $\approx$ $650\,\mathrm{km\,s}^{-1}$) on the southwestern side stands out clearly in the position-velocity diagram in Figure 4 from the motion of the higher-column density HI.

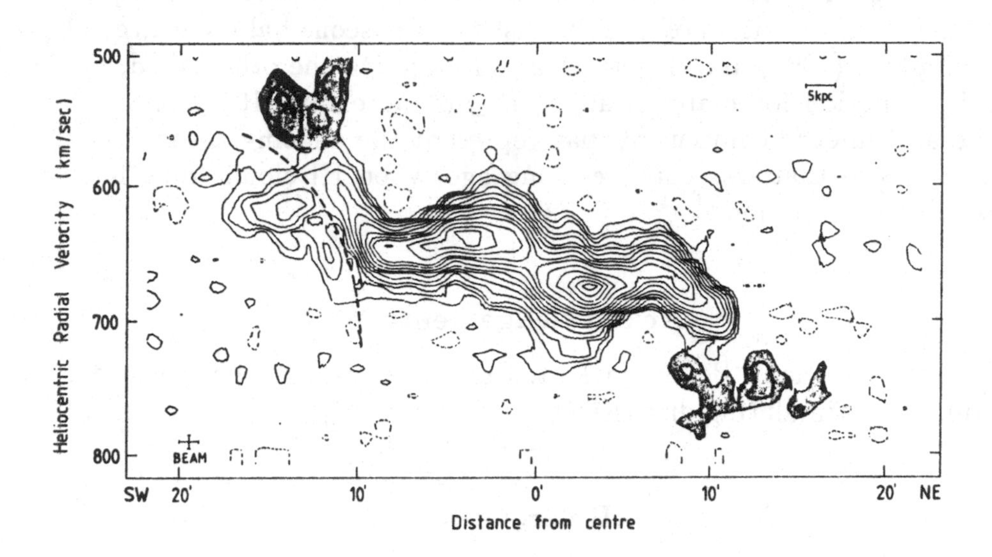

Fig. 4. Position-velocity diagram showing the high velocity regions (grey) at the extreme velocities (contours are -1, 1, 2, 3, 5, 8, 12, 17, 25, 35 mJy/beam area, followed by 45 to 145 mJy/beam area in steps of 20 mJy/beam area). Below and left of the dotted line, the asymmetric HI at the southwestern side is visible.

Discussion

The present HI observations of NGC 628 reveal an asymmetric HI distribution towards the southwest and the presence of two high-velocity-gas regions at large radii, symmetrically situated with respect to the centre. These deviations in the kinematics indicate that the very extended HI of NGC 628 has not settled into a well-behaved disk. A possible explanation is that the two high-velocity-gas complexes are tidal features induced by the infall and capture of an extragalactic object. This infalling material is now visible as the asymmetric, southwestern part and could be what remains of a large HI cloud or an HI-rich, irregular dwarf galaxy. Since intergalactic HI clouds are rare at present, the latter is more likely. In this scenario, the HI of the infalling dwarf is also tidally disrupted and becomes tail-shaped as it falls into the host galaxy. Since no optical counterpart is visible, the stellar content of the dwarf galaxy must also be disrupted and be dispersed below the optical detection limit. Thus, we may be witnessing a case of an apparently isolated, bright galaxy accreting an HI-rich dwarf. This kind of accretion would be important for the evolution of galactic disks and gaseous haloes as argued by Sancisi (1990) and Sancisi et al. (1990). On the other hand, the orbital period for material at 2.5 R_{Ho} is more than 10^9 years, and the asymmetric component may represent the ongoing assimilation of material that was captured long ago when HI clouds may have been more common.

Acknowledgements

We wish to thank Penny Sackett, Renzo Sancisi and Thijs van der Hulst for stimulating discussions.

References

Briggs, F.H. (1982). Astrophys. J. **259**, 544.

Kamphuis, J., van der Hulst, J.M. & Sancisi, R. (1990). In *The Interstellar medium in External Galaxies (Poster Session)*, NASA Conf. Proc. (eds. D.J. Hollenbach and H.A. Thronson, Jr.). Washington: NASA.

Sancisi, R. (1990). In *Windows on Galaxies*, Proceedings Erice Workshop (eds. G. Fabbiano, J. Gallagher and A. Renzini). Dordrecht: Kluwer.

Sancisi, R., Broeils, A., Kamphuis, J. & van der Hulst, T. (1990). In *Dynamics and Interactions of Galaxies* (ed. R. Wielen), p. 304. Berlin: Springer.

Shostak, G.S. & van der Kruit, P.C. (1984). Astron. Astrophys. **132**, 20.

van der Hulst, J.M. & Sancisi, R. (1988). Astron. J. **95**, 1354.

VLA HI observations of the radio galaxy Centaurus A

J. M. VAN DER HULST
Kapteyn Astronomical Institute, University of Groningen
J. H. VAN GORKOM
National Radio Astronomy Observatory and Columbia University
A. D. HASCHICK
Haystack Observatory
A. D. TUBBS
AT&T, Bell Labs, Morristown

Introduction

The elliptical galaxy NGC 5128 is the most nearby radio galaxy. The double lobed radio source associated with the galaxy has its axis almost perpendicular to the prominent and slightly warped dust lane (Kotanyi and Ekers, 1979). The existence of conflicting theoretical predictions regarding the origin of the dust lane, together with the unique opportunity to study the gas kinematics of a radio galaxy in detail prompted us to image NGC 5128 in neutral hydrogen with the Very Large Array (VLA). Here we summarize the results. For a more detailed description we refer to van Gorkom et al. (1990).

Observations and results

The HI was observed with the VLA with a spatial resolution of $33'' \times 77''$ and a velocity resolution of $25\,\mathrm{km\,s^{-1}}$. The sensitivity of the observations is 10 mJy/beam or 2.6 K brightness temperature. The HI appears largely associated with the prominent dust lane in Centaurus A and exhibits the same warped appearance. The HI velocities in the inner parts agree well with Hα emission line velocities (Bland et al. 1987) and indicate that the dust lane is a severely warped, rotating system. The HI in the outer parts, however, does not follow this kinematic pattern and is probably in highly non-circular orbits. This can be interpreted in terms of captured gas (a gas rich companion?) which has not yet settled in well behaved circular orbits, thus lending support to the hypothesis that the dust lane in Centaurus A is a result of a recent merger event.

HI absorption is seen against the nucleus and against the southern radio lobe. This indicates that the southern radio lobe is behind the dust and gas disk. Radio continuum depolarization was also found to be more significant in the direction of the southern radio lobe, confirming that the southern lobe is probably pointing away from the observer and shielded partly by the galaxy and the dust lane.

From the currently available data it seems likely that the central regions of NGC 5128 are devoid of any neutral gas. Absorption has been detected in the direction of the strong compact nucleus in HI, CO, OH and H_2CO. All species show absorption at the systemic velocity of the galaxy and faint absorption at velocities slightly below the systemic velocity. Thus far only the HI shows gas in absorption at velocities redshifted with respect to the systemic velocity. Since the redshifted HI absorption seen against the nuclear point source probably results from small clouds close to the center (van der Hulst et al. 1983), we tentatively conclude that the inner few hundred parsecs of Centaurus A are devoid of molecular gas. More complete and higher resolution CO data might be useful in verifying this.

The HI rotation curve is flat and is consistent with an elliptical galaxy model with a constant M/L at least out to 10 kpc. Using the rotation of the gas in the dust lane we estimate the mass out to $1.2r_e$ ($305''$) to be $2.5 \times 10^{11} M_\odot$ and find that the integrated mass to blue light ratio is 3.1 at that distance. The mass derived by assuming that the circular velocity as measured in HI remains constant out to the last measured X-ray point is considerably less than the value derived from X-ray data.

Acknowledgements

The National Radio Astronomy Observatory is operated by Associated Universities Inc. under a cooperative agreement with the National Science Foundation.

References

Bland, J., Taylor, K. & Atherton, P.D. (1987). Mon. Not. R. Astron. Soc. **228**, 595.
Kotanyi, C.G. & Ekers, R.D. (1979). Astron. Astrophys. **73**, L1.
van der Hulst, J.M., Golisch, W.F. & Haschick, A.D. (1983). Astrophys. J. **264**, L37.
van Gorkom, J.H., van der Hulst, J.M., Haschick, A.D. & Tubbs, A.D. (1990). Astron. J. **99**, 1781.

A geometric model for the dust-band of Centaurus A

RICHARD A. NICHOLSON
University of Manchester
KEITH TAYLOR
Anglo-Australian Observatory
JOSS BLAND
Rice University

Introduction

A detailed understanding of the morphology and kinematics of Centaurus A's dust-band is essential if this structure is to be used to constrain the shape and mass distribution of the host elliptical galaxy. Work towards formulating such a morphological/kinematic description (Taylor & Atherton 1983, Bland 1985, Nicholson et al. 1988, Nicholson 1989) concludes that many attributes of this structure can be explained with a thin warped disc geometry. As an extension to this work it is now demonstrated that the morphology and kinematics of the dust-band, as revealed through the comprehensive TAURUS data of the Hα ionized gas disc (Bland, Taylor and Atherton 1987, henceforth BTA), can be successfully explained with a thin warped disc geometry and a rotation curve consistent with that expected from an $r^{1/4}$ law mass distribution (Young 1976).

Apparent morphology of the dust-band

Figure 1a shows a schematic representation of Centaurus A's dust-band, a direct image of which is shown in Figure 1b. The two most prominent features are the 'thick' band of obscuration seen across the stellar nucleus and the symmetric clockwise twist in position angle (PA) with increasing radius. Closer inspection reveals the following features:

- the bisection of the apparent thick dust-band into two parallel dust lanes,
- the rounded feature seen just to the SW of the nucleus and its apparent merger with the less prominent SW dust lane,

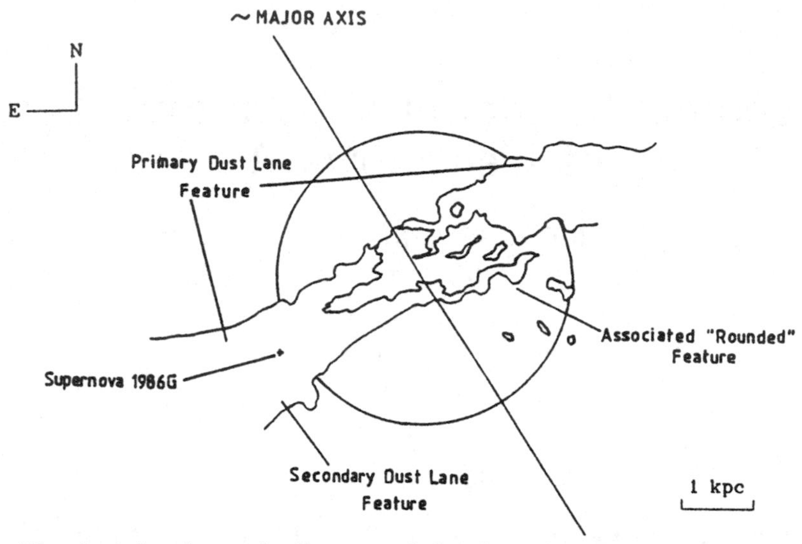

Fig. 1a. A schematic diagram of the dust-band of Centaurus A. Note how the thick dust-band bisects into two parallel dust-lanes to the NE and SW of the stellar nucleus. Also note the complexity of the dust-band morphology to the SW of the stellar nucleus.

- and the greater complexity and spread of the dust-band to the SW of the nucleus.

U,V,B photometry (Dufour et al. 1979) confirms that the most prominent reddening occurs along the main NE dust lane, with this component extending $\sim$ 12.5 kpc (assuming a distance of 5 Mpc) from the stellar nucleus. A second region of reddening, caused by the secondary dust lane to the SW of the stellar nucleus, is found to extend $\sim$ 2.5 kpc to either side of the nucleus before merging with the primary NE dust lane (Fig. 1a).

Morphology of the Hα ionized gas disc

A contoured TAURUS intensity map of Hα emission from Centaurus A's dust-band is shown in Figure 2 (BTA). Within the disc of diffuse Hα emission a ring ($r \sim 4.37$ kpc) of HII regions is easily identified at PA $\sim 125°$ and inclined $\sim 75°$ to the plane of the sky (Graham 1979). This ring is bisected along its major axis by two inner elongated bands of HII regions (BTA). Whereas the NE edge of the dust-band appears to be associated with the NE edge of the outer ring of HII regions (Graham 1979), the inner SW band of HII regions appears to be related to the secondary SW dust lane (Bland 1985, Nicholson 1989). Whilst this inner structure has been described as a hysteresis

Fig. 1b. A direct image of the Centaurus A system.

loop by BTA, it is also true that its appearance suggests a smaller radius ring ($r \sim 2.5$ kpc), at higher inclination ($\sim 85°$) to the plane of the sky and shifted in PA by $\sim 15°$ relative to the outer structure.

Kinematics of the Hα ionized gas disc

An iso-velocity contour map of the $2''$ spatial resolution TAURUS data is shown in Figure 3 (BTA). The most prominent features are the two pinched iso-velocity 'tongues' or 'spurs' that twist anticlockwise with increasing radius about the stellar nucleus. The velocity field displays a high degree of bi-symmetry at all radii and not, as has been implied in recent literature (Hayes et al. 1987), only in the central regions of the gas disc. However second order asymmetries do exist, with the iso-velocity tongues to the west of the nucleus appearing more elongated than the equivalent features to the east. The kinematic line-of-nodes (KLON) remains well defined from the outer edge of the ionized disc to within 0.5 kpc of the stellar nucleus. No evidence is found for the KLON twisting away from the dust-band at large radius to the NW of the nucleus, however differences of $\sim 30°$ are seen between the KLON and the dust-band ($r \sim 5.0$ kpc) SE of the nucleus.

The two bi-symmetric regions of line-splitting and broadening dis-

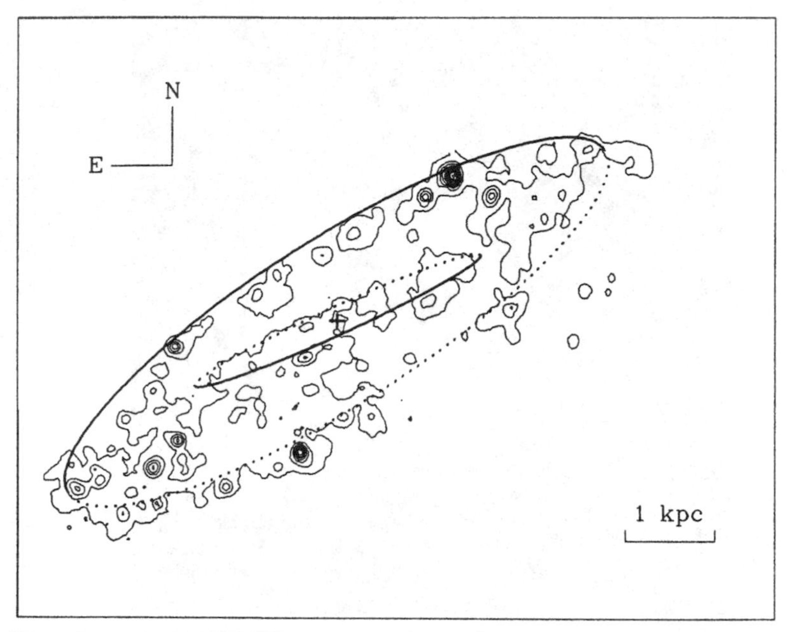

Fig. 2. A TAURUS contour intensity map of Hα emission associated with the dust-band of Centaurus A. Graham's ring of HII regions and a possible inner HII ring structure are clearly visible (broken lines represent the portions of each ring that are believed to lie behind the plane of the sky).

cussed by BTA are shown overlayed upon the iso-velocity contours in Figure 3. Average line-splitting values of $\sim 100\,\mathrm{km\,s^{-1}}$ are found throughout these regions with one area to the east of the nucleus having values as high as $140\,\mathrm{km\,s^{-1}}$. The two line-splitting regions are positioned $\sim 2.5 \pm 0.2$ kpc from, and symmetrically situated about, the nucleus at a PA of $\sim 110° \pm 5°$. It is important to note that the spatial coverage of the line-splitting regions only amounts to a few percent ($\sim 10\%$) of the ionized gas disc, and that these regions are spatially coincident with the twisted iso-velocity tongues. The spatial distribution of the these regions again hints at an anticlockwise twist with increasing radius.

Clear evidence is found for the evolution of the emission line profiles through these line-splitting regions, with the dominant emission line component on entering the line-splitting region becoming the secondary skewing feature on leaving. This switch in dominance is a strong indication that the line-splitting regions are related to a *localized* warp or fold in the gas disc geometry (Nicholson 1989).

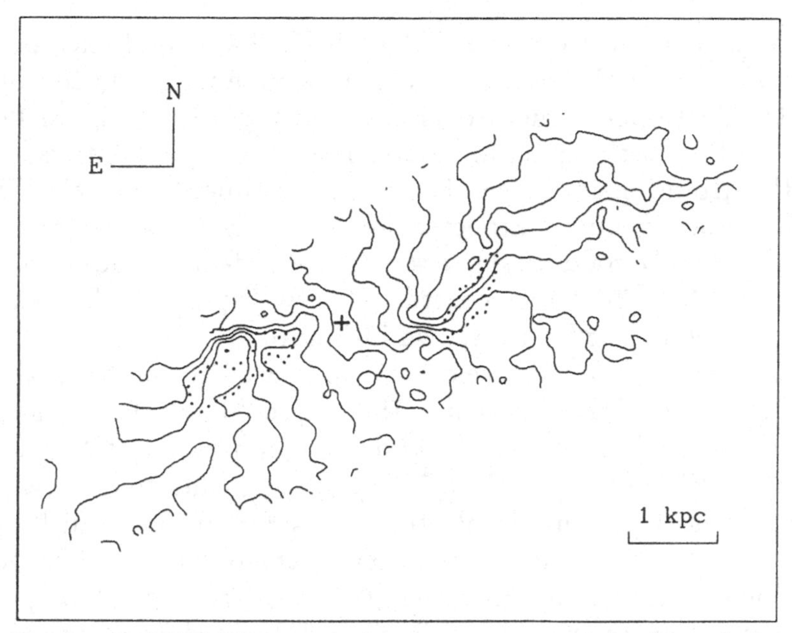

Fig. 3. A TAURUS iso-velocity contour map of the Hα velocity field. Regions with line profile widths greater than 60 km s^{-1} are enclosed by the broken lines.

Constraining the dust-band geometry

The geometric form of a structure, assumed to consist of material in simple closed circular orbits, is usually determined by fitting a series of concentric annuli to its velocity field (Rogstad 1976, Bland 1985). However, a potentially serious problem is caused by the inability to accurately determine the structure's velocity field from data with insufficient spectral resolution to resolve superimposed emission line features from different regions of the warped structure (Nicholson 1989). Further complications are caused by the fact that the number of velocity samples available for constraining the geometric and kinematic parameters for each annulus decreases as $1/r$, resulting in poorer fitting near the nucleus of the system. A detailed discussion of the results obtained using such ring-fitting techniques on the Centaurus A TAURUS data is given by Bland (1985).

If the mass distribution of the host elliptical is known, it is a simple process to recover the geometry of the dust-band, given the velocity gradient along the locus defined by the intersection of the structure with the plane of the sky (α_r). Assuming that the dust-band has no component of radial velocity, and that the material orbits the stellar nucleus in closed circular orbits, α_r is identical to the KLON. The angle α_r as derived from the TAURUS Hα data, and extrapolated out 12.5 kpc from the stellar nucleus in a manner suggested by the projected dust-band structure, is shown in Figure 4a. Figure 4a also shows the PAs of the inner and outer Hα structures identified in the TAURUS Hα intensity maps. The velocity gradient along the KLON and the rotation curve expected for a $r^{1/4}$ law mass distribution (Bailey & MacDonald 1981), assuming an effective radius of 305″ (Dufour et al. 1979) and a central velocity dispersion of 150 km s^{-1} (Wilkinson et al. 1986), are shown in Figure 4b.

Whilst the TAURUS Hα velocity map is used to constrain the *relative* inclination of the dust-band, the projected dust-band morphology must be used to deduce the *absolute* inclination of the structure to the plane of the sky. As the primary dust lane lies to the NE of the stellar nucleus this must imply that the structure at radii between 2.5 kpc $< r <$ 6.25 kpc also lies in front of the plane of the sky to the NE of the stellar nucleus. However at smaller radii ($r <$ 2.5 kpc) the front face of the structure must lie SW of the nucleus to explain the secondary SW dust lane. Finally, the dust-band's definite NE edge and ill-defined SW edge argues for the structure twisting through edge-on at large radius ($r >$ 6.25 kpc) so that it again lies to the SW of the stellar nucleus (Wilkinson et al. 1986, Nicholson 1989). The absolute inclination versus radius function (β_r) suggested by the velocity gradient along the KLON (assuming an $r^{1/4}$ mass distribution) and the projected dust-band morphology are shown in Figure 4c. Figure 4c also shows the inclinations of the inner and outer Hα structures identified in the TAURUS Hα intensity maps.

It is interesting to note that both $\alpha(r)$ and $\beta(r)$ imply that a discontinuity may exist in the warped disc structure at a radius around $\sim$ 2.5 kpc. This, together with the two ring structure implied by the TAURUS Hα intensity map, may suggest that the dust-band consists of two components: an inner 2.5 kpc warped disc, and an outer extended warped annulus of material.

Resultant geometric model

The velocity field and regions of line-splitting/broadening predicted for the 6.25 kpc Hα ionized gas disc are shown in Figure 5. Regions of line splitting are symmetric about the nucleus of the system with a maximum value of 130 km s^{-1}. The velocity field reproduces the prominent iso-velocity tongues that twist in a clockwise manner with increasing radius. The primary effect of differential dust obscuration

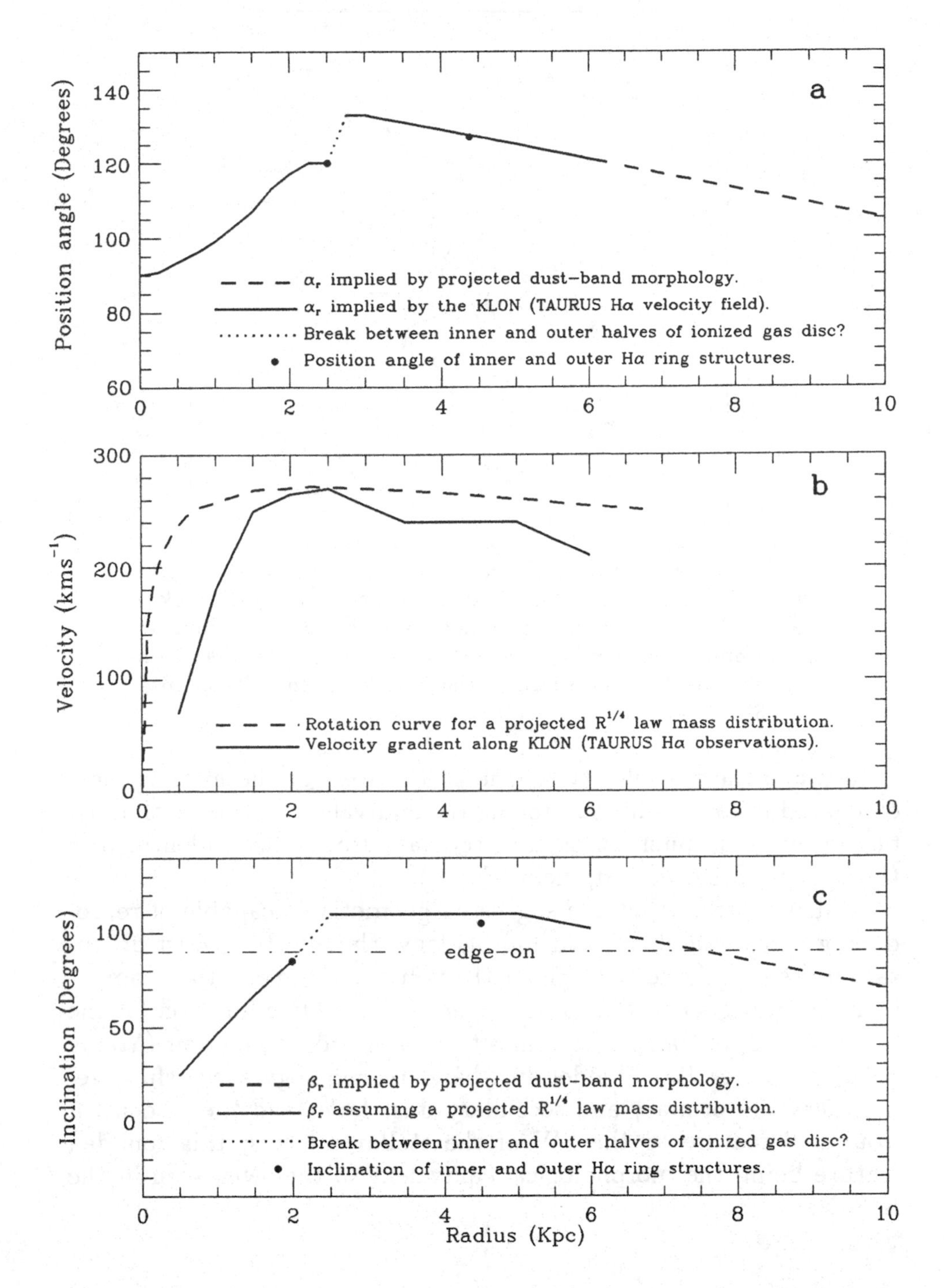

Fig. 4. Geometric and kinematic parameters for the dust-band structure derived from TAURUS Hα observations of the ionized gas disk and direct imaging of the Centaurus A system.

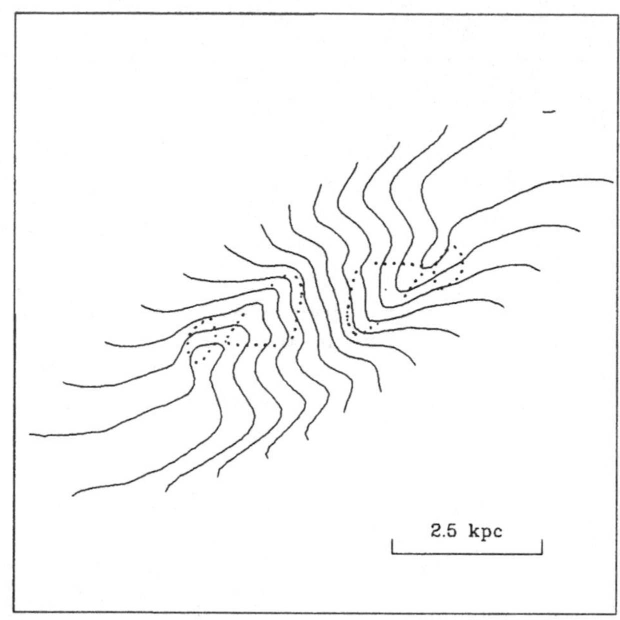

Fig. 5. Resultant model Hα velocity field. Regions where line profile widths are greater than 60 km s^{-1} are enclosed by the broken lines. The inclusion of differential dust obscuration results in second-order asymmetries between the iso-velocity tongues. Such features have been identified in the TAURUS Hα velocity map (Fig. 3).

is to make the iso-velocity tongue to the west of the nucleus more elongated in form, whilst fattening the equivalent feature to the east of the nucleus. Such an asymmetry has already been identified in the TAURUS Hα velocity field (Fig. 2).

To demonstrate that the suggested geometry is capable of reproducing a realistic dust-band morphology, the effects of dust obscuration are calculated throughout the volume of the elliptical component assuming a constant obscuration versus radius function for the warped disc and an $r^{1/4}$ luminosity versus radius function (Young 1976). The resultant model dust-band morphology successfully reproduces the primary and secondary dust lanes and the associated 'rounded' feature to the SW of the stellar nucleus; this rounded feature being the morphological equivalent to the twist seen in the KLON.

Discussion

Whilst it has been demonstrated that all the prominent kinematic

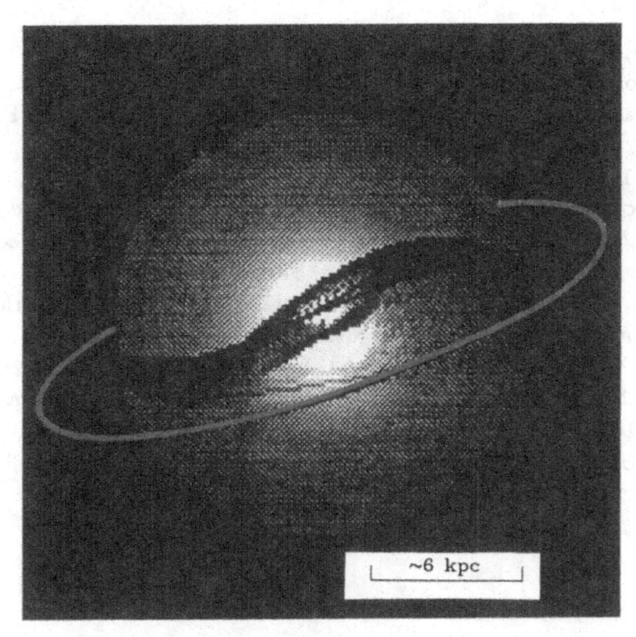

Fig. 6. Dust-band morphology predicted by the thin warped disk model. This model qualitatively reproduces all the features identified in direct images of the dust-band structure (Fig. 1).

features can be qualitatively reproduced with a thin warped disc geometry, how significant are radial and other non-circular motions? Bland (1985) concludes that no evidence is found for any radial contraction/expansion velocity component within the ionized gas disc ($r < 6.25\,\mathrm{kpc}$), this being supported by the apparent relationship between the KLON and the dust-band/ionized gas disc morphology (Nicholson 1989). Evidence for significant radial motions in the outer HI structure should also be viewed with some caution, as at large radius the suggested geometric model predicts a line-of-sight averaged velocity field whose KLON appear orthogonal to the dust-band structure (Nicholson 1989). Whilst the qualitative success of the thin warped disc model with circular orbits may imply that elliptical streaming motions, if present, are only second order perturbations to the Hα velocity field, such conclusions can only be verified after a more detailed investigation of the geometric nature of the dust-band structure.

References

Bailey, M.E. & MacDonald, J. (1981). Mon. Not. R. Astron. Soc. **194**, 195.

Bland, J. (1985). D. Phil. dissertation, Sussex University.

Bland, J., Taylor, K. & Atherton, P.D. (1987). Mon. Not. R. Astron. Soc. **228**, 595.

Dufour, R.J., van den Bergh, S., Harvel, C.A., Martins, D.H., Schiffer, F.H., Talbort, R.J., Talent, D.L. & Wells, D.C. (1979). Astron. J. **84**, 284.

Graham, J.A. (1979). Astrophys. J. **232**, 60.

Hayes, J.J.E, Schommer, R.A & Williams, T.B. (1987). In *Structure and Dynamics of Elliptical Galaxies*, IAU Symp. 127 (ed. T. de Zeeuw), p. 419. Dordrecht: Reidel.

Nicholson, R.A. (1989). D. Phil. dissertation, Sussex University.

Nicholson, R.A., Taylor, K., Sparke, L.S. & Bland J. (1988). In it The World of Galaxies (ed. H.G. Corwin & L. Bottinelli), p. 393. Berlin: Springer-Verlag.

Rogstad, D.H., Wright, M.C.H. & Lockhart, I.A. (1976). Astrophys. J. **204**, 703.

Taylor, K. & Atherton, P.D. (1983). In *Internal Kinematics and Dynamics of Galaxies*, IAU Symp. 100 (ed. E. Athanassoula), p. 331. Dordrecht: Reidel.

Wilkinson, A., Sharples, R.M., Fosbury, R.A.E. & Wallace, P.T. (1986). Mon. Not. R. Astron. Soc. **218**, 297.

Young, P.J. (1976). Astron. J. **81**, 807.

The circumgalactic ring of gas in Leo

STEPHEN E. SCHNEIDER
University of Massachusetts

In the course of observations subsequent to its discovery, the intergalactic gas in the M96 Group in Leo has finally revealed itself to be distributed in the form of an enormous ring of material surrounding the central two galaxies of the group (Schneider et al. 1989). As such, it is the largest circumgalactic ring of material ever found. It may represent an extreme of the sorts of rings found surrounding some early-type galaxies (van Driel 1987), or perhaps primordial debris that has never accreted onto any of the group's galaxies. Since the gas is located well outside of the optical disks of any of the galaxies and contains no stars down to very low limits, the ring also provides an unusually "clean" opportunity for studying the dynamics of a ring of gas in the dynamical environment of a few galaxies' haloes.

In this contribution, I will review the properties of this giant ring and briefly discuss some searches for intergalactic gas surrounding other early-type galaxies. These searches have yielded one other interesting case for comparison to the Leo ring.

Orbit of the ring

The ring of gas in Leo spans more than a degree on the sky and the original limited mapping (Schneider et al. 1983) hardly began to suggest its overall shape. Nevertheless, shortly after the discovery, Briggs (1983) pointed out that the velocity field across the southern part of the ring was consistent with an orbit around the group's central galaxies. Later mapping (Schneider 1985) located features that were indeed appropriate to a ring orbit, and deeper integrations (Schneider et al. 1989) demonstrated that gas practically forms a continuous ring (Figure 1).

Even before the final mapping, it was found that not only the sky

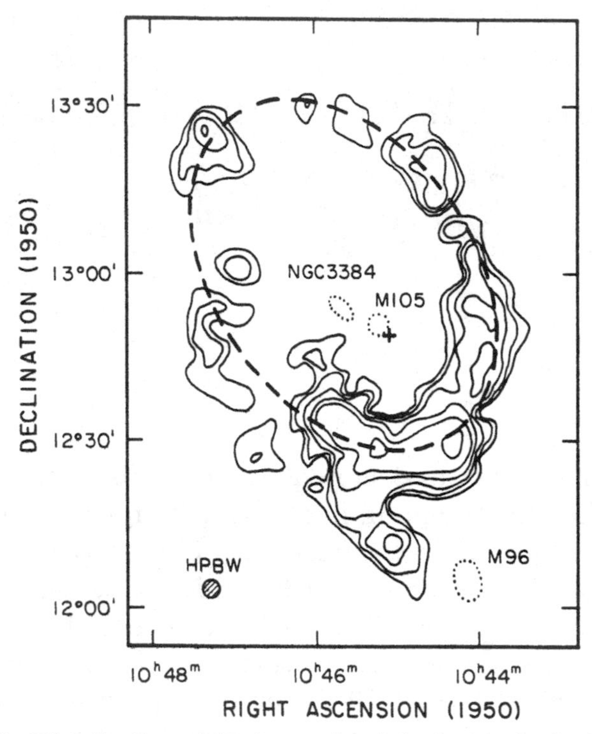

Fig. 1. Distribution of the intergalactic hydrogen in the M96 Group based on Arecibo observations. Contours of the integrated HI flux are shown at levels of 2, 4, 8, 16, 32, and 64 $\times 10^{18}$ cm^{-2}. The best-fit orbit model is shown by a dashed ellipse, and the predicted position of the orbit's focus is shown by a cross. The optical dimensions of nearby galaxies are marked with dotted lines.

distribution of gas, but its velocity pattern as well, could be quite successfully modeled as lying along the path of a single elliptical Keplerian orbit. Independent of any prior assumptions about the locations of nearby galaxies, the focus of the orbit was found to be near the central two galaxies in the group: M105, a giant elliptical, and NGC 3384, a lenticular. Further, the derived center-of-mass velocity matched the the centroid of the pair (assuming both have the same mass-to-light ratio). The success of the model is surprising given the probably complex gravitational environment inside any group of galaxies, but conversely this may suggest that interactions with galaxies other than the central two have been unimportant.

The modeling procedure used was to first fit an ellipse to the observed distribution on the plane of the sky, then to assign "Keplerian angles" (eccentric anomalies) to each point around the ring, and finally to solve for the observed radial velocity in terms of the

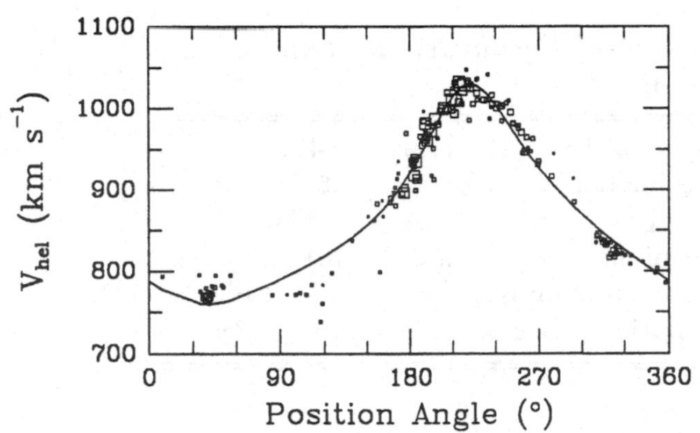

Fig. 2. Keplerian orbit model for the radial velocities around the intergalactic ring. Squares (plotted with sizes in proportion to the detected flux) show observed velocities, and the line is the best-fit model.

Keplerian angle, E:

$$v_{rad} = \left[\sqrt{\frac{GM(1-e^2)}{a(1-e^2 sin^2 E_\omega)}} \sin i \right] \frac{\cos(E - E_\omega)}{1 - e\cos E} + v_{com}$$

where M is the central mass, e is the eccentricity, a is the semi-major axis, i is the inclination, v_{com} is the center-of-mass velocity of the system, and E_ω is the Keplerian angle between the line of apsides and the line of nodes. The Keplerian angle is determined by applying an affine transformation to the ellipse to make it into a circle. This angle can be shown to be independent of any projection up to a simple additive constant, making it especially convenient for handling orbits that may be non-circular. One other free parameter is then the zero-point of E, which corresponds to the angle of the line of apsides.

The equation can be solved with a non-linear least squares algorithm for the term in brackets, E_ω, e, v_{com}, and the zero-point of E. Given the observed shape of the ellipse along with e and E_ω, it is in turn possible to estimate i, and for a given distance, M and a. The path of the orbit and the position of the predicted focus are shown in Figure 1. The fit to the observed radial velocities is shown in Figure 2, and the derived parameters are summarized in Table 1, based on a model fit to the complete set of Arecibo observations. The focus is closer to M105 than to NGC 3384, but the angle between the line of nodes and the line of apsides was not very well constrained, and the position could be shifted toward NGC 3384 without increasing the chi-square error very much. The results are

Table 1. *Parameters of the ring orbit.*

eccentricity	0.402
inclination	42°.5
v_{com}	$841\,\mathrm{km\,s^{-1}}$
semi-major axis	101 kpc
dynamical mass	$5.6 \times 10^{11} M_{\odot}$
rotation period	4.1×10^{9} yr

also in excellent agreement with the original model which was fit to only the few points of brightest emission around the ring.

The ring proves to be on a moderately eccentric orbit some 200 kpc across, whose pattern of radial velocities independently identifies the region nearest the M105/NGC 3384 pair as the pericenter of the orbit. Non-circular orbits are highly sensitive to the shape of the force law affecting them, and the existence of an elliptical orbit requires that the force law be $\propto r^{-2}$. Therefore the central galaxies must not have significant halos extending out beyond ~ 60 kpc (the pericenter radius of the ring). A limited halo extent is consistent with the inferred dynamical mass of $6 \times 10^{11} M_{\odot}$, which is only twice the *internal* dynamical mass determined for M105 and NGC 3384, giving an overall value of $M/L = 25$ for the pair. This is one of the best-constrained estimates of the total mass-to-light ratio for galaxies out to many times their optical radii, but clearly it would be useful to apply some more sophisticated dynamical models including hydrodynamics and possible interactions with other galaxies.
Another surprising result of the model is the implied orbital period of some 4×10^{9} years. The fact that gas is found at points all around the ring (although most of it is clumped toward the southern side), indicates therefore some minimum lifetime for the ring that may approach a Hubble time. If the feature is not primordial but originated in the injection of gas from a tidal encounter, it would presumably take more than one orbital period to spread the gas around into the form of a ring. Perhaps gas-dynamical models can set constraints on this timescale.

High-resolution observations show that some of the gas is relatively diffuse, while some is confined to virially stable clumps (Schneider et al. 1986). The situation is further complicated in that the degree of clumpiness appears to be different in different parts of the ring.

Therefore, it is not clear whether the continuous-gas or ballistic gas-clump models discussed at this workshop would better describe the intergalactic gas dynamics.

Rings into rings

The intergalactic gas in the ring gives a strong visual impression that it is interacting with a third galaxy exterior to it. A filament of gas extends from the southern portion of the intergalactic ring toward the (R)Sab spiral M96. One possible scenario would have M96 passing by the intergalactic gas, and tidally drawing some of the gas toward itself. Since the tidal feature should trail the direction of rotation of the ring, the angle of the southern extension implies that the ring must be rotating clockwise on the sky. The concentration of gas in the southern portion of the ring could be a result of this interaction, and the poor velocity fit and the disrupted appearance further clockwise around the ring (at P.A.$\sim 100°$) might then be explained as gas that has been disturbed by its slightly earlier passage by M96.

The fate of the gas actually stripped from the ring is uncertain, but there is tantalizing evidence that it is the source for the external ring surrounding M96 itself. M96 has a faint outer optical ring whose major axis is tilted by $\sim 60°$ relative to the inner bright disk suggesting that they do not share the same orbital plane. VLA observations of M96 show that 90% of the galaxy's neutral hydrogen resides in the outer ring as well (Schneider 1989). More detailed optical studies should be undertaken of M96 to determine if the inner disk does in fact rotate in a distinct plane from the outer ring, and whether the colors of the outer ring indicate an unusual star formation history.

The circumstances of the intergalactic ring appear to be ideal for numerical modeling, since self-gravity appears to be unimportant, and thus the intergalactic gas acts like test particles in the haloes of the neighboring galaxies. Such models might also show how the extension of gas toward M96 could have formed, and by how much the gravitational influence of M96 may have perturbed the mass estimates.

Based on restricted three-body simulations (Schneider 1989) I have argued that M96 cannot have remained as close to the ring as it is currently for more than a brief period. This model treated the M105/NGC 3384 pair as a single point with M96 acting as a perturber to an initially stable ring of massless particles. In the simulations, if M96 remained within $\sim$ 2 ring radii for any extended period of time, the ring became thoroughly disrupted within one orbit. Putting M96 on a more eccentric orbit or one more inclined to the ring plane left the ring intact except for some minor tidal features like the southern extension seen in the actual data. A more sophisticated model might include halo distributions around the galaxies to determine their effect. It would also be interesting to determine how the time-varying dipole field of the central binary system might affect the ring of gas, and to determine if it might permit for some transfer of orbital angular momentum from the binary orbit to the ring.

Rings around early-type galaxies

The intergalactic ring in Leo has spurred a variety of searches for similar types of objects. With an accidental discovery like that of the Leo ring, the problem is to figure out why we should have known it was there in the first place, but in this case hindsight has not proved to be very illuminating! Since the M96 group is the nearest group in de Vaucouleurs' (1975a) catalog dominated by early-type galaxies, one hypothesis was that such rings might be found in groups with similar-type galaxies. Extensive maps made of seven such groups proved unsuccessful. Another idea was that intergalactic gas might be associated with galaxies having outer rings like M96—but again no luck.

An alternative, observational, approach was suggested when it was found that some of the earlier HI spectra of M96 and NGC 3384 showed unrecognized features caused by the intergalactic ring. Large-beam 21 cm observations collect emission from the regions surrounding a galaxy as well as the galaxy itself. A literature search was therefore conducted to look for discrepancies between small- and large-beam observations of galaxies. Follow-up Arecibo observations of 14 cases showed that most of these discrepancies were produced by mundane problems—typographical errors, poor signal-to-noise ratios, sidelobe contamination, or bad pointing—but a few were more interesting. In several cases, faint companion dwarfs or tidal features were the source of the extra emission. In a few instances, extended disks of hydrogen were found surrounding the galaxies, and in the one case of the SB0/a galaxy NGC 5701, an extremely large disk of hydrogen was discovered, extending to $\sim$5 times its optical radius.

NGC 5701 was subsequently mapped at the VLA, and the gas is

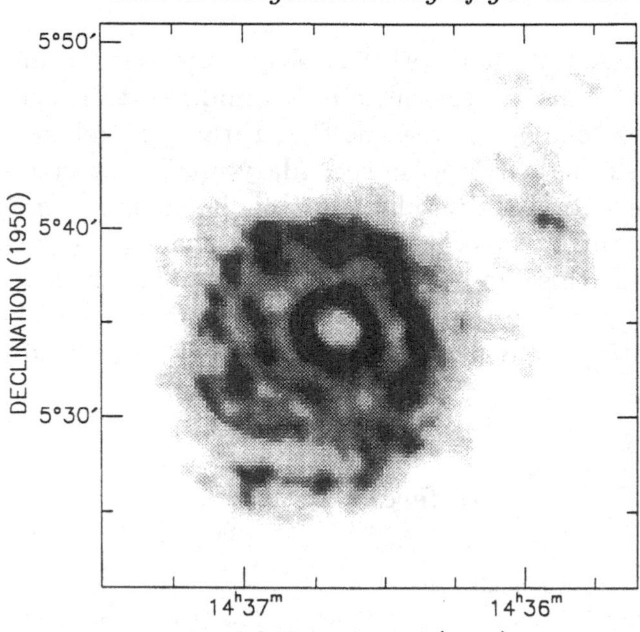

Fig. 3. VLA observation of the distribution of hydrogen around NGC 5701. The gray scale image has been "unsharp masked" to reveal details of the gas distribution. Optically, the galaxy extends only to the innermost ring seen in the HI distribution.

found covering a region up to 25′ (~ 150 kpc) across. In Figure 3, a gray-scale representation of the VLA HI map is shown. This map has been treated much like an unsharp masked photograph to highlight the variations across the disk. Optically the galaxy has a 4′.5 diameter, ending at the innermost bright ring in the HI map. A small detached cloud with a hydrogen mass of a few $10^8 M_\odot$ is present to the northwest. No optical counterparts have been found for any of the outer HI features.

The disk around NGC 5701 is clearly different from the ring in Leo, but it is also similar in its association with an early-type galaxy and the presence of small cloudlets of HI. If dissipative processes were present in a circumgalactic ring of gas, it might end up looking like NGC 5701. On the other hand, if NGC 5701 eventually clears out the central region of its HI disk, it would look very much like the ring in Leo.

It is interesting that both gaseous and optical outer rings appear to be more common around early-type galaxies (de Vaucouleurs 1975b; van Driel 1987). That rings and extended gas disks are common where disks are uncommon suggests either that disks "eat" rings or that early-type galaxies formed differently. There are good dynamical reasons discussed at this workshop why rings may be unstable in the presence of a disk potential, but I would like to suggest the alternative: maybe one of the reasons that early-type galaxies have rings and not disks is because the secondary-infall material at the time of galaxy formation had too high an angular momentum to settle down rapidly into a typical disk. I think the Leo ring can be best understood as this type of high angular momentum accretion, and perhaps many of the questions we ask about outer disks and rings may ultimately require that we look all the way back to initial conditions.

References

Briggs, F. (1983). Private communication.

de Vaucouleurs, G. (1975a). In *Stars and Stellar Systems, Vol. 9, Galaxies and the Universe* (ed. A. Sandage, M. Sandage, and J. Kristian), p. 557. Chicago: University of Chicago Press.

de Vaucouleurs, G. (1975b). Astrophys. J., Suppl. Ser. **29**, 193.

Schneider, S.E. (1985). Astrophys. J., Lett. **288**, L33.

Schneider, S.E. (1989). Astrophys. J. **343**, 94.

Schneider, S.E., Helou, G., Salpeter, E.E. & Terzian, Y. (1983). Astrophys. J., Lett. **273**, L1.

Schneider, S.E., Salpeter, E.E. & Terzian, Y. (1986). Astron. J. **91**, 13.

Schneider, S.E., Skrutskie, M.F., Hacking, P.B., Young, J.S., Dickman, R.L., Claussen, M.J., Salpeter, E.E., Houck, J.R., Terzian, Y., Lewis, B.M. & Shure, M.A. (1989). Astron. J. **97**, 666.

van Driel, W. (1987). PhD Thesis, University of Groningen.

Using gas kinematics to measure M/L in elliptical galaxies

TIM DE ZEEUW
California Institute of Technology & Sterrewacht Leiden

Abstract

The kinematics of cold gas in elliptical galaxies can be used to determine M/L as a function of radius, but this requires fitting the observed velocity fields with realistic triaxial models in which the gas occupies non–circular closed orbits. The derived M/L profiles may differ significantly from that obtained on the basis of circular motion in spherical potentials.

Introduction

In order to calculate the mass–to–light ratio M/L as a function of radius r in a galaxy, one needs to determine both the luminosity distribution $L(r)$ and the mass distribution $M(r)$. The former is usually obtained by measuring the projected surface brightness. By assuming a shape for the galaxy, one can then deduce $L(r)$ by deprojection. The mass distribution can be derived by measuring the kinematics of a set of tracer objects. If their orbits are known, the observed velocities can be used to find the potential in which the tracer objects orbit, from which the mass distribution which generates this potential follows. This again requires assumptions about geometry.

Deriving M/L in this manner is a relatively straightforward exercise for spiral galaxies. We have excellent reasons to believe that to good accuracy the luminous parts of these systems are flat circular disks, so that derivation of $L(r)$ is easy; the inclination of the disk follows from its apparent flattening on the sky. Moreover, there is an ideal kinematic tracer, cold HI, which must occupy the circular orbits in the equatorial plane. The observed line–of–sight HI velocities hence immediately give the circular velocity as a function of radius—the rotation curve—from which the mass distribution follows. The result is well-known: the derived rotation curves generally stay flat out to many optical radii, which means a density which falls off proportional to $1/r^2$. The light distribution decreases exponentially fast, however, so M/L increases outwards, indicating the presence of massive halos of unseen material in spiral galaxies (e.g., van Albada et al. 1985).

For elliptical galaxies the situation is not so straightforward. They are slowly rotating triaxial systems, so that two viewing angles are needed to specify the direction of observation, and two axis ratios need to be found for the determination of the intrinsic shape. Thus, instead of simply measuring an inclination, which is all that is needed for spirals, a total of four parameters must be obtained to fix the geometry for ellipticals. Furthermore, different morphologies are possible for a disk of settled cold gas (Steiman–Cameron & Durisen 1982, Merritt & de Zeeuw 1983), and it is less straightforward to derive a density distribution from the observed gas kinematics.

Velocity Fields in Triaxial Potentials

The stable simple closed orbits available to cold gas in a triaxial galaxy are not circular, but are approximately elliptic. In systems with a stationary figure, such closed orbits exist in two of the three principal planes, so that settled cold gas will be found either perpendicular to the short axis or perpendicular to the long axis of the system (Heiligman & Schwarzschild 1979). When projected on the sky, there will generally be a significant misalignment between the kinematic axes of the gas and the photometric axes of the stellar light. This results from the fact that for a triaxial light distribution the projected principal axes generally do not coincide with the major and minor axis of the projected surface brightness distribution (Stark 1977). Furthermore, the gas morphology will appear to be twisted and misaligned, due to the variation of surface density caused by the velocity change along an elliptic orbit (de Zeeuw & Franx 1989).

As a simple example, consider the logarithmic potential with zero core radius, defined as

$$V(x,y,z) = v_c^2 \ln\Big(\frac{x^2}{a^2} + \frac{y^2}{b^2} + \frac{z^2}{c^2}\Big), \qquad a \geq b \geq c > 0,$$

where (x, y, z) are Cartesian coordinates, a can be regarded as the unit of length, b/a and c/a are the axis ratios of the ellipsoidal equipotential surfaces, and v_c sets the velocity scale. The associated

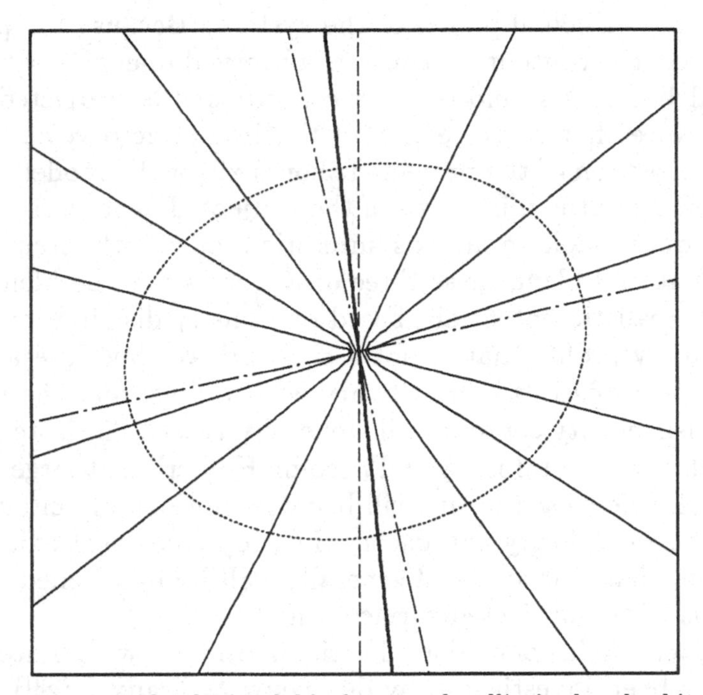

Fig. 1. Projected velocity field of gas on the elliptic closed orbits in the (x, y)–plane of Binney's logarithmic potential, with b/a=0.95 and c/a=0.9, for a direction of viewing defined by choosing the usual spherical polar angles equal to θ=60° and ϕ=45°. The associated density distribution has axis ratios 1:0.87:0.73. Contours of constant radial velocity are straight lines, and are drawn at linear intervals. The thick line is the zero velocity contour. The oval is a contour of constant projected surface density of the model. Its two principal axes are indicated by dot–dashed lines. The projected short axis of the density distribution itself is denoted by the short dashed line.

density is proportional to $1/r^2$, and hence has infinite total mass. Further properties of the density distribution, including its axis ratios, can be found in Binney (1981). Using epicyclic theory (e.g., Gerhard & Vietri 1986), it is easy to calculate the properties of the simple closed orbits in the principal planes. In this approximation the orbits are ellipses with axis ratio $2b/(a+b)$ in the (x, y)–plane and $2c/(a+c)$ in the (y, z)–plane. The former are elongated in the y–direction, and the latter are elongated in the z–direction. For $b = a$ the model is oblate, and the simple closed orbits in the (x, y)–plane are circular. The associated rotation curve is flat.

Figure 1 shows the velocity field for gas on the elliptic orbits around the short axis, as it would be observed from a viewing direction not in one of the principal planes of the system. Because the potential is scale–free, the contours of constant observed line–of–sight velocity are radial lines. Also plotted is a contour of the projected surface density. Note that the direction of the line of zero velocity differs from the direction of the projected short axis of the model, which in turn differs from the minor axis of the projected density, as expected for a triaxial model. In an axisymmetric model with circular orbits in the equatorial plane these three directions would be identical.

In more realistic potentials, based on density distributions with a radial density profile that is not a single power law, the axis ratio of the elliptic closed orbits will vary with radius, and the resulting observed isovelocity contours will generally twist. The potential of a model with finite total mass will become Keplerian at large radii, so that the elliptic closed orbits will become more nearly circular, and the line of zero velocity approaches the projected short axis at large radii. Note that this axis will generally still be misaligned from the minor axis of the projected surface brightness.

The velocity fields of cold gas in triaxial models with finite density cores have been investigated by de Zeeuw & Franx (1989). In this case the elliptic closed orbits become more and more elongated going inwards, so that there will be increasing shear between neighboring orbits near the center and the closed orbit description of the gas motion eventually breaks down at small radii. When the density profile diverges as a power of $1/r$ at small radii, i.e., has a cusp, then the orbits reach a limiting central axis ratio. Its value depends on the degree of asphericity of the potential, and on the slope of the profile, with shallower slopes resulting in more elongated orbits. The potentials relevant for elliptical galaxies are not far from spherical, and hence the epicyclic approximation is adequate for treating most systems with cusps. A detailed discussion of such systems is given by Lees and de Zeeuw (1991).

When the figure of the galaxy tumbles, e.g., about the short axis, cold gas will settle either on roughly elliptic orbits in the equatorial plane or on the family of anomalous orbits, which do not lie in any of the principal planes, but form a stable configuration which is intrinsically warped (Heisler, Merritt & Schwarzschild 1982; Magnenat 1982). The observed properties of such gas disks have not been modeled in any detail, but it is evident that a wide variety of morphologies and velocity fields can be obtained.

Elliptical galaxies with gas disks

The E3 galaxy NGC 5077 has a disk of ionized gas along its apparent minor axis, and gas velocities have been measured along seven position angles out to about an effective (half-light) radius (DeMoulin-Ulrich et al. 1984; Bertola et al. 1991). A simple model with the gas on circular orbits in a spherical potential fits the observed gas kinematics, but implies that inside half an effective radius M/L declines sharply toward the center, similar to what has been reported for the E4 galaxy NGC 7097 by Caldwell, Kirshner & Richstone (1986). However, constant M/L triaxial mass models with non-rotating figures and the gas in either of the two preferred planes can be constructed which are consistent with the morphology and kinematics of both the gas and the stars. In particular, they reproduce the observed misalignment of 23° between the line of zero velocity of the gas and the apparent major axis of the galaxy as being due to a projection effect. A constant M/L oblate model with the gas in a warped polar ring of the kind discussed by Sparke (1986) can be fitted to the data also.

Without higher resolution two-dimensional velocity measurements, we cannot decide which of these models is the correct one for NGC 5077. The results show, however, that there is no reason to suspect that M/L varies strongly over the luminous part of elliptical galaxies, and furthermore, that it is important to use the proper geometry when deriving density profiles, and hence M/L, from gas kinematics. This is true also for triaxial bulges of spiral galaxies (e.g., Gerhard & Vietri 1990).

Two-dimensional velocity information is available for the emission line gas associated with the inner dust lane of Centaurus A (Bland, Taylor & Atherton 1987). Ad hoc kinematic models have been constructed with the gas always on circular orbits whose inclination is adjusted separately at every radius. These models reproduce both the observed gas kinematics and the morphology of the dust lane, but require considerable inclination variations along the dust lane (Nicholson 1991). It should be noted, however, that Centaurus A is a flattened galaxy, with the inner dust lane falling along the projected minor axis. Even if the galaxy were oblate rather than triaxial then still the simple closed orbits available to the gas along the minor axis would not be circular but elliptic. Hence, rather than ascribing all observed deviations from purely planar circular motion to inclination variations of precisely circular orbits, it is more likely that the twisted velocity contours are caused in part by non–circular motions. This was investigated by Quillen et al. (1991), who obtained CO (2–1) line profiles along the inner dust lane of Centaurus A with the Caltech Submillimeter Observatory on Mauna Kea. The line profiles are invariably very broad, and are indeed inconsistent with planar circular motion. However, they can be fitted quite well with elliptic orbits in the flattened gravitational potential of the galaxy, calculated with a constant M/L. The spatial resolution of the CO observations is poor, due to the 30″ beam of the CSO, but it is clear that non–circular motions are of importance in the inner dust lane of Centaurus A.

A small number of elliptical galaxies are known with appreciable amounts of cold HI gas, which in some cases extends to many effective radii (van Gorkom 1991). These are ideal cases for studying the behavior of M/L with radius. Traditionally, HI velocity fields have been fitted with planar circular motion or with tilted circular ring models. It is clear from the examples mentioned above that while this may be correct for spirals, this is not necessarily the correct approach for modeling the HI kinematics in ellipticals. Indeed, Lees (1991) has fitted both circular and elliptic orbit models to the observed velocity field of the elliptical galaxy NGC 4278, and finds not only that the triaxial model with elliptic orbits fits better, but also that the derived M/L variation is significantly less than in the circular orbit model. This shows explicitly that the effects of triaxiality can be important even at large radii. It is clearly worthwhile to do a similar investigation of other elliptical galaxies with HI at large radii, such as NGC 5666 (Lake et al. 1987) and IC 2006 (Schweizer et al. 1989).

Teuben (1991) has demonstrated how the signature of non–circular motions can be detected by using standard circular orbit fitting routines. It is of considerable interest to apply his technique to the HI velocity fields of spirals. Detection of non–circular motions might well give interesting information on the intrinsic shape of the dark halos of these galaxies.

Conclusions

Projection effects and dynamical effects are important for the observed kinematics of cold gas in triaxial galaxies, and result in twisted velocity fields with misaligned kinematic axes. If the gas has had time to settle on the simple closed orbits, then high–resolution observations, combined with detailed modeling, can put significant constraints on the intrinsic shape of the galaxy, and on the variation of M/L with radius.

Acknowledgements

It is a pleasure to acknowledge stimulating discussions with Joanna Lees, Alice Quillen, Peter Teuben, and Jacqueline van Gorkom. This research was supported in part by NSF Grant AST 88-15132.

References

Bertola, F., Bettoni, D., Danziger, I.J., Sadler, E.M., Sparke, L.S. & de Zeeuw, P.T. (1991). Astrophys. J., in press.

Binney, J.J. (1981). Mon. Not. R. Astron. Soc. **196**, 455.

Bland, J., Taylor, K. & Atherton, P.D. (1987). Mon. Not. R. Astron. Soc. **228**, 595.

Caldwell, N., Kirshner, R.P. & Richstone, D.O. (1986). Astrophys. J. **305**, 136.

DeMoulin-Ulrich, M.-H., Butcher, H.R. & Boksenberg, A. (1984). Astrophys. J. **285**, 527.

de Zeeuw, P.T. & Franx, M. (1989). Astrophys. J. **343**, 617.

Gerhard, O.E. & Vietri, M. (1986). Mon. Not. R. Astron. Soc. **223**, 377.

Gerhard, O.E. & Vietri, M. (1990). In *Dynamics and Interactions of Galaxies* (ed. R. Wielen), p. 342. Berlin: Springer.

Heiligman, G. & Schwarzschild, M. (1979). Astrophys. J. **233**, 872.

Heisler, J., Merritt, D.R. & Schwarzschild, M. (1982). Astrophys. J. **258**, 490.

Lake, G., Schommer, R.A. & van Gorkom, J.H. (1987). Astrophys. J. **314**, 57.

Lees, J.F. (1991). This volume.

Lees, J.F. & de Zeeuw, P.T. (1991). In preparation.

Magnenat, P. (1982). Astron. Astrophys. **108**, 89.

Merritt, D.R. & de Zeeuw, P.T. (1983). Astrophys. J., Lett. **267**, L23.

Nicholson, R.A. (1991). This volume.

Quillen, A.C., Phillips, T.G., Phinney, E.S. & de Zeeuw, P.T. (1991). In preparation.

Schweizer, F., van Gorkom, J.H. & Seitzer, P. (1989). Astrophys. J. **338**, 770.

Sparke, L.S. (1986). Mon. Not. R. Astron. Soc. **219**, 657.

Stark, A.A. (1977). Astrophys. J. **213**, 368.

Steiman-Cameron, T.Y. & Durisen, R.H. (1982). Astrophys. J., Lett. **263**, L63.

Teuben, P.J. (1991). This volume.

van Albada, T.S., Bahcall, J.N., Begeman, K. & Sancisi, R. (1985). Astrophys. J. **295**, 305.

van Gorkom, J.H. (1991). In *Morphological and Physical Classification of Galaxies* (eds G. Busarello, M. Capaccioli & G. Longo). Dordrecht: Kluwer, in press.

Velocity fields of disks in triaxial potentials

PETER TEUBEN
University of Maryland

Abstract

Velocity fields of disks in triaxial halos are modeled and discussed. Residual velocities from a tilted-ring model fit along a galactic circle show a unique $\sin 3\theta$ signature, and provide a sensitive measure of the triaxiality of a potential. to some degree of the dark halo). the outer parts of disk galaxies.

Introduction

A large number of disk galaxies exhibit "flat" rotation curves, with the consequence that large amounts of dark matter are implied to be present in assumed spherically symmetric halos. This possibility of dark matter has not uniformly pleased astronomers and there have been a number of attempts to explain flat rotation curves without the necessity of a dark halo. Examples can be found in, e.g., Milgrom (1983), Valentijn (1990) and Braun (1990).

Oval distortions in the inner parts of galaxies are quite commonly observed (e.g., Bosma 1978, 1981). Blitz and Spergel (1990) studied the global asymmetries present in the longitude-velocity diagram of our Galaxy and interpreted it in the terms of a triaxial mass distribution (see also Gerhard and Vietri 1986). They proved that the triaxiality could not be in the outer regions of the Galaxy, and came to the conclusion that our Galaxy has a fairly strong central oval component, possibly in the form of a bar.

In addition elliptical galaxies are now widely thought to be triaxial, which may extend to dark halos. Primordial galaxy formation (see, e.g., Frenk et al. 1988) is also likely to result in triaxial halos. The collapse calculations reported by Katz give halos with axis ratio X:Y:Z of order 10:9:6.

It is well known that in disks with circular orbits the morphological and kinematical axes coincide and that the kinematical major and minor axes are perpendicular to one another (for a nice pictorial view see Fig. 1 in van der Kruit and Allen 1978). It is also well known (see, e.g., de Zeeuw, this volume) that both these observable aspects break down for oval* orbits, except in certain degenerate viewing angles.

Most disk galaxies show a varying position angle and inclination of the kinematical axes in the outer parts, which we call a *kinematical warp*. This can be contrasted with a *direct* or *morphological* warp, where we have direct evidence of a warp (Bosma, 1978). Some kinematical features can possibly be reproduced with an appropriate change of eccentricity and orientation of the orbits as a function of radius, without the need to deviate from the galactic plane. The presence of bars and oval distortions in disks and now the increasing evidence that halos are triaxial warrant such an exploration. This is not to say that warps do not exist, but one has to be able to disentangle geometry of the disk from the shape of the orbits. A nice example of such an exploration in M31 has recently been presented by Braun (1990).

Warps are a much more natural outcome if halos are triaxial, as we have seen during several talks in this conference. Bosma showed us that the vast majority of galaxies are warped to some degree. This is generally not correlated with the environment.

Oval orbits

In order to describe streaming on oval orbits a few cases have been considered:

1) closed elliptical orbits (harmonic potential),
2) closed orbits in a squashed logarithmic potential, and
3) closed orbits in the vertical plane of a Miyamoto disk;

they are subtly different in their orbit shape, or "ovalness", and the rate of change of velocity along the oval. The orbits in case 1) can be generated analytically, and are exact ellipses in position and velocity space, whereas in cases 2) and 3) the orbits had to be generated using a periodic orbit-finding algorithm (see, e.g., Teuben & Sanders

* From here on we shall use the term *oval* to denote anything non-circular, non-expanding but bi-symmetric. This includes the frequently used term *elliptical*.

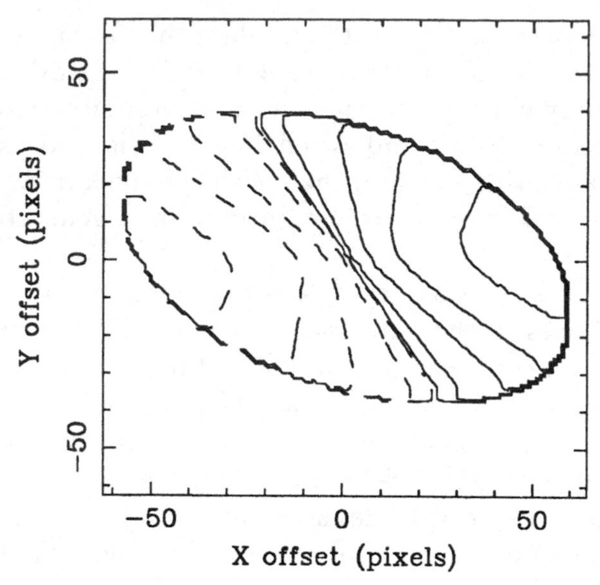

Fig. 1. Velocity field of projected set of epicyclic orbits (see text). Model with $Q = 0.6$, $i = 60°$, $\theta_0 = -60°$.

1985). For the set of projected elliptical orbits various quantities can be derived analytically (see Appendix).

In Figure 1 a contour diagram of the velocity field of a set of elliptical orbits at an intermediate viewing angle is plotted. The line of nodes is along the X axis. Note that the kinematic major axis also falls along the line of nodes in this particular model; this is generally not the case.

The numerical orbits in 2) and 3) deviate only a little from elliptical (e.g., see Fig. 2 in Binney and Spergel's 1982 paper on Fourier analysis of orbits). This means that when the major axis of an orbit is aligned with or perpendicular to the line of nodes, the velocity field is, for all practical purposes, indistinguishable from axisymmetric, though at a different interpreted inclination. Morphologically one would obtain

$$\cos i_{morph} = Q \cos i \tag{1}$$

for orbits along the line of nodes and

$$\cos i_{morph} = \frac{Q}{\cos i} \qquad Q < \cos i$$

and

$$= \frac{\cos i}{Q} \qquad Q > \cos i$$

for orbits perpendicular to the line of nodes in these two cases. Q

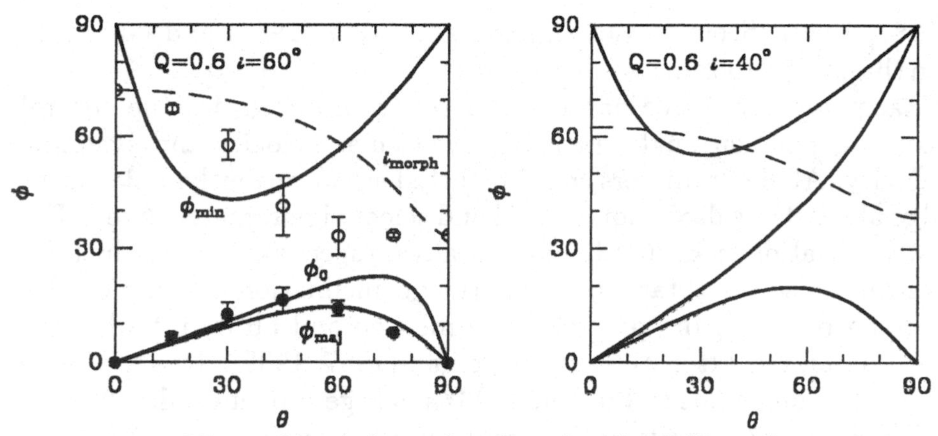

Fig. 2. Observed ϕ_0 and i as function of angle of ellipse, θ_0, away from line of nodes. Model with $Q = 0.6$; a) $i = 60°$ and b) $i = 40°$.

is the intrinsic axis ratio of the orbits. It can be shown that the *kinematic inclination* (see next section) is the same as the *morphological inclination* for elliptical orbits and a flat rotation curve. When the position angle of the ellipse is not 0° or 90° away from the line of nodes this is not the case, as can be seen in Figure 2.

Figure 2 shows how the apparent morphological and kinematical axes vary as functions of the position angle of the ellipse in the galactic plane away from the line of nodes. Two viewing angles are considered. The solid and open points are obtained from a least squares fit, as will be discussed in the next section.

At this point it is perhaps interesting to note that Rubin (1983) commented that all galaxies with *kinematical warps* known until then became more edge-on with increasing radius. Triaxial potentials could perhaps be a more natural explanation for this phenomenon.

The situation with changing position angle is more random. Assuming trailing spiral arms, the twist in the velocity field can also be interpreted as large scale inflow or outflow. In a sample of about a dozen galaxies one would interpret an equal amount with inflow and outflow.

Tilted-Ring Model Fit

To derive a rotation curve and other kinematic parameters from an observed two-dimensional velocity field the tilted-ring fit method is commonly used. This method was initially developed by Warner et al. (1973), and later refined by Bosma (1978, 1981) and Begeman

(1987, 1989). See also van Moorsel and Wells (1985) and Guhathakurtha et al. (1988).

Each ring has 6 free parameters: two coordinates for the central position, position angle, inclination, circular velocity and systemic velocity. All 6 parameters could be fitted simultaneously, but knowledge about the galaxy can be used and selected parameters fixed. For low inclination ($i < 40°$) it may be advantageous to fix the inclination, because the rotational velocity and inclination become tightly coupled in the fitting process. The method and all its pitfalls have been described extensively by Begeman (1987, 1989). It is available under the name ROTCUR in the GIPSY package (Allen et al. 1985).

The least-squares fitting is done to the function

$$v_{obs}(x,y) = V_{sys} + V_{rot}(R)\cos\theta\sin i \tag{2}$$

with

$$\cos\theta = \frac{-(x-X_0)\sin\phi_0 + (y-Y_0)cos\phi_0}{R} \tag{3a}$$

and

$$\sin\theta = \frac{-(x-X_0)cos\phi_0 - (y-Y_0)sin\phi_0}{R\cos i} \tag{3b}$$

Here $v_{obs}(x,y)$ denotes the observed radial velocity at (rectangular) sky coordinates x and y, V_{sys} the systemic velocity, V_{rot} the rotational velocity, i the inclination angle, θ the azimuthal distance from the major axis and R the galactocentric distance; θ and R are measured in the plane of the galaxy. the ring the parameters $V_{sys}, V_{rot}, i, \phi_0, X_0$ and Y_0. position angle ϕ_0 of the major axis is defined as the angle, sky and the major axis of the receding half of the galaxy. If the inclination is derived from the velocity field, we call it the *kinematic inclination* (i_{kin}).

Errors in the kinematic parameters are easily recognized, as the most common ones have a very systematic pattern in the residual velocity field (see van der Kruit and Allen 1978). An error in the position angle shows up as as a $\sin\theta$ residual along a galactic circle, whereas an error in the inclination results in a $\cos 3\theta$ residual.

Figure 3 shows typical residuals along a galactic circle for an oval with $Q = 0.9$; a mildly triaxial case. The mean rotation velocity in this model is 80. The upper left panel shows the best least squares fit as obtained with ROTCUR, whereas the other three panels show residuals when one of the fit parameters was deliberately offset by a small amount. All data within 40° on either side of the minor axis ($\theta = \pm 90°$) were excluded from the fit.

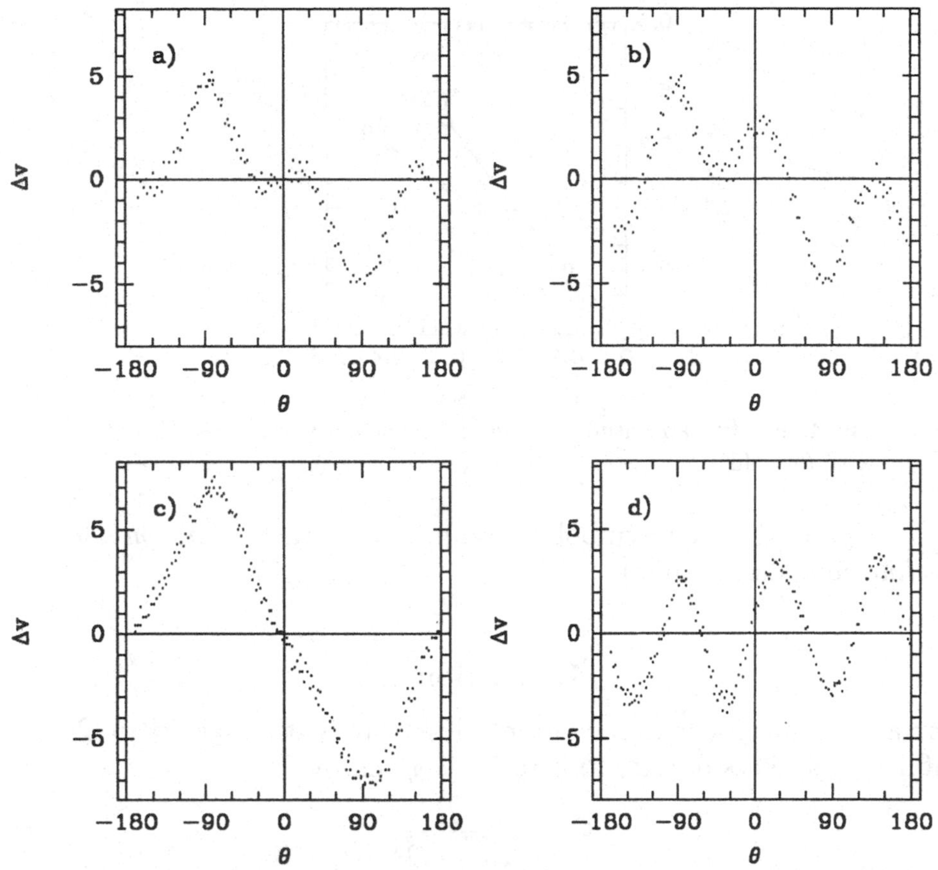

Fig. 3. Velocity residuals along a galactic circle (−180°–180°). a) best fit i and ϕ_0; b) $i+3$; c) ϕ_0-3; d) ϕ_0+3.

What is clearly visible are the $O(\sin\theta - \sin 3\theta)$ residuals in the best fit, which have a unique signature and are different from the ones introduced by errors in the known parameters, such as i and ϕ_0.

Figure 4 shows how the amplitude of this residual depends on the eccentricity of the orbits. The magnitude of the residuals also depends slightly on the inclination of the disk. For any given velocity field and set of residual velocity plots one can then put an upper limit on the shape of the orbits.

For an extreme oval ($Q = 0.6$) and a number of relative orientations (θ_0) the velocity field was fitted using ROTCUR as if it were axisymmetric. The kinematic position angle and inclination were derived and added to Figure 2a. The best-fit kinematic major axis (ϕ_{maj}, filled circles) falls more or less on the curve derived for an ellipse with constant velocity (bottom solid curve). The *kinematic inclination*

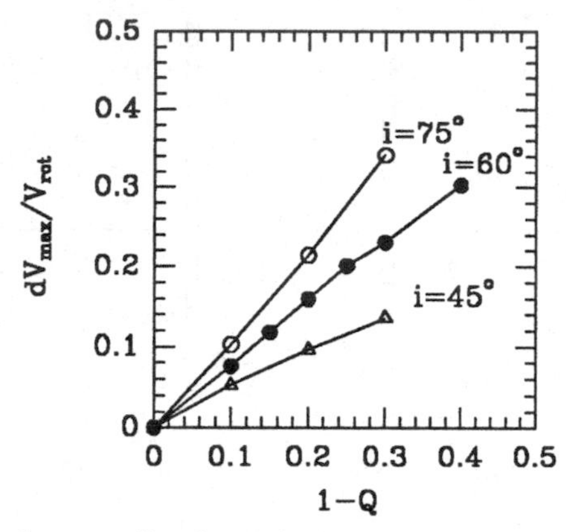

Fig. 4. Relative amplitude of $\sin\theta - \sin 3\theta$ velocity residuals. Model with $\theta_0 = 45°$.

is also plotted (open symbols); compare it with the *morphological inclination* (dashed line).

Discussion

For a squashed logarithmic potential with potential axis ratio Q_p at infinity, the mass density axis ratio, is given by

$$Q_d = \frac{Q_p^2}{\sqrt{2 - Q_p^2}} \tag{4}$$

The orbits in this potential have an axis ratio Q_o very nearly Q_p, perpendicular to the mass distribution (orbits do not become circular at infinity owing to the infinite total mass of the logarithmic potential), see also Gerhard and Vietri (1986).

If an analysis of the velocity field, as described above, results in an upper limit of 10% in the velocity residuals, a lower limit on the axis ratio of the orbits can be read off in Figure 4 as $Q_p \approx 0.92$. The corresponding axis ratio of the density in the midplane would then be $Q_d \approx 0.79$.

On the other hand, if Katz finds that the density axis ratio in his collapsed halo is $Q_d = 0.9$, the axis ratio of the potential (again using eq. (4)) would be 0.965 (which is also that of the orbits) and hence the typical $\sin 3\theta$ velocity residuals would be 3%, for favorable viewing angles. This would be difficult to detect with most current data sets, but not impossible.

We conclude that the *existing* tilted-ring fit provides a sensitive method to investigate residuals along apparent galactic circles. The velocity residuals for oval orbits show systematic $\sin 3\theta$ deviations, with an underlying $\sin\theta$ curvature due to an expected wrong fit of the position angle. A small modification to the fitting procedure should make it possible to estimate the 7th kinematic parameter, Q_o. How this relates to the axis ratio of the potential or that of the density is a totally different story. A rotating potential introduces resonances, which can increase the eccentricity of the orbits, especially near the Lindblad resonances. A number of interesting observed galaxies, plus the appropriate statistical treatment will be discussed in a forthcoming paper by Teuben (1991).

Acknowledgements

I thank enlightening discussions with and constructive comments from Leo Blitz, Tim de Zeeuw, Lee Mundy, Bikram Phookun, and Martin Schwarzschild. I also wish to thank all those of my colleagues who created the GIPSY and NEMO packages, without which this work would not have been so easy.

References

Allen, R.J., Ekers, R.D. & Terlouw, J.P. (1985). In *Data Analysis in Astronomy* (eds. V. di Gesú et al.). London: Plenum.

Begeman, K. G. (1987). PhD Thesis, University of Groningen.

Begeman, K. G. (1989). Astron. Astrophys. **223**, 47.

Binney, J.J. & Spergel, D. (1982). Astrophys. J. **252**, 308.

Blitz, L. & Spergel, D. (1990). Astrophys. J., in press.

Bosma, A. (1978). PhD Thesis, University of Groningen.

Bosma, A. (1981). Astron. J. **86**, 1791.

Braun, R. (1990). preprint.

de Zeeuw, P.T. (1991). This volume.

Frenk, C.S., White, S.D.M., Davis, M. & Efstathiou, G. (1988). Astrophys. J. **327**, 507.

Gerhard, O.E. & Vietri, M. (1986). Mon. Not. R. Astron. Soc. **223**, 377.

Guhathakurtha, P., van Gorkom, J.H., Kotanyi, C.G. & Balkowsky, C. (1988). Astron. J. **96**, 851.

Rubin. V. (1983). In *Internal Kinematics and Dynamics of Galaxies*, IAU Symp. 100 (ed. E. Athanassoula), discussion in Bosma's contribution, p. 21. Dordrecht: Reidel.

Milgrom, M. (1983). Astrophys. J. **270**, 371.

Teuben, P.J. & Sanders, R.H. (1985). Mon. Not. R. Astron. Soc. **212**, 257.

Teuben, P.J. (1991). In preparation.

Warner, P.J., Wright, M.C.H. & Baldwin, J.E. (1973). Mon. Not. R. Astron. Soc. **163**, 163.

Valentijn, E. (1990). Nature **346**, 153.
van der Kruit, P.C. & Allen, R.J. (1978). Annu. Rev. Astron. Astrophys. **16**, 103.
van Moorsel, G.A. & Wells, D.C. (1985). Astron. J. **90**, 1038.

Appendix

In this Appendix we will derive a few handy relationships for exact elliptical orbits seen in projection. Consider an ellipse with axis ratio Q and major axis at a position angle θ_0 from the X axis.

Defining the X axis as being the line of nodes, an inclination i for the ellipse yields an apparent major axis position angle ϕ_0 (away from the line of nodes) and apparent axis ratio q in the projected ellipse. Intuitively one expects ϕ_0 to be always smaller than θ_0; one finds

$$\tan 2\phi_0 = \frac{2B\cos i}{A\cos^2 i - C} \tag{A1}$$

and

$$q^2 = \frac{(A\cos^2 i + C)\cos 2\phi_0 + (A\cos^2 i - C)}{(A\cos^2 i + C)\cos 2\phi_0 - (A\cos^2 i - C)}, \tag{A2}$$

where we have used the coefficients

$$\begin{aligned} A &= Q^2\cos^2\theta_0 + \sin^2\theta_0, \\ B &= (1-Q^2)\sin\theta_0\cos\theta_0, \\ C &= Q^2\sin^2\theta_0 + \cos^2\theta_0 \end{aligned} \tag{A3}$$

from the equation of the unprojected ellipse, $Ax^2 - 2Bxy + Cy^2 = 1$. When the term *morphological inclination* is used, we have used the apparent axis ratio, q, and assumed it was a circle in projection:

$$\cos i_{morph} = q\,. \tag{A2b}$$

The position angle of the kinematical minor axis can be derived by a simple geometrical construction: the tangent along the projected ellipse w.r.t. the X axis defines the point where the radial velocity is zero. The same argument can be applied to finding the kinematic major axis, but since this defines where the maximum velocity is achieved, we also need to make an assumption for how the velocity along the ellipse varies. We shall consider two cases.

First of all we write the projected ellipse also as $ax^2 - 2bxy + cy^2 = 1$ (note the lower case coefficients). Taking the tangent ($y = constant$) along the ellipse the angle of the kinematical minor axis away from the true minor axis (the Y axis) is found to be

$$\tan\Delta\phi_{min} = -\frac{b}{a} = -\frac{\sin\phi_0\cos\phi_0(1-q^2)}{q^2\sin^2\phi_0 + \cos^2\phi_0}\,. \tag{A4}$$

which is 0 for $\theta_0 = 0°$ and $\theta_0 = 90°$. If we assume the velocity along the ellipse to be constant, the other tangent (x = *constant*) defines where the kinematic major axis crosses the ellipse. One then finds that the angle of the kinematical major axis away from the true major axis (the X axis) is given by

$$\tan \Delta\phi_{maj} = \frac{b}{c} = \frac{\sin\phi_0 \cos\phi_0 (1 - q^2)}{q^2 \cos^2\phi_0 + \sin^2\phi_0} . \tag{A5}$$

Since the velocity along an ellipse generally increases towards the minor axis, the angle derived in eq. (A5) will generally be smaller. In fact, for epicyclic orbits it can be shown that the kinematic major axis is always along the line of nodes, i.e., $\Delta\phi_{maj} = 0$, which is perhaps a surprising result.

Figure 2 shows all the relevant angles as a function of θ_0 for a case where $Q > \cos i$ (left diagram) and $Q < \cos i$ (right diagram).

Modeling the atomic gas in NGC 4278

JOANNA F. LEES
Princeton University Observatory

Abstract

Using VLA data on the extended atomic gas disk in the elliptical galaxy NGC 4278, we calculate kinematic models for the velocity distribution. Axisymmetric models cannot reproduce a number of important features in the velocity field, such as the kinking of the zero-velocity contour near the nucleus. A triaxial model is presented that fits this and other peculiarities very well.

Introduction

Extended disks and rings of atomic gas have now been detected and mapped in a number of elliptical galaxies (van Gorkom et al. 1986; Lake, Schommer and van Gorkom 1987; Kim et al. 1988; Schweizer, van Gorkom and Seitzer 1989). The kinematics of the gas in these systems allows a unique opportunity to probe the three-dimensional mass distribution of an elliptical galaxy.

NGC 4278 is an optically "normal" E1 galaxy located in a small group at a distance of 16.4 Mpc. It has an optical heliocentric velocity of 630 km s^{-1}, and an absolute magnitude $M_B = -20$. A few weak dust patches are evident near the nucleus, as well as ionized gas extending to a radius of over five kiloparsecs with a LINER-type emission-line spectrum (Osterbrock 1960; Demoulin-Ulrich et al. 1984). It also is a fairly bright (about 500 mJy at 21 cm) compact flat-spectrum radio source. The extended HI disk was previously mapped at Westerbork by Raimond et al. (1981).

NGC 4283, an E0 galaxy at a projected separation of 3.′4 and an absolute magnitude of $M_B = -18$, and NGC 4286, an S0 galaxy at a distance of 8′ and $M_B = -16$, are both close neighbours, and the Sab galaxy, NGC 4274, is also located about 20′ north.

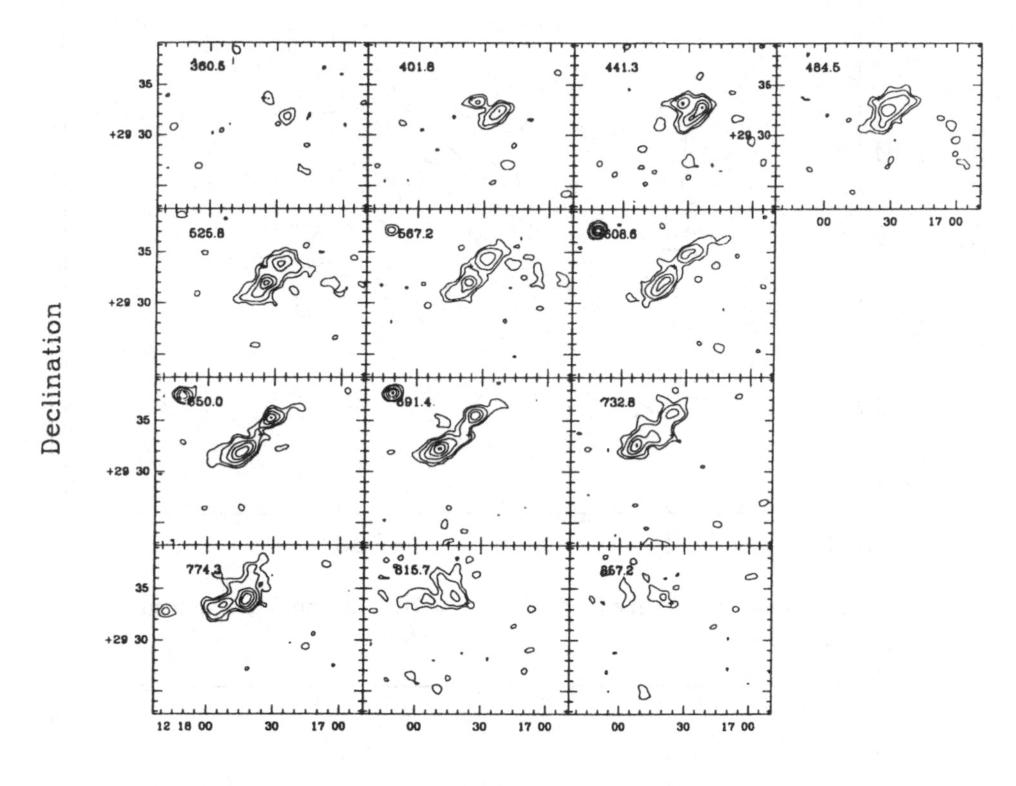

Fig. 1. Channel maps showing emission from NGC 4278 (center), and NGC 4286 (upper left). The velocities are shown in the top left of each channel, and the contours are 0.0005, 0.001, 0.002, ..., 0.005 Jy beam^{-1}.

Observations

The HI observations were made on 3–5 April 1987 with the VLA (Lees, van Gorkom and Knapp 1990). The spatial and velocity resolutions were $64'' \times 55''$ and $41.4\,\mathrm{km\,s^{-1}}$ respectively.

The final channel maps showing emission from NGC 4278 and NGC 4286 are displayed in Figure 1. As is evident from the velocity field of Figure 2, the emission from NGC 4278 is that of an inclined, rotating disk. The projected rotational velocity is about $220\,\mathrm{km\,s^{-1}}$, and the rotation curve appears to rise to the largest detectable radius of more than thirty kiloparsecs, twice the optical light radius, implying an increasing mass-to-light ratio in the outer parts of the galaxy.

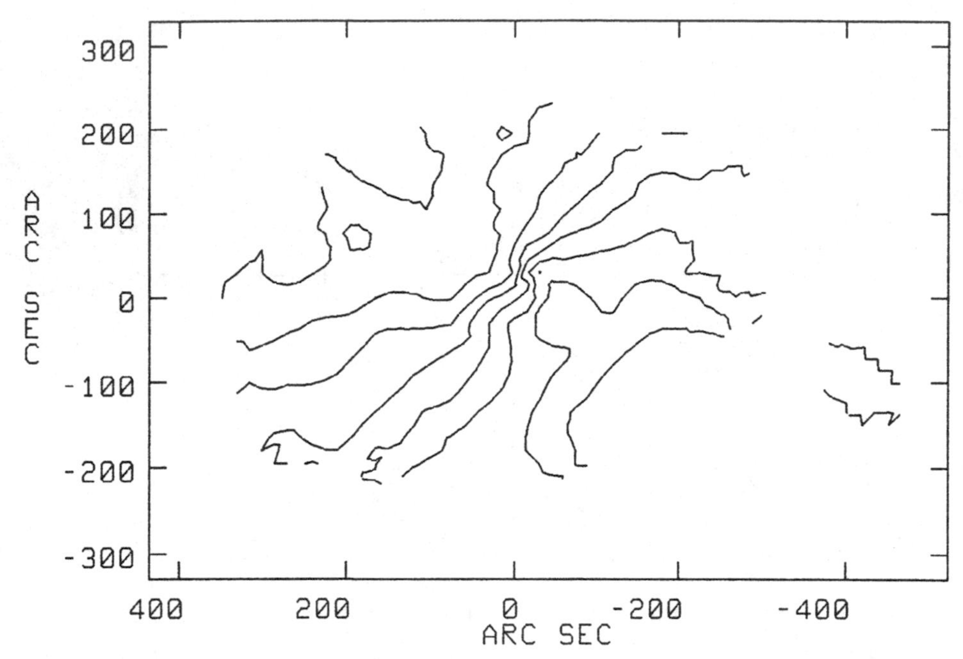

Fig. 2. Observed velocity field of NGC 4278. Contours are spaced by 50 km s^{-1}.

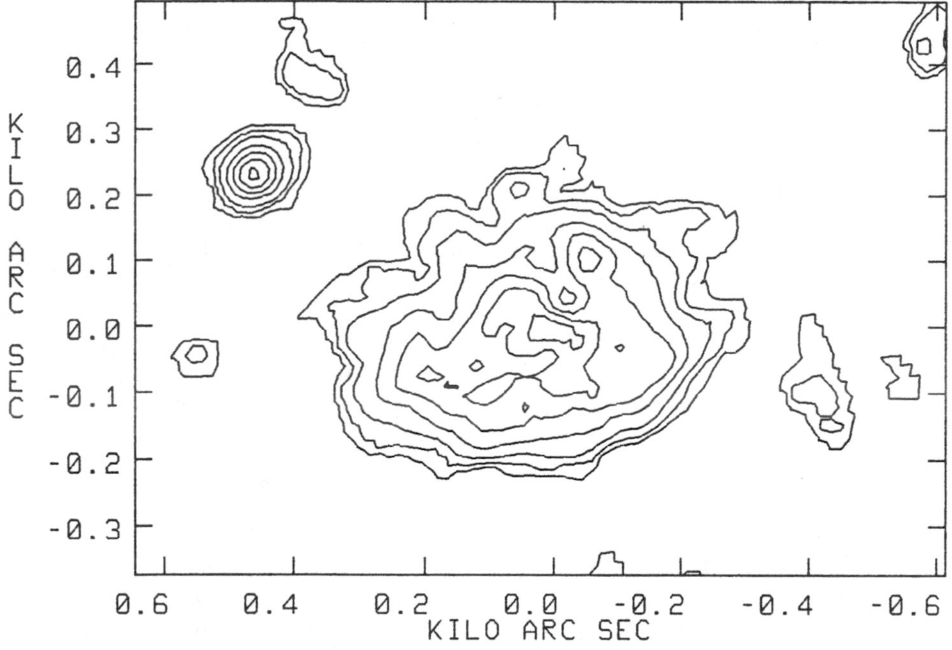

Fig. 3. Total HI surface density in NGC 4278 and NGC 4286. The contours are 20, 50, 100, 200, ..., 500 Jy beam^{-1} m s^{-1}.

The total HI surface density is displayed in Figure 3, where it can be seen that the HI distribution is quite clumpy, and in fact extended perpendicularly to the kinematic major axis. Some very extended emission to the southeast, as much as 50 kpc from the nucleus, is evident in the channel maps at 480–570 km s^{-1}.

Axisymmetric mass models

As a first fit to the data, we assumed that the gas was on circular orbits in a plane with constant inclination angle, i, and with a rotation curve of the form

$$v(R) = \frac{v_{max}\left(\frac{R}{R_{max}}\right)}{\left(\frac{1}{3}+\frac{2}{3}\left(\frac{R}{R_{max}}\right)^p\right)^{\frac{3}{2p}}}.$$

The free parameters are the central position of the galaxy, (x_c, y_c); the position angle of the gas kinematic major axis, $\phi_{\rm HI}$; the inclination of the gas disk, i; the galaxy's systematic velocity, $v_{\rm HI}$; the maximum rotational velocity, v_{max}; the turnover radius of the rotation curve, R_{max}; the rotation curve index in the above formula, p; and the maximum gas surface density, $N_{\rm HI}$. For simplicity, we assumed a constant surface density for the disk (since the gas is so clumpy anyway), and arbitrarily truncated it at a radius of five arcminutes. We then did a non-linear least-squares fit to the data cube using the subroutine CUBIT, written by J.A. Irwin and W.D. Cotton, in the Astronomical Image Processing System (AIPS) software package, using the Levenberg-Marquardt algorithm. This program simultaneously models the HI density and velocity distributions in the entire data cube (rather than just the first and second moments).

The final parameters of the fit are displayed in Table 1. Most of the parameters were found to be quite insensitive to the exact initial guess, the precise form of the rotation curve, the form of the density distribution, or the area of the disk that was included in the fit. However, the inclination angle is quite poorly constrained by the data, with the final value varying between 20° and 50°. The central velocity and position agree well with the optical values, and the kinematic major axis position angle is equal to that of the ionized gas of 40° (Demoulin-Ulrich et al. 1984), but is 30° away from the stellar kinematic major axis at 13° (Davies and Birkinshaw 1988). The rotation curve does not begin to fall until over 3′ from the nucleus, whereas the effective radius of the optical light is only 30″. The

Table 1. *Best-fit parameters for the axisymmetric model.*

Central position, (x_c, y_c)	$12^h17^m40^s$ $29°32'26''$
Major axis position angle, ϕ_{HI}	46°
Inclination, i	$\sim 33°$
Systemic velocity, v_{HI}	$636\,km\,s^{-1}$
Rotational velocity, v_{max}	$\sim 450\,km\,s^{-1}$
Turnover radius, R_{max}	202″
Rotation curve index, p	0.6

mass-to-light ratio increases by a factor of 25 from the nucleus to the outer edge of the HI disk.

The model channel maps and velocity field, smoothed to the resolution of the data, are shown in Figures 4 and 5 respectively. Both show similar overall features to Figures 1 and 2. However, a few differences should be noted. Firstly, since the data are much less extended along the major axis, the emission is systematically too strong in the model at very high and low relative velocities. Another significant difference can be identified in the twisting of the velocity contours of the data near the nucleus of the galaxy, so that the observed kinematic major axis of the atomic gas varies with radius, from almost 70° in the nucleus to 40° at larger radii. This cannot be reconciled with an axisymmetric, unwarped model for the gas disk.

Thus, there are four major problems with an equilibrium axisymmetric mass model for NGC 4278. One is the twist in the zero-velocity contour near the nucleus; secondly, the atomic gas is extended along the kinematic minor axis of the HI; third, there is a 30° misalignment between the projected angular momentum vectors of the gas and stars; and, finally, is the fact that the stellar isophotes twist by a full 20° with radius (Peletier 1989).

Triaxial mass models

One solution to the above discrepancies between the data and the model is to choose a triaxial potential, which can in principle solve all four points raised in the previous section (de Zeeuw and Franx 1989). To this end, we chose to use a modified version of the potential that Hernquist (1990) pointed out gives a very good agreement with the

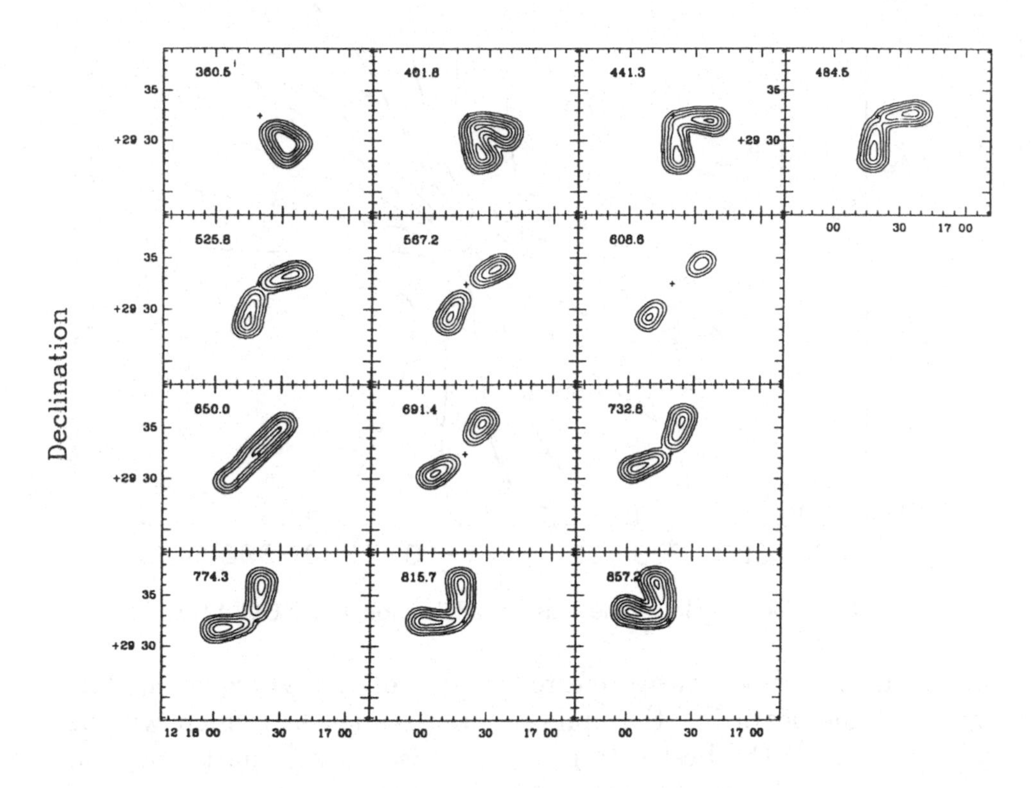

Fig. 4. Channel maps produced from the axisymmetric model of NGC 4278 of Table 1.

de Vaucouleurs law when projected. The potential is of the form

$$V(r,\theta,\phi) = u(r) - v(r)P_2^0(\theta,\phi) + w(r)P_2^2(\theta,\phi),$$

where P_2^0 and P_2^2 are the spherical harmonic functions, $u(r)$ is the spherical Hernquist potential,

$$u(r) = -\frac{GM}{r + r_0},$$

and $v(r)$ and $w(r)$ are of the form

$$v(r) = -\frac{GMr_1r}{(r+r_2)^3}, \qquad w(r) = -\frac{GMr_3r}{(r+r_4)^3}.$$

Here, M is the total galaxy mass, and r_0,...,r_4 are constants to be fitted to the data (see Lees and de Zeeuw, in preparation, for general properties of the kinematics of cold gas in potentials of this form). To fit the gas kinematics, we then assume that the gas lies in closed

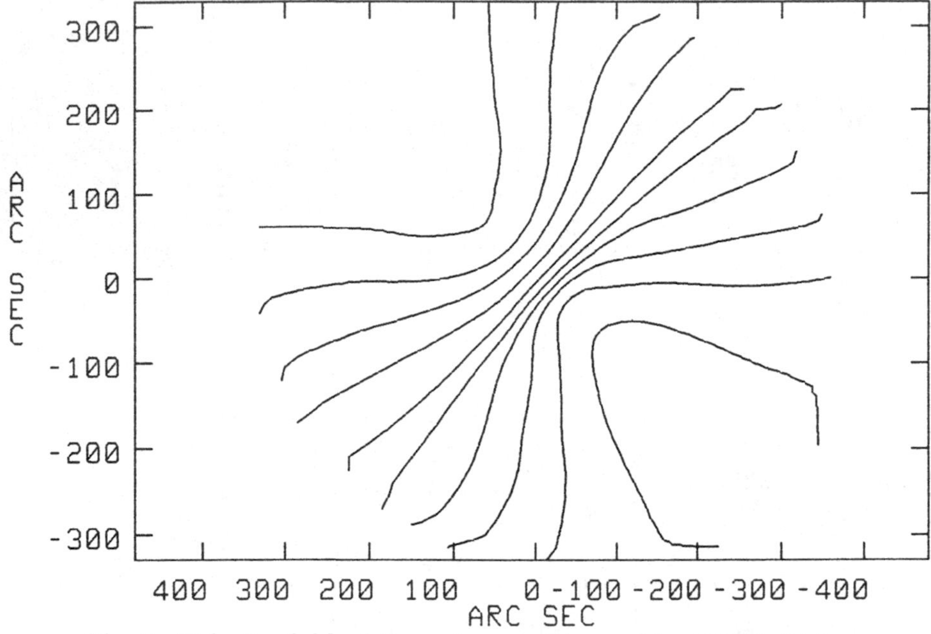

Fig. 5. Velocity field of the axisymmetric model of NGC 4278.

orbits in one of the two preferred planes of the galaxy, calculate orbital velocities using the epicycle approximation (which will be valid so long as the best-fitting potential is not very flattened, and can be checked later by numerical orbit integration; see for example Gerhard and Vietri 1986), and then project to compare with the data. As with the axisymmetric model discussed in the previous section, we performed a non-linear least-squares fit to the entire data cube. In this case, the free parameters are (x_c, y_c), $\phi_{\rm HI}$, $v_{\rm HI}$, $N_{\rm HI}$ (these parameters all being the same as in the axisymmetric case), as well as a scaling velocity, v_{scale} (related to the total mass M, and similar to v_{max} in the axisymmetric model); the scale length for the mass model, r_0 (again, similar to R_{max} from before); the viewing angles, i (the inclination, as in the previous section) and ϕ; and finally the axis ratios of the density distribution at the center, p_0 and q_0, and at large radii, p_∞ and q_∞ (which can be expressed in terms of r_1,...,r_4). Thus, we have four more parameters than in the axisymmetric case. (We could make this less if we were to constrain the axis ratios at large radii, either as being equal to the axis ratios at the center — although such a model may not be self-consistent (e.g., , Miralda-Escudé and Schwarzschild 1989) — or as being axisymmetric or spherical).

Table 2. *Best-fit parameters for the triaxial model.*

Central position, (x_c, y_c)	$12^h17^m40^s$
	$29°32'26''$
Major axis position angle, $\phi_{\rm HI}$	43°
Systemic velocity, $v_{\rm HI}$	$636\,{\rm km\,s^{-1}}$
Mass scale length, r_0	$5.0'$
Viewing angles, i	46°
ϕ	44°
Axis ratios at center, p_0	0.86
q_0	0.64
Axis ratios at infinity, p_∞	0.90
q_∞	0.85

The final parameters for the fit to the HI are shown in Table 2. Model channel maps are displayed in Figure 6, and the velocity field in Figure 7. It is clear that the triaxial model is indeed able to reproduce most of the features of the data that could not be explained with the axisymmetric model. The main point is that the velocity contours twist in the nuclear regions, in a manner identical to that seen in the data. This feature is also apparent in other ellipticals, such as NGC 5666 (Lake et al. 1987; van Gorkom 1990, private communication). To agree with the extent of this twisting, however, we need quite a large scale length for the mass distribution: $R_{1/2\ {\rm mass}} \sim 1.8 r_0 \sim 9'$, 18 times the half-light radius. This is consistent with the observation of a rising rotation curve mentioned previously. However, in this case, the mass-to-light ratio only varies by a factor of six over the HI disk, four times less than for the axisymmetric model. We also note that there is a 10° offset between the major axis of the gas and the stellar isophotes, and that the projected axis ratio, $a'/b' \sim 0.85$, is quite similar to that observed, without any *explicit* fit to this parameter.

In conclusion, then, this triaxial model is able to reproduce the gross velocity features of the atomic gas disk in NGC 4278 quite well, and also some of the optical ones. It should be possible to use the constraints imposed by these other observables in the fit, such as the apparent axis ratio of the stellar isophotes, the position angle of the major axis of the isophotes, but we have not yet included these. We also hope to further constrain the velocity field in the central regions, where the HI resolution is poor, using recent observations

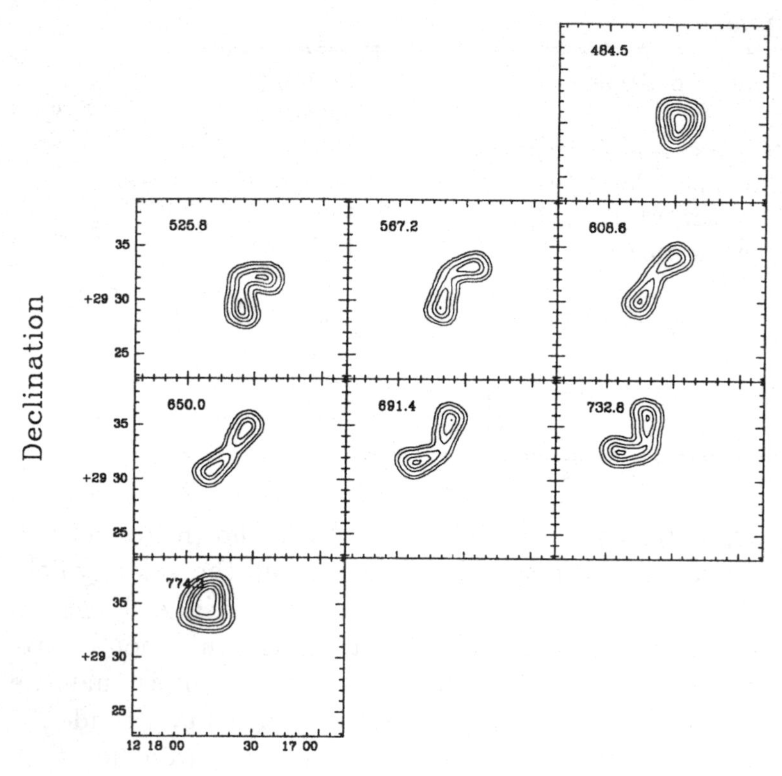

Fig. 6. Channel maps produced from the triaxial model of NGC 4278 of Table 2.

of the ionized gas (Lees 1991, in preparation), and to apply this technique to other ellipticals with extended atomic disks.

Acknowledgements

The author would like to thank Jacqueline van Gorkom and Jill Knapp for the use of the HI data, and Tim de Zeeuw for his useful suggestions and help with the models.

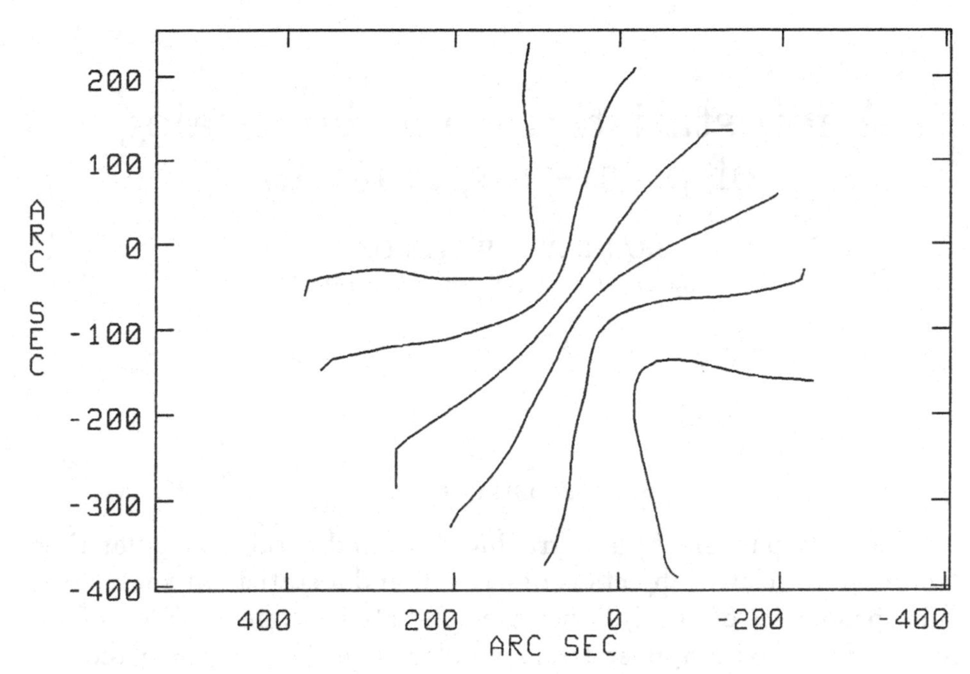

Fig. 7. Velocity field of the triaxial model of NGC 4278.

References

Davies, R.L. & Birkinshaw, M. (1988). Astrophys. J., Suppl. Ser. **68**, 409.

Demoulin-Ulrich, M.-H., Butcher, H.R. & Boksenberg, A. (1984). Astrophys. J. **285**, 527.

de Zeeuw, T. & Franx, M. (1989). Astrophys. J. **343**, 617.

Gerhard, O.E. & Vietri, M (1986). Mon. Not. R. Astron. Soc. **223**, 377.

Hernquist, L. (1990). Astrophys. J. **356**, 359.

Kim, D.-W., Guhathakurta, P., van Gorkom, J.H., Jura, M. & Knapp, G.R. (1988). Astrophys. J. **330**, 684.

Lake, G., Schommer, R.A. & van Gorkom, J.H. (1987). Astrophys. J. **314**, 57.

Lees, J.F., van Gorkom, J.H. & Knapp, G.R. (1990). Bull. Am. Astron. Soc. **21**, 1178.

Miralda-Escudé, J. & Schwarzschild, M. (1989). Astrophys. J. **339**, 752.

Osterbrock, D.E. (1960). Astrophys. J. **132**, 325.

Peletier, R. (1989). PhD Thesis, University of Groningen.

Raimond, E., Faber, S.M., Gallagher, J.S. & Knapp, G.R. (1981). Astrophys. J. **246**, 708.

Schweizer, F., van Gorkom, J.H. & Seitzer, P. (1989). Astrophys. J. **338**, 770.

van Gorkom, J.H., Knapp, G.R., Raimond, E., Faber, S.M. & Gallagher, J.S. (1986). Astron. J. **91**, 791.

A few statistics from the catalog of polar-ring galaxies

BRADLEY C. WHITMORE
Space Telescope Science Institute

Abstract

A recently published photographic atlas and catalog of polar-ring galaxies and related objects is described, and a statistical analysis of some properties of the galaxies are tabulated. Roughly 0.5% of all nearby S0 galaxies appear to have polar rings. When corrected for various selection effects (e.g., non-optimal viewing orientation, possible dimming or limited lifetime of the ring) the percentage increases to about 5% of all S0 galaxies which now have, or have had in the past, a polar ring bright enough to make it into the catalog. The angle between the major axis of the two components shows a strong preference for being nearly perpendicular. However, deviations of 10–30° degrees are not uncommon. Polar rings show a variety of shapes; roughly half are narrow like a wedding ring while the other half are extended annuli with the central region cut out. Roughly two-thirds of the rings show some obvious warping; of these, roughly half have an integral sign type warp, which is typical of most warped disk galaxies, while the other half show a peculiar banana-shaped warp. A fair correlation of the asymmetry of the ring with the presence of outer debris, and a weak correlation of the angle between the ring and disk with the ring asymmetry, offer the possibility of determining the approximate age of a polar-ring galaxy based on its morphological appearance.

I. Introduction

The discovery of polar-ring galaxies has provided an unexpected but very welcome opportunity to study galaxies in new ways. These systems generally consist of a normal S0 galaxy with a ring of gas, dust and young stars aligned roughly in a perpendicular orientation with respect to the major axis of the disk. This unique geometry allows us several new insights into the structure and evolution of galaxies.

1. The stability of polar orbits

Three models have been proposed to explain the apparent stability of polar features in polar-ring galaxies. Accretion, either from tidal capture of matter from a nearby system or the merger of a gas rich companion, is generally agreed to be the source of the ring material. This material is spread into an annular disk or ring over a few orbital periods. The galactic gravitational field, dissipation, and the self-gravity of the accreted material play a role in the subsequent evolution of the ring. The proposed models differ in their interpretation of the importance and details of each of these. The *preferred orientation model* (e.g., Steiman-Cameron and Durisen 1982, Steiman-Cameron 1990) proposes that the gravitational potentials of S0 galaxies are slightly triaxial in nature. This small departure from axisymmetry will lead a fraction of all dissipative disks accreted by S0s to settle into a stable polar orientation. In the *ring self-gravity model*, Sparke (1986, 1990) demonstrated that a sufficiently massive ring captured at a high initial inclination in an axisymmetric galaxy modifies the gravitational potential of the galaxy in such a manner that differential precession in the disk can be eliminated, and the system can be quasi-stable. The *statistical selection* model (Schweizer, Whitmore and Rubin 1983) argues that polar rings may not be equilibrium structures, but rather that their structure is an artifact of the initial capture orientation. The apparent ring stability then results from the long timescale for differential precession to destroy the near planar appearance of the rings which are captured at high inclination.

2. The shape of the dark halo

The existence of stars and gas in two nearly perpendicular planes provides the opportunity of mapping the shape of the gravitational potential in three dimensions. Based on the observations of three polar-ring systems, Whitmore, McElroy and Schweizer (1987a, hereafter WMS) find that the massive dark halos around these galaxies are nearly spherical. However, a more detailed set of mass models have been constructed for NGC 4650A by Sackett and Sparke (1990a, b), who find that a more flattened dark halo is also compatible with the data for this particular galaxy.

3. The merger rate

While it is now well established that interacting and merging galaxies exist, we do not know whether they comprise only an interesting sidelight or the fundamental mechanism for determining the structure and dynamics of most galaxies. In order to make this estimate for the particular types of interactions that can form a polar ring we need to know: 1) the fraction of S0 galaxies that currently appear to have polar rings, 2) the correction for non-optimal viewing orientations, and 3) the correction for possible dimming or limited lifetime of polar-rings. Whitmore et al. (1990; hereafter PRC for Polar Ring Catalog) find that about 5% of all S0 galaxies have, or have had in the past, a polar ring bright enough to make it into the catalog. Since it is likely that accretion events with the correct impact parameters to form a polar-ring galaxy are relatively rare, the number of accretion events for S0 galaxies in general is certainly much larger than this 5%.

4. Insight into how other galaxies formed

Several classes of objects have some characteristics which are similar to polar-ring galaxies (see §II). For example, many elliptical galaxies have dust-lanes oriented along their minor axis but have no extended luminous component associated with this material. Could the formation mechanism be the same in both cases but the resulting products appear different because of differences in the gravitational potential of S0 and elliptical galaxies? Another example is for spiral and S0 galaxies which have box-, peanut-, and X-shaped bulges. Whitmore and Bell (1988) have suggested that these systems might represent "failed" polar rings, where the impact was at an intermediate angle rather than being nearly perpendicular.

In the past, only four systems had been observed in enough detail to firmly establish their status as S0 galaxies with polar rings (NGC 2685: Schechter and Gunn 1978; A 0136-0801: Schweizer, Whitmore and Rubin 1983, WMS; NGC 4650A: Schechter, Ulrich and Boksenberg 1984, WMS; ESO 415-G26: WMS). Four other galaxies were listed as good candidates by Schweizer, Whitmore and Rubin (1983) along with 14 possibly related galaxies. Observations reported in PRC show that two of these good candidates (UGC 7576 and UGC 9796 [= II ZW 73]) are also S0 galaxies with polar rings. A third candidate from this list, AM 2020-504, turns out to have an elliptical galaxy as the central component (see §IIIa). The other candidates and related objects have not been observed in enough detail to make this determination.

The PRC provides a large enough sample to examine some of the statistical properties of polar-ring galaxies. What fraction of polar rings are narrow rings and what fraction are more extended annuli? Are the warps similar to those seen in normal galaxies? Is the central component always an S0 galaxy? Is the angle between the two components random, or is there a preference for being nearly perpendicular? These are some of the questions we will address in this article.

II. The Polar Ring Catalog

The PRC is divided into four categories, based on the certainty of the identification. This identification is difficult because a polar-ring galaxy is only obvious if it is seen in a certain orientation on the sky. A graphic illustration of this problem is shown in Figure 1, in which a flat disk and an orthogonal ring, whose inner radius is the same size as the outer radius of the disk, are shown from a variety of viewing orientations. The angle α is the amount the ring has been rotated, while the angle β is the amount the disk was subsequently rotated. Finally, the model is silhouetted to depict a two-dimensional observation on the sky. Only when both ring and disk are seen relatively edge-on is the identification obvious. If the projection angle on the sky is not optimal, a polar-ring galaxy can masquerade as an Sa galaxy (e.g., upper left of Fig. 1) or a barred "theta" galaxy (e.g., lower left of Fig. 1). These cases of mistaken identity could generally be distinguished by detailed kinematic and photometric observations. This figure also indicates that for every polar-ring galaxy we identify, there are one or two others that have been missed because of their orientation on the sky.

The criteria used to categorize the galaxies in the PRC are described below.

Category A - Kinematically Confirmed (6 objects)

1. Spectroscopic evidence must exist for two nearly orthogonal angular momentum vectors with similar amplitudes.

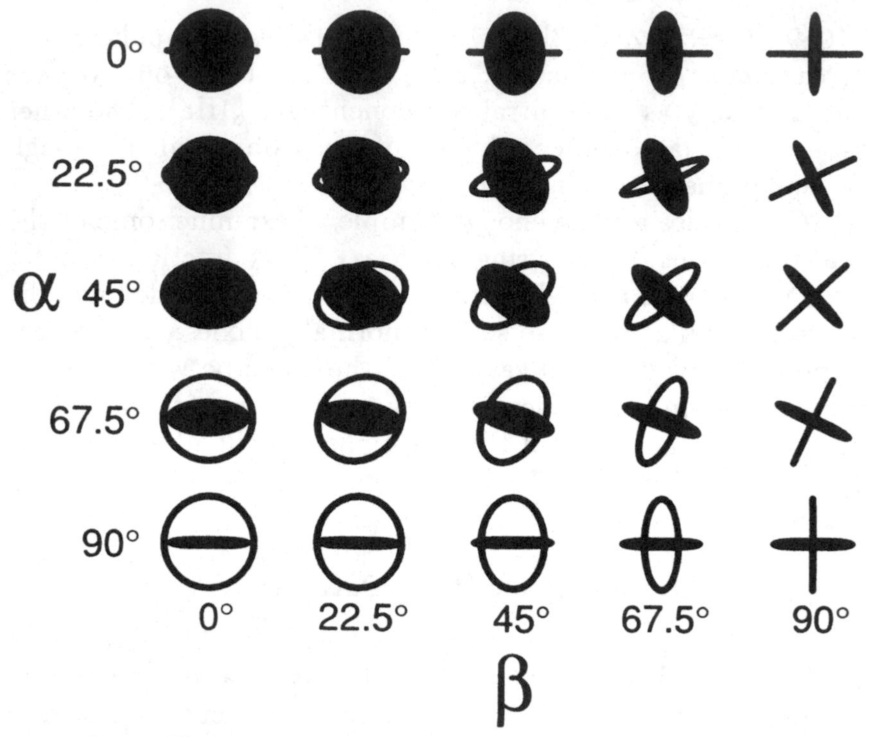

Fig. 1. The appearance of a polar-ring galaxy as seen from a variety of viewing orientations. Note that less than half of the viewing orientations allow easy identification of the system as a polar-ring galaxy.

2. The system must be in an equilibrium or quasi-equilibrium configuration, as shown by the fact that:
   - both components must have similar systemic velocities, and
   - the centers of the two components must be nearly aligned.
3. The ring must be comparable in size to the inner component, must be luminous (i.e., dust-lane ellipticals are not included), and it must be nearly planar (i.e., the chaotic appearance of M 82 would be ruled out).

Category B - Good Candidates (27 objects)

As shown in Figure 1, a polar-ring galaxy can be difficult to distinguish when viewed from certain projection angles, and can look like a fairly normal galaxy (i.e., an edge-on spiral galaxy with an abnormally spherical bulge, or a barred galaxy with ring-like arms). Category B includes systems that are oriented in such a way that we can be fairly sure the system is actually a polar-ring galaxy based on morphological appearance, but kinematic observations have not yet been made to confirm the classification.

1. The projected major axis of the two components must be nearly orthogonal.
2. The centers of the two components must be nearly aligned.
3. The ring must be comparable in size to the inner component, must be luminous, and it must be nearly planar.

Category C - Possible Candidates (73 objects)

The criteria for this category are necessarily more ambiguous. In some cases an interaction or merger is clearly occurring involving material aligned near the minor axis of the other component. The question is whether the resulting product will be a polar ring. In other cases a galaxy may appear to be quite normal with only very subtle clues suggesting that it might actually be a polar ring seen from an orientation which is not advantageous.

Category D - Systems Possibly Related to Polar-Ring Galaxies (42 objects and 8 related classes)

While it is unlikely that these galaxies are polar-ring galaxies, or will ever become polar-ring galaxies, they share some characteristics with them implying that they may have had similar evolutionary histories. In addition to galaxies from our search, we also list classes of galaxies that might be related to polar-ring galaxies, or may be mistaken for polar-ring galaxies. These include:

- dust-lane ellipticals
- box, peanut and X-shaped bulges
- "Mayall-type" objects
- "smoke-ring" galaxies
- superpositions
- "theta" galaxies
- "Hoag-type" galaxies
- kinematically related objects

See the PRC for photographs of examples of galaxies in these classes, and for a discussion of how some of them might be related to polar rings.

III. A few statistics from the Polar Ring Catalog

a) Morphological type of the central component

The central component of all the polar-ring systems in Categories A and B appear to be diffuse systems; either S0 disks or ellipticals. In no case is the central component a spiral galaxy. This can probably be explained by the need for one, or both, of the two components to be free of gas since the orbits intersect. In principle, both components could be gas free. The fact that the outer component contains gas in all seven of the well-observed systems indicates that gaseous dissipation is probably required to form the ring. If the outer ring must necessarily be gaseous in order to settle to a disk, then the inner component cannot be gaseous because of the intersecting orbits.

In six of the seven well-studied polar-ring galaxies the central component has turned out to be a rapidly rotating S0 disk. The exception is AM 2020-504 which has a slowly rotating elliptical for a central component (Whitmore 1987b; note, however, that even this galaxy has a rapidly rotating system in the inner few arcseconds of the central elliptical). The component aligned with the minor axis of dust-lane ellipticals rarely shows an extended luminous component. Perhaps this difference is just due to a difference in the gravitational potential of the two types of systems. For example, the presumably more flattened potential of the S0 systems may be able to stabilize the ring at greater radii than in an elliptical galaxy. The presence of the ring in AM 2020-504 may indicate that this elliptical galaxy is flatter than a typical elliptical.

Another interesting point is that none of the central components appear to have a bar-like structure (Galletta 1990). Perhaps the degree of asymmetry the bar would produce in the gravitational potential is enough to disrupt the formation of a ring in these cases. Computer modelling of this situation should be able to answer this question.

b) Narrow ring or wide annulus?

Polar rings appear to come in two flavors; 1) an annulus (i.e., a disk-like component with the central region cut out), and 2) a narrow ring which is not extended in radius. Figure 2 shows two galaxies which are examples of annular systems (NGC 4650A in the upper left and A0136-0801 in the lower left) and two galaxies which are examples of narrow ring systems (ESO 415-G26 in the upper right and AM 2020-504 in the lower right). In many cases the polar ring component is too edge-on to make this distinction obvious. Table 1 includes a summary of various characteristics for Category A and B galaxies from the PRC. Column 2 is a determination of what type of ring the galaxies have. In the 16 cases where a clear determination can be made, seven are annular rings while nine are narrow rings.

There are no obvious correlations between the width of the polar ring and other properties. For example, while ESO 415-G26 and NGC 2685 have narrow polar-rings and appear to be young systems with chaotic distributions of luminosity and HI, IC 1689 and AM 2020-504 also have narrow rings but appear to be very symmetric.

c) The angle between the ring and the disk

The distribution of angles between the ring and disk components may provide insight into the formation mechanism responsible for making polar-ring galaxies. For example, the statistical selection hypothesis (see §I) would predict a relatively flat distribution, with a gradual tendency for the more oblique angles to be less prevalent since they would decay fastest due to differential precession. The observed distribution is shown in Figure 3. We find a strong tendency for the components to be nearly orthogonal (i.e., 0 degrees in Fig. 3). The distribution for the intermediate angle between 10 and 35 degrees is essentially flat. No conclusions can be drawn beyond 35 degrees since the selection effect of including only objects which are more nearly orthogonal than parallel begins to remove these objects from the sample.

These results appear to support the preferred orientation model (see §I and Steiman-Cameron 1990) since it predicts that the systems should evolve toward the orthogonal orientation. It also suggests the possibility of two separate distributions, with the more nearly orthogonal systems being intrinsically different than the systems with intermediate orientations (e.g., perhaps they are older, and have therefore had more time to settle to the preferred plane). There is a weak correlation of the angle between the ring and disk with the asymmetry of the rings which supports this idea. The more nearly perpendicular rings are more symmetric than the oblique rings. The correlation coefficient is 0.33 with 27 galaxies in the sample. The probability of coming from a random distribution is 8%.

A possible psychological effect to consider would be that in cases that are difficult to measure because one component is relatively face

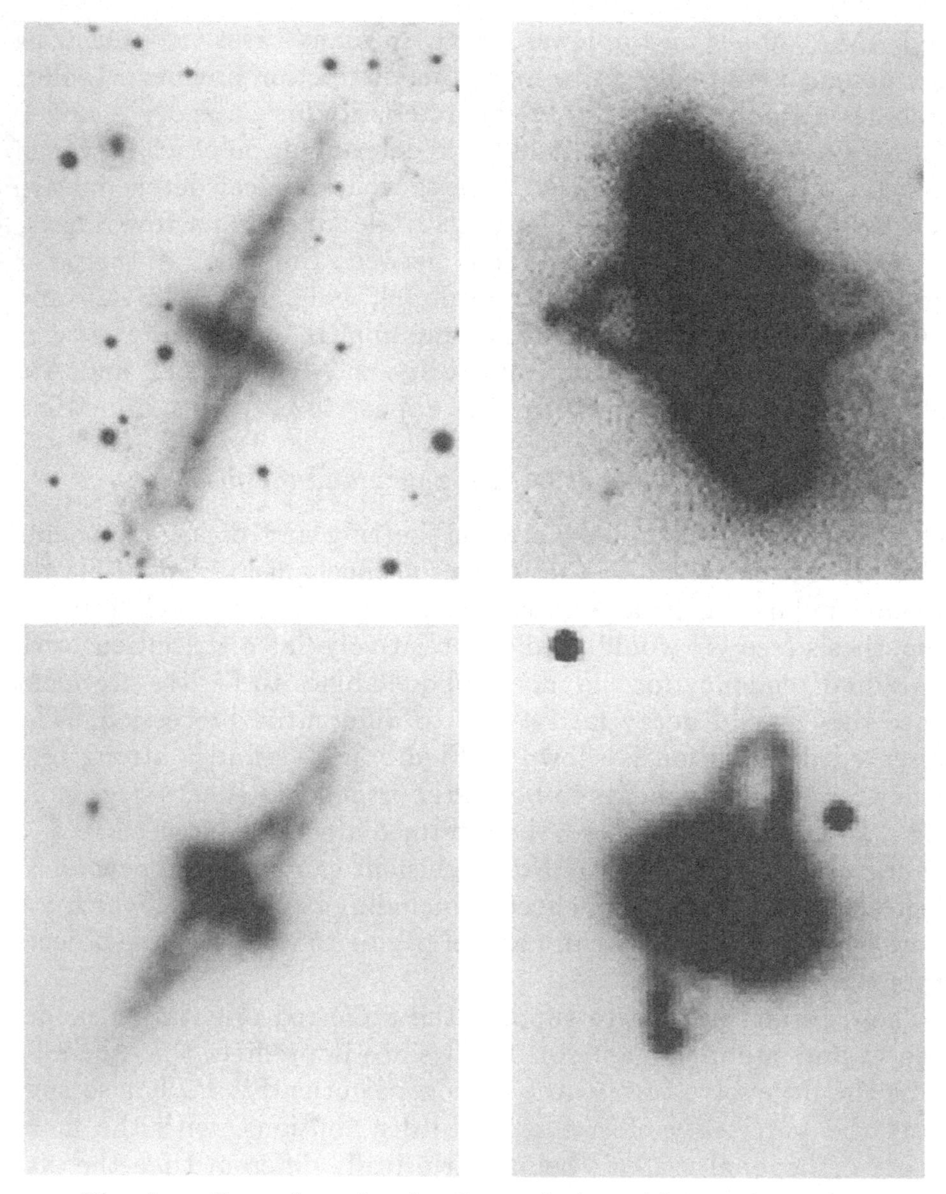

Fig. 2. Examples of polar-ring galaxies with annular rings (NGC 4650A in the upper left and A0136-0801 in the lower left) and with narrow rings (ESO 415-G26 in the upper right and AM 2020-504 in the lower right).

on, the eye may tend to measure a more perpendicular angle. This does not appear to be a problem, however, since the distribution of angles for poor quality determinations (i.e., quality rating of 3 in Table 1) is the same as the total distribution.

Table 1. *Some characteristics of polar-ring galaxies.*

PRC name, other id	Type	Angle (deg)	Type of warp	Ring symmetry	Outer debris
Category A – KINEMATICALLY CONFIRMED					
A-1, A 0136-0801	annulus(1)	7 (1)	integral-inner(1)	1 (1)	1 (1)
A-2, ESO 415-G26	narrow (1)	16 (1)	—	2 (2)	3 (1)
A-3, NGC 2685	narrow (1)	7 (2)	—	3 (1)	3 (1)
A-4, UGC 7576	—	12 (2)	banana (2)	1 (2)	1 (2)
A-5, NGC 4650A	annulus(3)	12 (2)	integral-outer(1)	1 (2)	2 (1)
A-6, UGC 9796, IIZw73	annulus(1)	26 (2)	straight (2)	2 (2)	1 (2)
Category B - GOOD CANDIDATES					
B-1, IC 51	narrow (3)	2 (2)	—	2 (3)	3 (2)
B-2, A 0113-5442	annulus(3)	10 (3)	banana (2)	2 (2)	—
B-3, IC 1689	narrow (2)	2 (2)	—	1 (2)	1 (2)
B-4, A 0336-4905	—	13 (2)	—	1 (2)	3 (2)
B-5, A 0351-5458	narrow (1)	32 (2)	—	2 (2)	—
B-6, AM 0442-622	—	34 (2)	—	—	—
B-7, Abell 548-17	—	7 (2)	banana (3)	2 (3)	—
B-8, AM 0623-371	—	4 (2)	—	1 (2)	—
B-9, UGC 5119	—	6 (2)	—	2 (2)	—
B-10, A 0950-2234	—	3 (2)	—	2 (2)	—
B-11, UGC 5600	—	3 (2)	—	—	3 (1)
B-12, ESO 503-G17	—	9 (1)	—	1 (2)	—
B-13, Abell 1631-14	annulus(3)	30 (3)	—	3 (2)	—
B-14, Abell 1644-105	narrow (2)	2 (3)	—	1 (1)	1 (2)
B-15, A 1256-1710	—	7 (3)	—	—	—
B-16, NGC 5122	annulus(3)	1 (1)	straight (2)	1 (2)	—
B-17, UGC 9562	—	18 (2)	straight (2)	2 (2)	2 (2)
B-18, AM 1934-563	annulus(3)	30 (2)	integral-outer(1)	—	—
B-19, AM 2020-504	narrow (1)	1 (1)	—	1 (1)	1 (1)
B-20, A 2135-2132	—	3 (2)	—	1 (2)	2 (2)
B-21, ESO 603-G21	—	1 (2)	integral-inner(2)	2 (2)	2 (2)
B-22, A 2329-4102	narrow (2)	36 (2)	—	2 (2)	—
B-23, A 2330-3751	—	8 (2)	—	—	—
B-24, A 2333-1637	narrow (3)	43 (3)	—	2 (2)	—
B-25, A 2349-3927	—	5 (2)	straight (2)	3 (2)	3 (2)
B-26, A 2350-4042	—	4 (2)	banana (2)	2 (2)	—
B-27, ESO 293-IG17	—	1 (2)	—	—	—

NOTES: Column 1: Name from PRC, other identification; Column 2: type of ring (annulus means a disk with the central region cut out; narrow means the ring is not extended in radius); Column 3: angular distance away from perpendicularity of the disk and ring; Column 4: type of warp (integral means the two sides bend in opposite directions [inner or outer refers to which part of the ring is more perpendicular to the disk], banana means the two sides bend in the same direction); Column 5: how symmetric the ring is with respect to the plane of the disk (1=very symmetric to 3=very asymmetric); Column 6: amount of material not associated with either the ring or disk (1=no debris to 3=extensive debris). Quality ratings are included in parenthesis (1=very good to 3=uncertain).

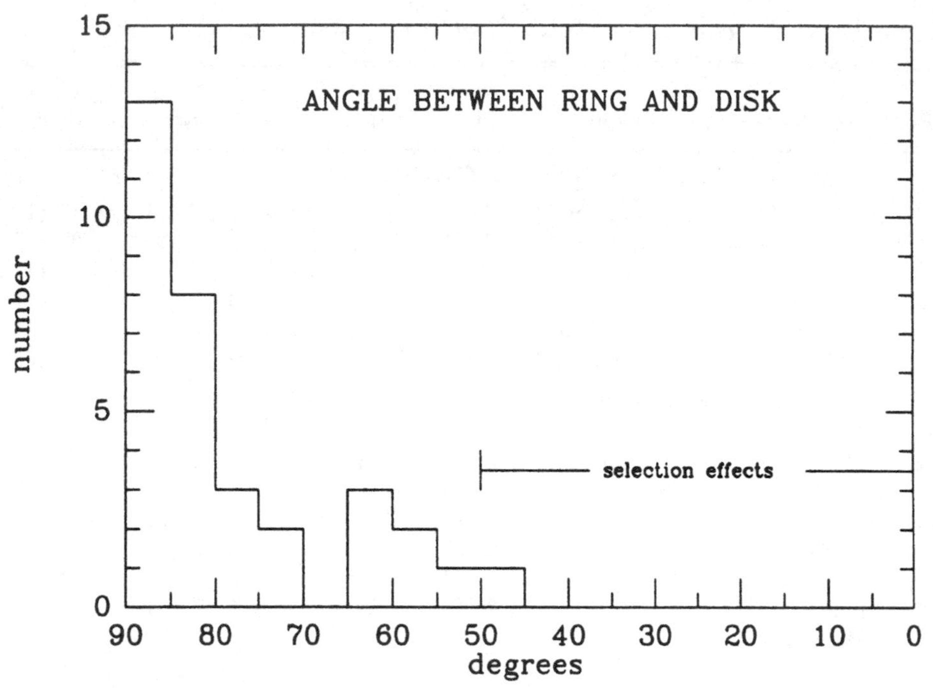

Fig. 3. The distribution of galaxies from category A and B of the PRC as a function of the angle between the ring and the perpendicular to the disk.

d) Are the polar rings warped or straight?

The nature of the warps which are seen in many polar rings may also provide clues about the formation mechanism (e.g., see Lake and Norman, 1983). In addition, Sparke (1986) has argued that self-gravity can hold a polar ring together against the effects of differential precession. Detailed models have been constructed for the self-gravity model in a few cases. These provide reasonable fits to the observed warp (Sparke 1990).

Of the 12 cases of category A and B galaxies where a clear determination can be made (i.e., after eliminating the narrow-ringed systems), only four appear to be essentially straight. Four more appear to be shaped like an integral sign, with the two sides warping in opposite directions. Of these four, two appear to be more nearly aligned along the minor axis at small radii (i.e., "integral-inner" designation in Table 1), while two are better aligned with the minor axis at large radii. Most surprising is the fact that four systems appear to have banana-shaped warps, with both sides bending in the same direction. UGC 7576 is a good example of this type of warp. Spiral galaxies that are warped invariably show integral-shaped warps rather than banana-shaped warps.

e) Ring symmetry and the presence of outer debris

Table 1 also includes an estimate of the degree of asymmetry of the ring and the amount of outer debris around the galaxy which does not appear to be associated with either the ring or disk component. The degree of ring asymmetry was estimated by determining whether the ring was brighter or more extended on one side of the disk than the other. The determination of the amount of outer debris was especially difficult since it depends sensitively on the type of "plate material" available. We have therefore chosen not to report the quality 3 determinations because most of them were for objects where either CCD images or SRC plates were not available.

The presence of asymmetric rings and/or outer debris may indicate that the system is relatively young, and has not had a chance to settle into an equilibrium. The fact that the two parameters show a fair correlation with each other (correlation coefficient = 0.61 for 15 objects; 1.5 % chance of being from a random distribution) provides some evidence for this interpretation. The weak correlation between the angle and the asymmetry of the ring (§IIIc) suggests that all three parameters may provide information about the age of polar-ring galaxies.

IV. Future work

On the theoretical side, probably the most important work would be the development of computer models that show how the rings originated. Is our simple model of the accretion of a gas rich dwarf or material from a passing galaxy viable? Can the differences between narrow rings and annuli be understood in terms of the impact parameters or the type of material that is accreted? Are polar rings stable or transient phenomena? How are polar rings related to dust-lane ellipticals and other similar objects? It should now be possible to directly address these questions using the latest generation of computer modeling techniques.

On the observational side, one of the most important observations would be improved measurements to determine the shape of the dark halos with more precision. Equally important would be observations that allow us to age date the formation of the ring (see WMS for some examples of how this might be done), with the goal of placing the various polar-rings in the proper chronological sequence in order to see how these systems evolve.

I wish to thank my co-authors in the development of the PRC; Ray Lucas, Douglas McElroy, Tom Steiman-Cameron, Penny Sackett and Rob Olling. Ray Lucas also helped determine some of the values in Table 1. I also wish to thank Dana Berry for helping to put together Figure 1.

References

Galletta, G. (1990). Private communication.

Lake, G. & Norman, C. (1983). Astrophys. J. **270**, 51.

Sackett, P.D. & Sparke, L.S. (1990a). In *Dynamics and Interactions of Galaxies* (ed. R. Wielen), p. 367. Berlin: Springer.

Sackett, P.D. & Sparke, L.S. (1990b). Astrophys. J. **361**, 408.

Schechter, P.L. & Gunn, J.E. (1978). Astron. J. **83**, 1360.

Schechter, P.L., Ulrich, M.-H. & Boksenberg, A. (1984). Astrophys. J. **277**, 526.

Schweizer, F., Whitmore, B.C. & Rubin, V.C. (1983). Astron. J. **88**, 909.

Sparke, L. (1986). Mon. Not. R. Astron. Soc. **219**, 657.

Sparke, L. (1990). In *Dynamics and Interactions of Galaxies* (ed. R. Wielen), p. 338. Berlin: Springer.

Steiman-Cameron, T.Y. & Durisen, R.H. (1982). Astrophys. J., Lett. **263**, L51.

Steiman-Cameron, T.Y. (1990). Astrophys. J. , in press.

Whitmore, B.C. & Bell, M. (1988). *Ap. J.*, **324**, 741.

Whitmore, B.C., Lucas, R. A., McElroy, D. B., Steiman-Cameron, T. Y., Sackett, P. D. & Olling, R. P. (1990). Astron. J. **100**, 1489 (PRC)

Whitmore, B.C., McElroy, D. & Schweizer, F. (1987a). Astrophys. J. **314**, 439. (WMS)

Whitmore, B.C., McElroy, D. & Schweizer, F. (1987b). In *Structure and Dynamics of Elliptical Galaxies*, IAU Symp. 127 (ed. T. de Zeeuw), p. 412. Dordrecht: Reidel.

Kinematics of polar-ring galaxies

PENNY D. SACKETT
University of Pittsburgh

Abstract

A review of polar-ring kinematics is presented. Implications for the nature and origin of polar-ring galaxies, the shape of their potentials, and possibilities for further study are presented.

What is a Polar-Ring Galaxy?

Given its recent popularity (as evidenced during this meeting) it is tempting to declare NGC 4650A the morphological prototype for the class of polar-ring galaxies. Roughly speaking, a galaxy that looks like NGC 4650A is a spindle-shaped galaxy surrounded by an outer ring or annulus of luminous material that is so severely tilted so as to be nearly perpendicular to the major axis of the inner body. This morphological picture has been used as a guide in the tedious task of searching sky surveys and galaxy catalogs to identify polar-ring candidates (Schweizer, Whitmore & Rubin 1983, hereafter SWR, and Whitmore et al. 1990). Observational studies of these first polar-ring candidates led to the discovery that polar rings not only appear to encircle their central galaxies in a highly inclined plane, but also rotate in this plane with speeds comparable to those found in the central galaxies.

In constructing a catalog of 157 polar-ring galaxies, candidates and related objects (hereafter PRC), Whitmore and colleagues (1990) were guided by a definition based on the kinematical structure observed in most of the early prototypes. Their definition (see PRC and Whitmore, this volume) labels a system a "true" polar-ring galaxy if it appears to be in quasi-equilibrium and has a luminous, planar, highly-inclined ring that is known to rotate over the poles of its comparably-sized central disk, with comparable angular momentum. Based on this definition, 6 galaxies are confirmed in the PRC as "true" (Class A) polar-ring galaxies, 27 are listed as probable (Class B) polar-ring candidates, and 73 are listed as possible (Class C) polar-ring galaxies; 51 other related objects (Class D) are also included. Only a few of the most promising polar-ring candidates have been studied kinematically.

The definition used in the PRC serves to separate dust-lane ellipticals from polar-ring galaxies, weeds out interacting galaxies with chaotic plumes and tails, and eliminates galaxies with *non-luminous* rings of neutral hydrogen on highly inclined orbits (such as those found by van Driel, 1987). This distinction, while pragmatically useful in defining a class free of interlopers, is also somewhat artificial. In this workshop, for example, we have heard NGC 3718 described both as a galaxy with a severe warp (Briggs) and as a badly precessing polar-ring galaxy (Sparke). Cen A, another favorite, may be considered a dust-lane elliptical or a polar-ring galaxy, according to taste. It is quite possible that these various classes of objects form a continuum, a continuum that we must understand if we wish to explain the differences as well as the similarities between the classes.

General Characteristics of Polar-Ring Galaxies

Listed in Tables 1 and 2 are some general morphological, photometric, and kinematic characteristics for all nine Class A or B polar-ring galaxies for which published rotation speeds in the ring and/or disk are available. A discussion of these characteristics follows. Throughout this work, *all* distance-dependent quantities are calculated using $H_0 = 75\,\mathrm{km\,s^{-1}\,Mpc^{-1}}$.

Polar-ring galaxies are difficult objects to study, in part because they are rare (approximately 0.5% of all S0 galaxies have observable polar rings, PRC). Their rarity implies that most catalogued polar-ring galaxies are relatively distant and subtend small angles (roughly one arcminute) on the sky. The central bodies of the polar-ring galaxies listed in Table 1 are small compared to normal S0 disks; the absolute diameters of the central components are on the order of 10 kiloparsec. The size of the optical rings varies greatly, from slightly smaller than the central disks to nearly four times as large. Polar-ring galaxies tend to be rather faint, with blue luminosities on the order of $5 \times 10^9 L_{\odot,B}$, which exacerbates the difficulty of their small size. In some cases, furthermore, the rings themselves can be especially faint; this combined with unfavorable viewing angles allows many to escape classification as polar-ring galaxies (cf. PRC).

Morphological details of polar rings vary. Polar rings may have relatively small radial extent or be quite extended. They may be thin bands or broad annuli; some of the annuli are warped, others are not (see Whitmore, this volume). Comparison of columns 6 and 7 of Table 1 shows that those rings with similar inner and outer radii (thin bands, labeled R) tend to hug their central disks, but polar annuli (labeled A) can extend to nearly four times the central disk size in the optical, and the HI usually extends even further. Thin, wedding-band polar rings are not seen far from the central disk.

Kinematical data are given in Table 2. The speeds listed for the disk and ring are the maximum observed speeds along the major axis of each component; no corrections have been applied (for inclination, for example) and in some cases the maximum recorded speed is still on the rising part of the rotation curve (marked with a * in Table 2). In cases for which speeds can be measured over a large radial extent, the (Newtonian) gravitational field of the luminous matter alone is not sufficient to explain the flattening of the rotation curves observed at large radii: dark matter is needed (SWR, Whitmore, McElroy & Schweizer 1987a, hereafter WMS, and Sackett & Sparke 1990, hereafter SS). Rotation speeds in the disk and ring components are comparable, which has implications for the shape of the potential (see below). The central velocity dispersion of the disk is low compared to disk rotation speeds, implying that the central galaxy of polar-ring systems is a rotationally-supported, inclined S0, not an elliptical. The one exception to this rule that has been observed to date is AM 2020-504 (Whitmore, McElroy & Schweizer 1987b and PRC), which has a large velocity dispersion and exhibits rapid rotation in the extreme inner disk (about $100\,\mathrm{km\,s^{-1}}$ at $3''$), but slow rotation further out ($20\,\mathrm{km\,s^{-1}}$ at $10''$).

Polar rings are rich in gas and dust. Where the ring passes in front of the central galaxy, it dims the disk luminosity profile; often the dust can be observed by visual inspection alone. With one exception, polar-ring galaxies that have been observed in the 21cm line of neutral hydrogen have been shown to contain gas masses on the order of $10^9 M_\odot$ or larger (see Table 2). Three other polar-ring candidates that have been observed in neutral hydrogen but whose kinematics are unknown (so that they do not appear in Tables 1 and 2), are: IC 51, $M_{HI} \approx 1.3 \times 10^9 M_\odot$ (Richter and Huchtmeier, private communication), IC 1689, $M_{HI} \leq 0.53 \times 10^9 M_\odot$, (van Gorkom, Schechter

Table 1. *Optical Properties of Polar-Ring Galaxies.*

Galaxy Name (1)	m_B (2)	L_B $10^9 L_{\odot,B}$ (3)	$D_c \times D_r$ $''$ (4)	$D_c \times D_r$ kpc (5)	D_r/D_c (6)	Ring Type (7)	Ring Angle (8)
A0136-0801	15.82	4.0	24 × 66	8.6 × 24	2.8	A	7
ESO 415-G26	14.60	7.9	60 × 54	17 × 15	0.82	R	16
NGC 2685	12.1	3.6	144 × 108	8.8 × 6.6	0.75	R	7
UGC 7576	15.85	6.2	30 × 102	14 × 46	3.4	?	12
NGC 4650A	14.27	4.6	48 × 144	9.1 × 27	3.0	A	12
UGC 9796	16.04	3.4	24 × 90	8.8 × 33	3.75	A	26
UGC 9562	14.2	1.3	36 × 60	3.4 × 5.7	1.7	?	18
AM 2020-504	14.5	10.8	42 × 42	13 × 13	1.0	R	1
ESO 603-G21	15.50	1.8	30 × 60	6.2 × 12	2.0	?	1

NOTES: Column 1: Galaxy; Column 2: Apparent blue magnitude (from the PRC); Column 3: Approximate blue luminosity in 10^9 solar blue units (no corrections for extinction have been applied); Column 4: Approximate diameters of central and ring components in arcseconds (estimated by author from optical photographs); Column 5: Approximate diameters of central and ring components in kpc; Column 6: Ratio of ring to central galaxy size; Column 7: Apparent angle in degrees that polar ring makes with polar axis (from Whitmore, this volume); Column 8: Ring type: A = Annulus, R = Thin Ring (from Whitmore, this volume)

& Kristian 1987), and UGC 5600, $M_{HI} \approx 3.4 \times 10^9 M_\odot$ (inferred from values given in Huchtmeier & Richter 1989). A recently completed 21cm survey of polar-ring galaxies with the 140′ radio telescope at Green Bank (Richter, Sackett & Sparke) should add to our knowledge of the gas content of polar-ring systems.

The faintness and large HI mass of polar-ring galaxies gives them *very* high ratios of HI mass to blue optical luminosity (see Table 2), higher than those usually seen in late-type spirals and especially striking when compared to the gas content of normal S0 galaxies, which often have no detectable HI at all (cf. Wardle & Knapp 1986). As in the spiral arms of late-type disks, the gas and dust of polar rings are often accompanied by blue colors, $H\alpha$ emission, and bright, clumpy HII regions.

Synthesis observations have shown that the neutral hydrogen in polar-ring galaxies is spatially associated with the optical ring, not with the central body (Shane 1980, Schechter et al. 1984, van Gorkom, Schechter & Kristian 1987). The radial extent of the HI ring is often substantially larger than that of the optical ring. In cases for which

Table 2. *Kinematical Information for Polar-Ring Galaxies*

Galaxy Name (1)	v_{hel} km s^{-1} (2)	Dist Mpc. (3)	v_{max} Disk (4)	v_{max} Ring (5)	σ_0 km s^{-1} (6)	M_{HI} $10^9 M_\odot$ (7)	M_{HI}/L_B $M_\odot/L_{\odot,B}$ (8)
A0136-0801	5528	73.5	140[1]*	183[1]	67[2]	1.5[3]	0.38
ESO 415-G26	4625	59.1	170[2]	188[2]	127[2]	3.2[3]	0.41
NGC 2685	876	12.6	138[5]	107[4]	60[5]	1.2[4]	0.33
UGC 7576	7036	93.1	134[5]*	235[6]	116[5]	5.0[6]	0.81
NGC 4650A	2910	38.9	110[2]	123[2]	77[2]	4.6[3]	1.0
UGC 9796	5407	75.2	136[5]*	175[6]	73[5]	4.6[6]	1.4
UGC 9562	1242	19.7	?	110[7]	?	0.92[8]	0.71
AM 2020-504	5058	66.2	20[5]	241[5]	153[5]	?	?
ESO 603-G21	3150	42.6	111[5]	132[5]	?	?	?

NOTES: Column 1: Galaxy; Column 2: Heliocentric velocity in km s^{-1} (from the PRC); Column 3: Distance in Mpc (using $H_0 = 75\,\mathrm{km\,s^{-1}\,Mpc^{-1}}$ and correcting for motion of the Galaxy); Column 4: Maximum speed observed in disk in km s^{-1} (uncorrected for inclination), * indicates that rotation curve has not yet flattened, see text for a discussion of AM 2020-504; Column 5: Maximum speed observed in the ring in km s^{-1} (uncorrected for inclination); Column 6: Central velocity dispersion of central object in km s^{-1}; Column 7: Neutral hydrogen mass in 10^9 solar units; Column 8: Approximate ratio of HI mass to blue luminosity in solar units.

REFERENCES: 1. Schweizer, Whitmore & Rubin 1983; 2. Whitmore, McElroy & Schweizer 1987a; 3. van Gorkom, Schechter & Kristian 1987; 4. Shane 1980; 5. Whitmore et al. 1990; 6. Schechter et al. 1984; 7. Balkowski, Chamaraux & Weliachew 1978; 8. Gordon & Gottesman 1981.

kinematical data are available for both stars and gas, the rotation speeds are similar. Since the central disks of polar-ring systems are devoid of gas, kinematical information for the disks is gleaned from stellar absorption line data, making disk rotation speeds more uncertain than those in the rings.

The last columns of Tables 1 and 2 hint at a possible trend of ring angle with M_{HI}/L_B ratio. Galaxies with extended polar rings ($D_r/D_c \geq 1.5$) that are inclined significantly away from being perfectly polar seem to have the largest M_{HI}/L_B ratios. (IC 1689 has a tight polar ring inclined 2-3 degrees from polar and $M_{HI}/L_B \leq 0.09$; UGC 5600, which seems to have a quite polar inner ring and an outer ring at a different orientation, has $M_{HI}/L_B \approx 0.91$.) Sparke (1986) has shown that the self-gravity of polar rings can maintain stable orbits at intermediate angles if the ring mass is sufficiently high. A coherent, extended ring that warps toward the poles at large radii then results; the amount of warping decreases with increasing ring mass. This may provide an explanation of the possible trend of ring angle with gas-richness: to the extent that M_{HI}/L_B provides a crude measure of the ratio of masses in the ring and disk, high M_{HI}/L_B systems would be those in which the self-gravity of the ring was relatively strong, perhaps strong enough to maintain an extended polar ring at off-polar angles. If this hypothesis is correct (the statistics are obviously still quite poor!), polar-ring galaxies with extended, off-polar rings should be associated with large amounts of HI, especially if the rings are relatively planar (as is the case for UGC 7576). On this basis, Abell 1631-14, AM 1934-563, and in particular, ESO 503-G17, may be expected to have large M_{HI}/L_B ratios.

Double Ringed Systems

At least one clear case of a polar-ring-like system with *two* outer rings is known: ESO 474-G26. The rings of this galaxy are inclined at different angles to the roughly spherical-looking central object; the rotation speeds in each component are comparable (PRC). The nature and orientation of the central body have not yet been determined. The Helix Galaxy, NGC 2685, has a highly inclined, tight inner ring of gas and stars rotating over the poles of its central S0 disk. As shown by Shane (1980) it also has a much larger and more massive ring of neutral hydrogen in rotation at large radii, which seems to be associated with a faint, diffuse optical ring that can be seen on deep exposures. Shane reports that the major axis of this outer HI ring is approximately the same as that of the central S0, although the ring appears to be slightly more face-on. A third case of a double polar-ring object may be UGC 5600; like the Helix Galaxy it has a bright, tight inner ring which appears to go over the poles of the flattened central object, and a more diffuse, but knotty, outer optical ring with position angle and ellipticity close to that of the central body.

How do Polar Rings form?

The presence of two nearly perpendicular angular momentum vectors of comparable size argues against polar-ring galaxies being formed in a single protogalactic collapse. A more likely alternative is that during a tidal encounter, material is stripped from a neighboring galaxy, or perhaps even from the outskirts of the host galaxy itself, and that this material later settles onto polar orbits since those are the ones most stable against differential precession in an oblate potential. (Schweizer (1986) suggests that we may be witnessing this process taking place in the NGC 3808 system: an optical filament connects this spiral to its S0 companion, around which it seems to form a partial polar ring.) As a variation of this hypothesis, the host galaxy may be involved in the wholesale capture of a dwarf galaxy, whose stars and gas then become smeared out on the (relatively) stable polar orbits to form a ring. Are stars present at the formation of the ring or are they formed later in the gas-rich environment found in almost all polar rings? Active star formation is certainly taking place in many polar rings, as evidenced by HII regions and blue colors, but this does not rule out the possibility of an underlying older stellar population as well. Lack of dissipation in the stellar component, however, would probably prevent any stars present in the pre-ring material from evolving with the dissipational gas to form a common ring.

The hypothesis that polar-ring material originates from outside the parent body is not without its difficulties. We have seen (Rix & Katz and Quinn, this volume, and references therein) that forming a smooth, broad ring from the capture of a single dwarf object is not always easy. On the other hand, if the material is captured during a tidal interaction, it is difficult to understand how the small object usually found at the center of the final polar-ring galaxy is able to gravitationally "convince" substantial amounts of mass to leave the passing neighbor. As stressed earlier, large amounts of HI are associated with the rings, amounts comparable to the total gas content of many gas-rich spirals. In NGC 4650A, a maximum disk fit to the rotation curve of the central S0 gives an upper limit to the mass of the stellar disk that is equal to the mass in polar ring in gas alone (SS), implying that the much more extended ring of NGC 4650A actually contains *substantially* more angular momentum than its host S0! How did such a small, low-mass galaxy acquire such a large, massive ring? A dark matter halo can help only if the halo of the host galaxy is considerably more massive than that of the donating neighbor. This may argue for cannibalism of a gas-rich dwarf over tidal stripping as a formation mechanism for polar rings.

If an accretion event is responsible for the formation of polar-ring galaxies, how does a galaxy acquire *two* rings of different orientation?

It is very difficult to imagine how double polar-ring systems could have formed simultaneously from a single interaction, and a second capture event is likely to disturb the fairly fragile stability of the first ring. Sparke (this volume) suggests that double-ringed systems may form from a single accretion event if the accreted mass is spread over a large range in radius. In that case, self-gravity may cause the tidal material to evolve into two or more ring fragments of different radii and orientation, with the outer ring being the more polar of the two. Detailed observational studies of double-ringed systems are needed to determine whether the rings could be the remnants of the same accreted material; in NGC 2685 and UGC 5600, however, first appearances would suggest that the inner rings are the more polar, contrary to the otherwise attractive hypothesis of Sparke. Recent VLA observations in the 21cm line by Mahon (private communication 1990) may help to clarify the situation for NGC 2685.

Another nagging problem, raised by SWR, is that too many polar rings are too nearly polar, especially if the polar-ring material has been captured from an external source by a system with a dominant, spherical halo. Several solutions to this dilemma have been proposed: Sackett & Sparke suggest that the dark matter halo may not be spherical, and indeed does not appear to be so for NGC 4650A (SS); Steiman-Cameron & Durisen (1982) point out that adding a small amount of triaxiality to the gravitational potential can create a conical zone of moderate opening angle for capture onto stable polar orbits; and Sparke (1986) reminds us that self-gravity can hold polar rings together against differential precession even in an oblate potential, but only if the ring is fairly close to polar, massive, and warps toward the poles. These explanations are not mutually exclusive. An alternative explanation is that polar rings are not stable, but that the most polar ones have the longest lifetimes and thus are the ones we are most likely to observe (SWR).

Shape of the Gravitational Potential

The shape and stability of polar-ring orbits are critically dependent on the form of the underlying gravitational potential. The HI masses of the rings are so large in some cases that the polar rings themselves can significantly influence the shape of the gravitational potential, complicating calculations and necessitating a self-consistent treatment of the ring orbits (SS). Attempts to use ring and disk kinematics as a probe of the shape of the gravitational potential (under the

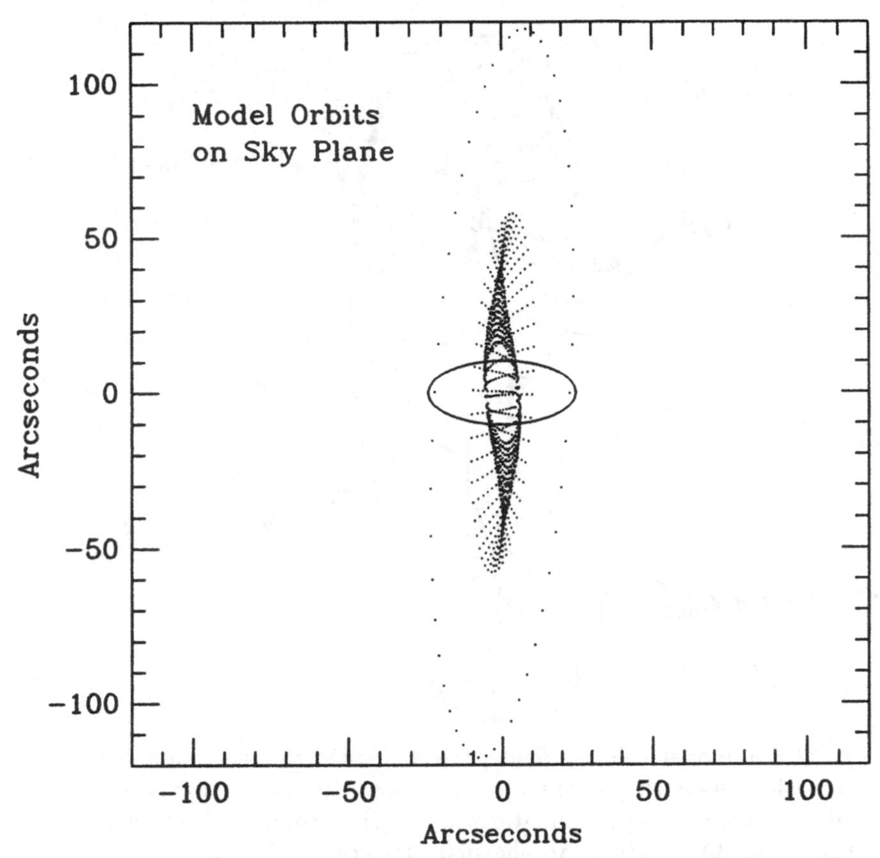

Fig. 1. The morphological appearance on the sky of a model polar ring consisting of perfectly polar, concentric, but non-coplanar orbits. Scale lengths and inclinations are appropriate to NGC 4650A; the central oval signifies the inner disk. In this model, the apparent twisting of the ring major axis is entirely an artifact of projection of the non-circular polar orbits onto the sky.

assumption that the rings are on closed orbits) have been made by SWR, WMS, and SS; their results do not place strong constraints on the shape of the dark halo. The earlier results of SWR and WMS, based on a comparison of the similar disk and ring rotation speeds at one galactocentric radius, suggested that the dark halos of three polar-ring galaxies were quite round. A recent analysis by SS of one of these systems (NGC 4650A), which fit predicted rotation curves from detailed mass modeling to the full extent of the observed curves, has indicated that a quite flat halo (E6) is also consistent with the data and can, in fact, provide a better fit. The method of comparing disk and ring rotation curves is promising, but more systems

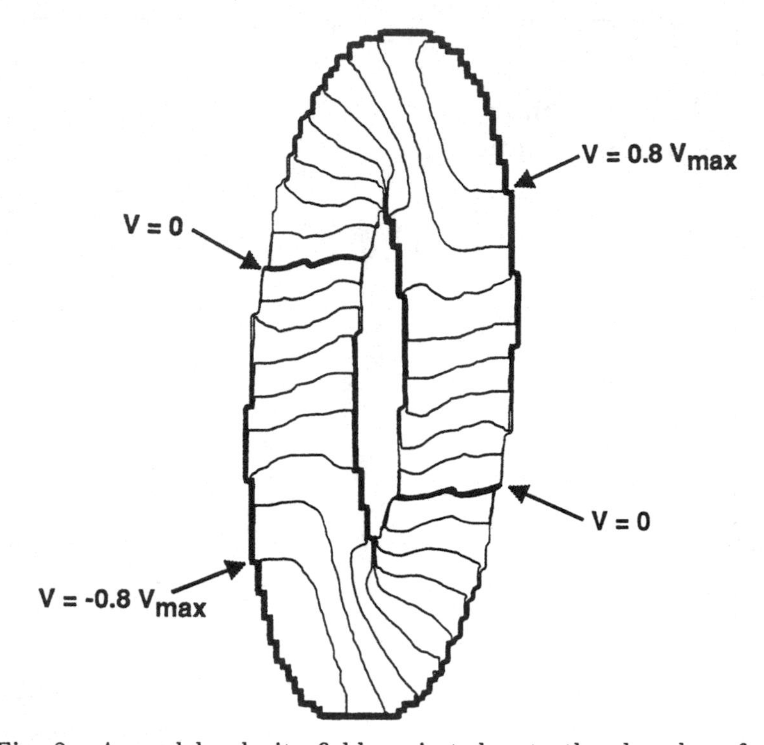

Fig. 2. A model velocity field projected onto the sky plane for a massless polar ring in the potential of a Miyamoto-Nagai disk, using scale lengths and inclinations appropriate to the disk and ring of NGC 4650A. Contours are shown in steps of 0.1 v_{max}, where v_{max} is the maximum line-of-sight velocity in the ring. Note that the zero velocity contour is not aligned with the morphological minor axis and that both the major and minor kinematic axes are badly twisted. The degree of misalignment could be used as a measure of the non-circularity of the polar orbits and thus the oblateness of the gravitational potential.

must be studied in detail and the rotation curves, especially those for the central components, must be extended to larger galactocentric radius.

Steiman-Cameron, Kormendy & Durisen (1990, see also Steiman-Cameron, this volume) have had remarkable success in reproducing the complicated dust-lane structure of NGC 4753 over a factor of 7 in radius, taking as a model a settling disk that has become severely twisted due to differential precession in a scale-free potential. Their extremely good fit implies that this system has a dark halo that retains a moderate flattening (E0.5–E3) to large radii.

If the gravitational potential is significantly flattened toward the plane of the inner stellar disk (which it must be at small radii where the dark halo presumably does not dominate), then polar orbits will not be circular, but elongated along the polar axis. When projected onto the sky plane, these non-circular ring orbits, if polar but not entirely co-planar, can create the appearance of not only a warp, but a twisting of the ring (see Figure 1). In principle, this morphological signal could be used as a diagnostic of the shape of the potential; in practice, this approach may be very difficult to implement. In any event, use of closed circular orbits to model the motion of polar gas, especially at small radii, should be treated with caution.

More promising as a probe of the potential shape may be the apparent twisting of the kinematic minor and major axis of the ring caused by the projection of the non-circular polar orbits onto the sky plane (see Figure 2). This misalignment of the kinematic and morphological axes is present even in an axisymmetric potential; it results from the non-circularity of polar orbits (see also Teuben, this volume). The axial ratio of the rotational speeds departs from unity much more strongly than the axial ratio of the orbit radii; this may make *kinematical* twisting a promising diagnostic tool. When the kinematical twisting is accompanied by a real warp, however, as may be the case in NGC 4650A, the two effects will be quite difficult to disentangle (Nicholson 1989).

Acknowledgements

It is a pleasure to thank Peter Teuben for help in preparing Figure 2 using NEMO, Linda Sparke and Frank Briggs for many stimulating and instructive discussions, and my colleagues on the polar-ring catalog project (PRC): Brad Whitmore, Ray Lucas, Doug McElroy, Tom Steiman-Cameron and Rob Olling, whose hard work produced much of the data given in Tables 1 and 2.

References

Balkowski, C., Chamaraux, P. & Weliachew, L. (1978). Astron. Astrophys. **69**, 263.

Gordon, M., & Gottesman, S. (1981). Astron. J. **86**, 161.

Huchtmeier, W. K. & Richter, O.-G. (1989). *A General Catalog of HI Observations of Galaxies* (Springer-Verlag, New York).

Nicholson, R. A. (1989). Ph.D. Thesis, Sussex University, U.K.

Sackett, P. D. & Sparke, L. S. (1990). Astrophys. J. **361**, 408. (SS)

Schechter, P.L., Sancisi, R., van Woerden, H. & Lynds, C.R. (1984). Mon. Not. R. Astron. Soc. **208**, 111.

Schweizer, F. (1986). Science **231**, 227.

Schweizer, F., Whitmore, B. C. & Rubin, V. C. (1983). Astron. J. **88**, 909. (SWR)

Shane, W. W. (1980). Astron. Astrophys. **82**, 314.

Sparke, L. S. (1986). Mon. Not. R. Astron. Soc. **219**, 657.

Steiman-Cameron, T. Y. & Durisen, R. H. (1982). Astrophys. J. **263**, L51.

Steiman-Cameron, T. Y., Kormendy, J., & Durisen, R. H. (1990). In preparation.

van Driel, W. (1987). PhD Thesis, University of Groningen, The Netherlands.

van Gorkom, J. H., Schechter, P. L. & Kristian, J., (1987). Astrophys. J. **314**, 457.

Wardle, M. & Knapp, G. R. (1986). Astron. J. **91**, 23.

Whitmore, B. C., McElroy, D. B. & Schweizer, F. (1987a). Astrophys. J. **314**, 439. (WMS)

Whitmore, B. C., McElroy, D. B. & Schweizer, F. (1987b). In *Structure and Dynamics of Elliptical Galaxies*, IAU Symp. 127 (ed. T. de Zeeuw), p. 413. Dordrecht: Reidel.

Whitmore, B. C., Lucas, R. A., McElroy, D. B., Steiman-Cameron, T. Y., Sackett, P. D., & Olling, R. P. (1990). Astron. J. **100**, 1489. (PRC)

Dynamics of polar rings

L. S. SPARKE
Washburn Observatory, University of Wisconsin-Madison

A polar ring is a ring of gas, dust and stars orbiting at a large angle to the equator of a flattened galaxy - usually an S0 galaxy, less often an elliptical. In some cases the inner galaxy is known to rotate rapidly about its short axis, consistent with the interpretation that it is a disk seen nearly edge-on; the ring, of course, rotates about its own, nearly orthogonal, minor axis. Whitmore et al. (1990) have compiled a 'Polar Ring Catalog' listing known and suspected ring systems, from which they estimate that about 5% of S0 galaxies are now polar ring galaxies or have been so in the past; the phenomenon is neither very common nor extremely rare. Presumably polar ring galaxies are formed in a two-stage process, with the ring material being accreted after the galaxy disk has been built up; but the details of this process are not at all clear.

Once a polar ring is in place, however, some features of the dynamics are quite straightforward. For example, the rings are not exactly orthogonal to the inner galaxy, but are more usually slightly displaced from the pole (Whitmore 1984). Polar rings are not narrow annuli, but are extended over a range in radius; this can be seen particularly clearly in the photograph of A1036-0801 presented by Schweizer et al. (1983)—the outer radius of the stellar disk is about three times the radius at the inner edge. The quadrupole moment of the central galaxy will cause an inclined ring to precess about the galaxy pole, at a rate which depends on distance from the galaxy center; simple estimates (e.g., Schechter et al. 1984b) suggest that differential precession will cause the ring to become noticeably twisted in much less than a Hubble time.

A model including differential precession *only* can explain most of the ring structure observed in NGC 3718. This galaxy, classified SBa (peculiar), was mapped in neutral hydrogen by Schwarz (1985).

He found a very complex gas distribution, which could be modelled as the warped and twisted disk shown at the top right of Figure 1. The gas orbits are obviously quite inclined to the optical disk, which appears near to face-on, while the HI disk passes from about 75° inclination, through edge-on, to $i = 65°$ on the other side. At the same time, the major axis of the gas ring twists on the plane of the sky.

To model the gas sheet in NGC 3718, I assumed (Sparke 1990) that the potential of the galaxy was given by a combination of a thin Kuzmin disk and a spherical unseen halo in which the rotation speed rises asymptotically to a constant value. The scale length of the stellar disk was set according to optical data, and the halo parameters chosen in a 'maximal disk' fit to the flat rotation curve which Schwarz (1985) inferred from his HI observations; the resulting fit implies $M/L_B = 8$ for the luminous part of the galaxy. In this potential, the precession of an orbit with any given tilt could be found. The calculation was started with the orbits of all the ring material aligned at 10° to the pole of the galaxy (representing the plane of the orbit from which the disk material was presumably captured), and precession was allowed to occur over several orbital times.

The structure resulting after 6 orbits at 250″ radius (about 3.5×10^9 years) is shown at the top left of Figure 1. The model is projected to look as much as possible like the observed galaxy, with the ring orbit at 160″ seen edge-on; this requires that the S0 disk is about 30° to face-on, with the major axis at position angle 90°. The panel below shows how good the match is; starting from a nearly-planar disk, differential precession seems adequate to explain the structure of this ring, at least outside 150″ radius. The prediction which this model makes for the orientation of the stellar disk can be checked by measuring the rotation and kinematic axes; work on this is in progress (Balcells, Schwarz and Sparke).

Schwarz (1985) found that, in a coordinate system defined by the plane of the *innermost gas orbits*, the gas ring has a shape which is warped by almost 90° but is nearly untwisted (to within about 15°) about the pole of that system. Because of this property, Steiman-Cameron (private communication) was able to model the ring as an equilibrium structure of tilted orbits in the potential of a tumbling triaxial halo. However, the pole of a system in which the gas ring appears untwisted does not fall close to the axis of the stellar disk, but points in some other, apparently arbitrary, direction. According to the present model, the beautiful 'restricted' warp results from the

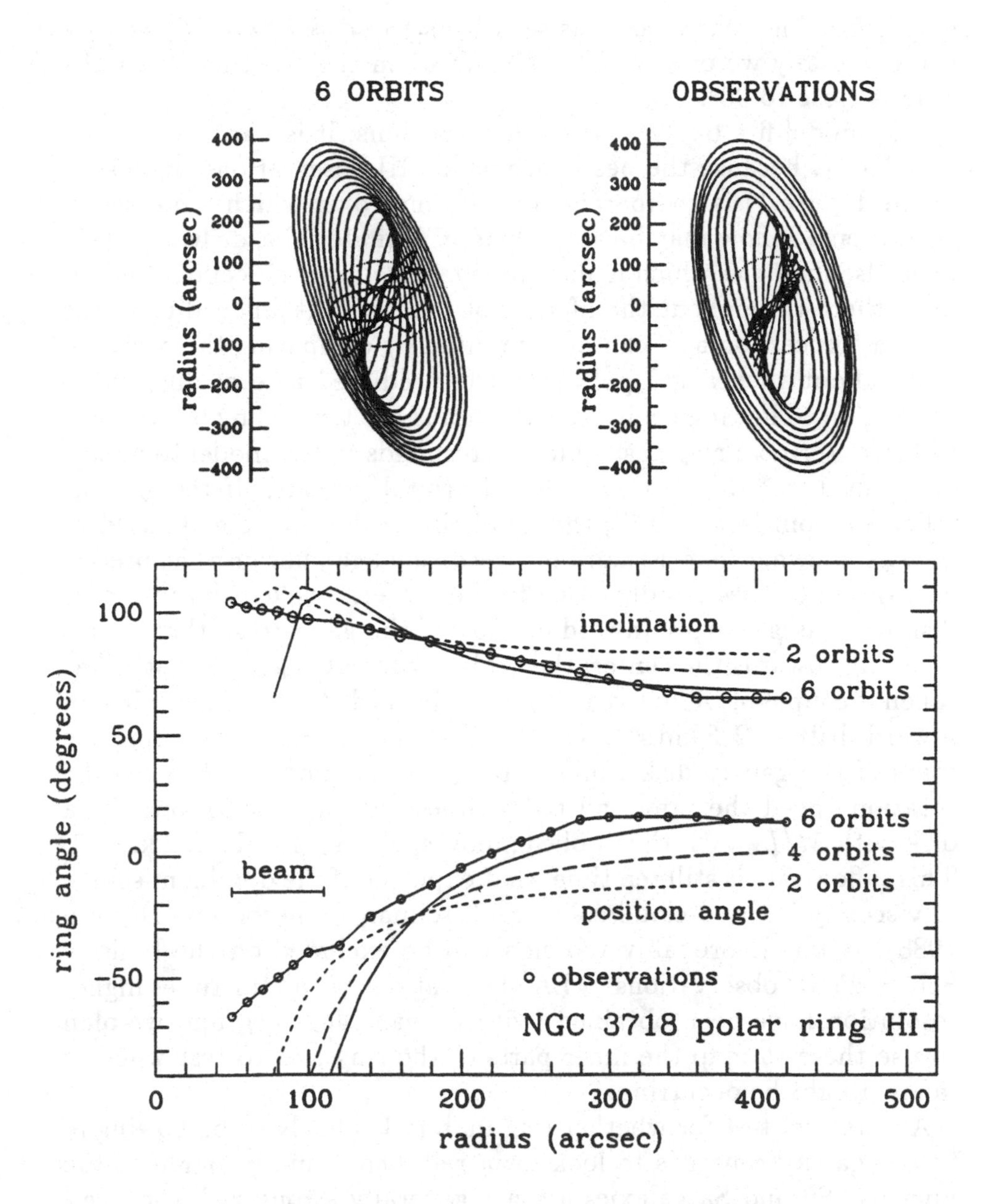

Fig. 1. A precessing model for the polar ring in NGC 3718. Top left: the model, projected to look like the observed ring; the dotted ellipse shows the plane of the S0 disk. Time is measured in units of the orbital period at 250″ radius. Top right: the ring structure fitted by Schwarz (1985) to his HI observations. Bottom: inclination (upper curves) and position angle of the major axis (lower) of the ring as a function of radius. The observations (circles) are compared with the model at various times; the beam size of the HI observations is shown on the left of the figure.

gas disk being nearly polar as seen from the *disk of the S0 galaxy* – coincidentally we observe it at a time when the twisting about the pole is close to 90°.

The model fits badly in the central regions; it is hard to tell just how badly, because the beam size of the HI observations (indicated in the Figure 1) is comparable to the range over which there is disagreement. Inflow may be at work here. If material is drifting slowly inwards, then gas which is now at say 100″ from the center has not been there throughout the lifetime of the ring. At any radius in the ring, a 'precession age' can be computed by comparing the observed twist about the galaxy pole with the predicted rate of differential precession. The ring orbits at 100″ radius are twisted by 62° relative to the outermost ring gas, which corresponds in the model to an age of about 1.2×10^9 years, or 5 local orbital periods. If the gas has fallen in from larger radii, the infall timescale can be estimated as being somewhat longer than the 'precession age', because the precession rate increases rapidly closer to the center. The density in the HI disk decreases sharply inward of 200″; if the gas within that radius is falling towards the center, then this argument suggests that it has taken a couple of billion years to move inwards to 100″, implying an inward drift at 2–3 $\mathrm{km\,s^{-1}}$. If the model is changed by reducing the mass of the galaxy disk while increasing the halo mass to keep the rotation speed the same, all these timescales become longer: for a disk with $M/L_B = 4$, the implied inflow speed is only about 1 $\mathrm{km\,s^{-1}}$. This inflow rate is still ten times that expected if the dominant source of viscosity is cloud-cloud collisions (Steiman-Cameron and Durisen 1988), but it is probably too small to be excluded on the basis of Schwarz's HI observations. VLA observations with four times higher resolution have been obtained (with G. van Moorsel), and we plan to use these to map the inner parts of the ring and to test whether inflow might be occurring.

An indirect test for whether ring material is likely to be flowing in to the galaxy center is to look for a radio continuum source at the nucleus. S0 and Sa galaxies are not generally strong radio sources, presumably because they lack the gas to fuel an active nucleus. But if material is flowing inward from the polar ring, that will provide a source of gas for accretion, and may trigger radio emission. Work is in progress (Richter, Sackett and Sparke) to survey galaxies in the Polar Ring Catalog radio continuum emission; we are indebted to Dr. P. Biermann for suggesting this.

But not all polar rings suffer such strong differential precession: for example, the ring in UGC 7576 is seen nearly edge-on and appears extremely flat, so that it cannot be twisted by more than a few tenths of a radian. The ring is about 12° from polar (Whitmore 1984), but appears smooth and relaxed, with little ionized gas (Mould et al. 1982), suggesting that it is fairly old. The self-gravity of a polar ring around an oblate S0 galaxy can hold it together and prevent twisting, if enough mass is present in the ring (Sparke 1986); in this case, the whole ring precesses uniformly about the pole of the S0 galaxy without change of shape. The polar rings which have been observed in neutral hydrogen are gas-rich (e.g., Schechter et al. 1984a, van Gorkom et al. 1987), some containing several billion solar masses of gas. This can be enough to hold the ring together in a stable equilibrium state.

In UGC 7576, the kinematics of the neutral hydrogen (Schechter et al. 1984a) suggests that the gas probably lies in a wide ring, with an outer radius around twice the radius of the inner edge—so that severe differential precession might be expected. Something is obviously acting to stabilize this ring. A stable self-gravitating model for the ring (Sparke 1990) is shown in the second figure. The material making up the ring is assumed to follow nearly circular orbits which precess slowly in the potential of the galaxy and of their own gravity. When precession is slow compared to the orbital motion, the ring can be treated as a collection of precessing concentric wires; this amounts to averaging over the fast orbital motion. Then the mutual gravitational attraction of material in different parts of the ring can be computed as the attraction between uniform circular wires.

Following the HI observations of Schechter et al. (1984a), the ring is taken to have constant density in the range $26'' < r < 49''$. The galaxy is modelled in the same way as for NGC 3718, with a Kuzmin disk and a spherical halo potential. Again, these were chosen to make the rotation curve nearly flat at the largest measured value of $250\,\mathrm{km\,s^{-1}}$ out to the edge of the ring, while putting as much mass in the disk as is reasonably possible, which yielded $M/L_B = 7.5$ for the stellar component. The polar ring was given a mass equal to that measured in gas, $7\times10^9 M_\odot$ (corrected to $H_0 = 75\,\mathrm{km\,s^{-1}\,Mpc^{-1}}$, and multiplied by a factor of 1.4 to allow for the helium content). The model is displayed first with the galaxy disk tipped by 30° away from edge-on as indicated by photometry (Whitmore 1984), and then with the galaxy edge-on to show the structure more clearly.

The ring exhibits no twisting about the galaxy pole, and a poleward warp is evident. This warping would be less if the ring mass were

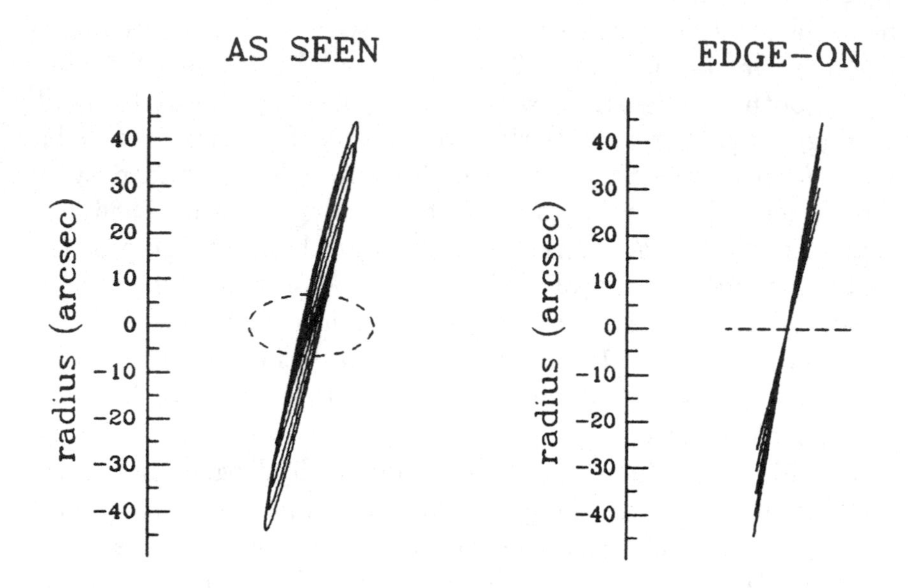

Fig. 2. A self-gravitating equilibrium model for the polar ring in UGC 7576; the dotted line shows the plane of the S0 disk. The model is displayed first in the observed orientation, then with the disk edge-on to show the structure better.

larger; if the ring mass is too small, no stable equilibria are found. For a given ring mass, the curvature is most critically dependent on the radial distribution of ring matter. Again, HI observations at high resolution would provide the best test of this model.

To test whether self-gravity is likely to be important for polar rings in general, Richter, Sackett and Sparke are currently engaged in a neutral hydrogen survey of objects in the Polar Ring Catalog. Since in those polar ring galaxies which have been mapped with an interferometer, the HI always lies in the polar ring and is not associated with the stellar disk, it is reasonable to assume that the mass of any HI detected in a single-dish observation gives a lower limit to the mass of the ring. We expect the masses to be in the region of $10^9 M_\odot$ if self-gravity is a significant force.

Even if self-gravity does not lock a polar ring into a steady state, it can still cause the characteristic warp up towards the pole. To see this, consider a very simple model for a polar ring, consisting just of two concentric circular wires; the inner one has radius r, mass m, and is tilted by an angle θ away from the equatorial plane of the S0 galaxy, while the outer wire has radius r', mass m', and tilt

θ'. The galaxy model is equally simple; the rotation curve is taken to be completely flat, V =constant, and the potential has constant oblateness η (see Sparke 1986). Then any motion of the two wires must conserve both the total energy E and the angular momentum J_z about the symmetry axis of the galaxy. These quantities are given by

$$E = \text{kinetic energy} - 3V^2\eta[m \cos^2\theta + m' \cos^2\theta']/8 + V_m(\theta, \theta') ,$$

where the middle term represents the potential energy of the wires in the galaxy potential, and $V_m(\theta, \theta')$ is the mutual potential energy; and

$$J_z = mrV \cos\theta + m'r'V \cos\theta'.$$

If the ring is light, then the term in V_m is small compared to the others, since it is quadratic rather than linear in the ring mass. Then any motion of the wires must satisfy

$$\Delta E = 0 \approx \Delta(\text{k.e.}) - 0.75V^2\eta m\Delta(\cos\theta)[\cos\theta - (r/r') \cos\theta'] .$$

If the ring is initially static and nearly coplanar ($\theta \approx \theta'$), then any motion requires $\Delta(\cos\theta) > 0$; the tilt of the inner wire must decrease. Thus the transfer of energy and angular momentum between the wires due to self-gravity causes the inner wire to be pushed up towards the pole, while the inner wire settles towards the equator.

This effect can be seen in the polar ring in NGC 4650A. Figure 3 shows the time-evolution of a polar ring in a galactic potential taken from the models of Sackett and Sparke (1990). In this paper, an attempt was made to constrain the flattening of the unseen halo of NGC 4650A by fitting simultaneously the rotation curve in the polar ring and that in the stellar disk. The galaxy model consisted of a bulge, an exponential disk and an oblate massive halo; it was found that halos having a shape between E0 and E6 could all fit the observations. Here, time-dependent calculations of the motion of an initially coplanar polar ring in the model potentials with an E0 (leftmost column of Figure 3), E3 (central column) and E6 (right column) halo are shown. The models for the galaxy potential are exactly the 'best-fit' models of Sackett and Sparke, except that the exponential disk is replaced by a Kuzmin disk for ease of calculation; for the E0 model the halo core radius $r_c = 20''$ and the asymptotic circular speed $V_H = 120\,\mathrm{km\,s^{-1}}$, while for the E3 model $r_c = 19''$, $V_H = 140\,\mathrm{km\,s^{-1}}$ and for the E6 model $r_c = 14''$, $V_H = 160\,\mathrm{km\,s^{-1}}$. The polar ring mass and density distribution were also taken from that work. The ring was initially taken to be coplanar and inclined at 10° to the pole of the central S0 disk.

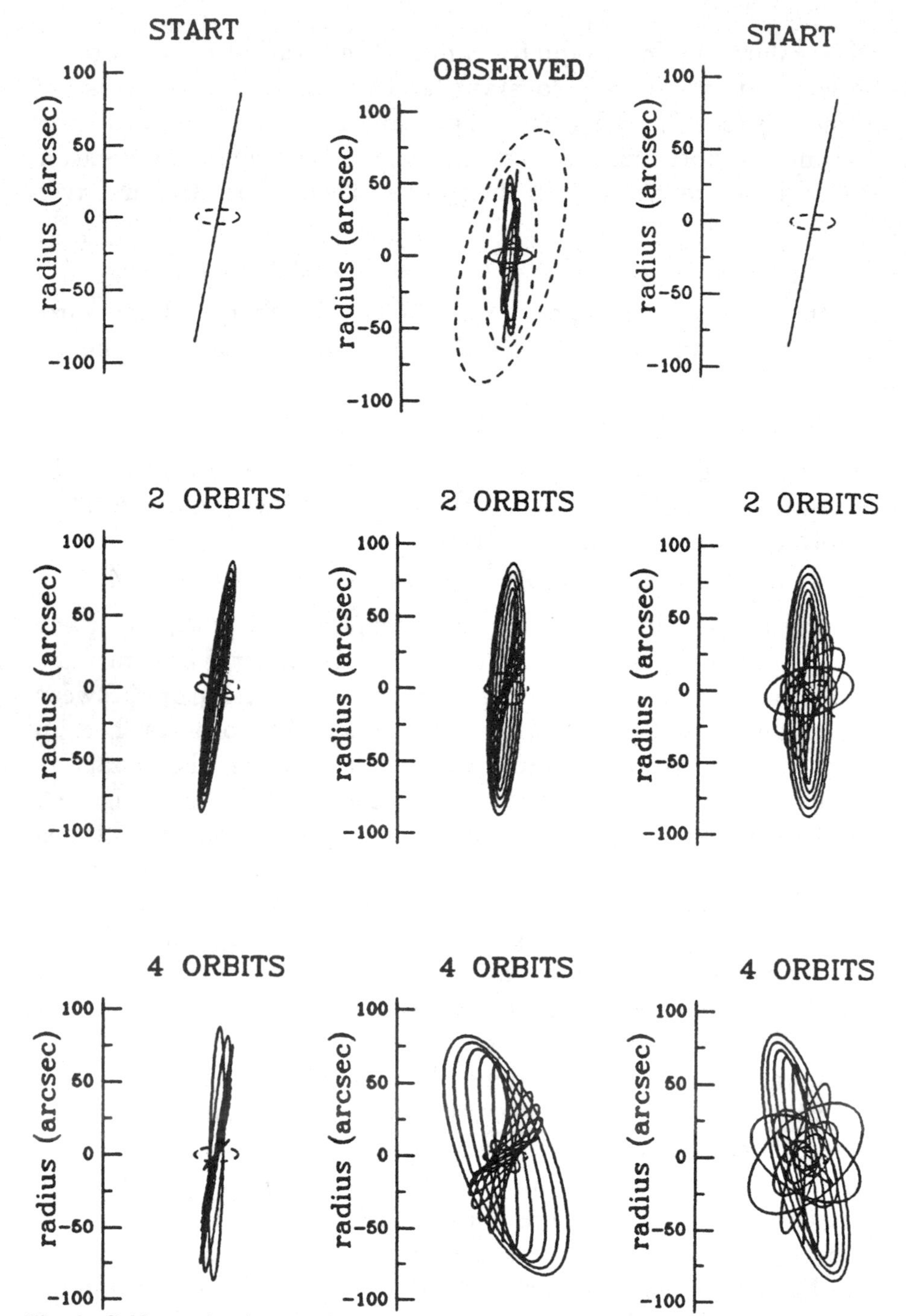

Fig. 3. Self-gravitating models for the polar ring in NGC 4650A, compared with the ring structure deduced from the $H\alpha$ velocity field (center top). The leftmost column shows the development of an initially planar ring in the potential of a disk and a spherical halo component, the middle column corresponds to an oblate E3 halo and the right column to an E6 halo.

These models have been tipped to match the observed orientation of the S0 disk, and compared with the tilted-ring structure deduced by Nicholson (1989) from a two-dimensional $H\alpha$ velocity field obtained with the Taurus imaging Fabry-Perot device. The outer, dashed lines represent extrapolations from the dust and HI structures. In the model with the spherical halo, the polar ring never becomes strongly twisted; after about 4 orbits at radius $40''$ (1.67×10^9 years) the model resembles the structure of the inner part of the observed ring, but it never twists enough in the outer parts to match the dust-lane orientation. The model with the E3 halo does better; after 2 orbits (8×10^8 years), the outermost part of the ring is not far from the orientation of the dust ring at $65''$, while the inner parts show the characteristic warp upwards to the pole. Although the model with the E6 halo can fit the measured rotation curves in the ring and the stellar disk, it cannot match the observed ring structure—the model undergoes far too much differential precession, and becomes completely disordered in the inner parts. Conclusions drawn on the basis of these models should be treated with some caution, since the thin Kuzmin disk has a slightly different potential from the finite-thickness exponential disk used by Sackett and Sparke (1990), and the resulting ring shapes depend on the distribution of mass within the ring, which is only poorly constrained by either direct observations or rotation-curve fitting. But with further work this method may prove a useful constraint on the flattening of unseen halos.

Eventually, an unstable ring like this will break up into a number of sub-rings which precess independently; the outer ring material is pushed into an orbit closer to the pole than the original ring plane, while inner gas sinks towards the equator of the S0 galaxy (Sparke 1986). Thus a single accreted gas cloud might give rise to a multiple ring system, such as that seen in ESO 474-G26 (Schweizer et al. 1983). However, dissipative processes act more rapidly at smaller radii and lower inclinations to remove ring gas; if the inner material disappears completely, the remaining ring will be thinner and in a more nearly polar orbit than that on which the gas was originally accreted. This might resolve a puzzle pointed out by Schweizer et al.(1983): why so many polar rings are so nearly polar, when the initial orbits of the captured material must have been much more uniform in angle.

It is not known how polar rings were formed. One possibility is that they are the remains of gas-rich dwarf galaxies which have been captured into high angular momentum orbits about the central galaxy, and torn apart by differential rotation. This interpretation is supported by the fact that at least one polar ring system (MCG – 5-7-1 = ESO 415-G26; van Gorkom et al. 1987) shows the faint shells and ripples which are usually taken as signatures of a merger. On the other hand, polar rings contain an amount of HI, several billion solar masses, which is much larger than a typical dwarf galaxy, and indeed is comparable to the entire HI mass of a large spiral. In NGC 4650A, the mass of the polar ring in HI alone is comparable to that of the luminous part of the central galaxy.

Another possibility is that polar rings represent the delayed inflow of primordial gas, in the same way that the disks of spiral galaxies might form over a Hubble time by the infall of external gas. If the incoming material did not share the rotation axis of the inner galaxy, a structure like a polar ring would result. The giant ring in Leo (Schneider et al. 1989), a large (200 kpc diameter) coherently rotating structure containing at least a billion solar masses of hydrogen, may be a larger-scale example of such a phenomenon.

One obvious test of the origin of polar rings is to measure the heavy-element abundances in the ring material. If the rings are indeed captured dwarf galaxies, abundances around a few tenths of solar might be expected. Extremely low abundances are likely to indicate that the the ring gas was primeval; since the rings are thin and spread-out, it is hard to see how they could retain gas lost by evolving stars, and any metals which are now present are probably a relic of a time when the ring material had a more compact form.

References

Mould, J. et al. (1982). Astrophys. J. **260**, L37.

Nicholson, R.A. (1989). PhD Thesis, University of Sussex.

Sackett, P.D. & Sparke, L.S. (1990). Astrophys. J. **361**, 408.

Schechter, P.L., Sancisi, R., van Woerden, H. & Lynds, C.R. (1984a). Mon. Not. R. Astron. Soc. **208**, 111.

Schechter, P.L., Ulrich, M.-H. & Boksenberg, A. (1984b). Astrophys. J. **277**, 531.

Schneider, S.E., Skrutskie, M.F., Hacking, P.B., Young, J.S., Dickman, R.L., Claussen, M.J., Salpeter, E.E., Houck, J.R., Terzian, Y., Lewis, B.M. & Shure, M.A. (1989). Astron. J. **97**, 666.

Schwarz, U.J. (1985). Astron. Astrophys. **142**, 273.

Schweizer, F., Whitmore, B.C. & Rubin, V.C. (1983). Astron. J. **88**, 909.

Sparke, L.S. (1990). In *Dynamics and Interactions of Galaxies* (ed. R. Wielen), p. 338. Berlin: Springer.

Sparke, L.S. (1986). Mon. Not. R. Astron. Soc. **219**, 657.

Steiman-Cameron, T.Y. & Durisen R.H. (1988). Astrophys. J. **325**, 26.

van Gorkom, J.H., Schechter, P.L. & Kristian, J. (1987). Astrophys. J. **314**, 457.

Whitmore, B.C. (1984). Astron. J. **89**, 618.

Whitmore, B.C., Lucas, R.A., McElroy, D.B., Steiman-Cameron, T.Y., Sackett, P.D., Olling, R.B. (1990). Astron. J. **100**, 1489.

Mergers and the structure of disk galaxies

LARS HERNQUIST
University of California, Santa Cruz

1. Introduction

There is a growing body of evidence that indicates that environmental effects play a central role in the evolution of galaxies. Transient encounters are likely responsible for unusual features seen in at least some disk galaxies. Using a restricted method, Toomre and Toomre (1972) demonstrated that the narrow "bridges" and "tails" frequently associated with nearby pairs of galaxies are a natural consequence of grazing collisions between disks. Toomre and Toomre were also able to reproduce the global morphology of M 51, strengthening the interpretation that grand–design spiral structure is driven by interactions of galaxies. Refined models show that self–gravity does not adversely affect either of these claims (e.g., Barnes 1988; Hernquist 1990a). Observations that interacting galaxies tend to be bluer than their isolated counterparts are most easily explained by enhanced rates of star formation during collisions (Larson and Tinsley 1978). Though controversial, there are hints that peculiar activity in the nuclei of some galaxies, such as Seyferts, may be triggered by transient encounters (Dahari 1984; Kennicutt and Keel 1984; see, however, MacKenty 1989, 1990; Fuentes-Williams and Stocke 1988).

There are also compelling reasons to believe that the accretion of intergalactic material and complete mergers are relevant to some aspects of galactic evolution. It has been suggested that disks in spirals (e.g., Gunn 1982), nuclear activity in some ellipticals (Gunn 1979), and "shells" around galaxies (Malin 1979) result from various types of late infall. Some ellipticals may have formed from mergers of comparable–mass disk galaxies (Toomre 1977) and many of the brightest infrared objects (e.g., Sanders et al. 1988) and radio sources (e.g., Heckman et al. 1986) appear to be galaxies that are either currently merging or have just recently done so.

Simple estimates suggest that mergers with less–massive companions can have a profound influence on the structure of disks in spiral galaxies. This review focuses on such events where the mass ratio between the companion and victim disk is quite small, say $\lesssim 10\%$. In more violent mergers, i.e., those involving more massive companions, disks are completely destroyed and the remnants are more akin to ellipticals rather than any kind of spiral galaxy (e.g., Barnes 1988).

Why are mergers between spiral galaxies and their satellites of interest? Kinematic evidence supports the view that most galaxies are surrounded by dark halos having masses significantly larger than the luminous masses of disks. Owing to dynamical friction, satellites' orbits decay and they may eventually be consumed by the larger galaxy. Applying Chandrasekhar's treatment of drag to orbital decay, Ostriker and Tremaine (1975) and later Tremaine (1981) showed that the mass accreted in satellites by an isothermal halo is roughly

$$M_{acc} \sim 4 \times 10^8 \left[(H_0 t) \ln \Lambda\right]^{0.6} \left[\frac{100\,\mathrm{km\,s^{-1}}}{\sigma}\right]^{0.6} \left(\frac{M}{L}\right)^{1.6} h^{-2}\, M_\odot, \quad (1)$$

where $H_0 t$ measures time in units of the Hubble time, $\ln \Lambda$ is a Coulomb logarithm, σ is the halo velocity dispersion, M/L is the constant mass–to–light ratio, and $h = H_0/100$. For values thought to be typical of galaxies like our own, $\ln \Lambda \sim 3$, $\sigma \sim 200\,\mathrm{km\,s^{-1}}$, and $M/L \sim 10$, equation (1) predicts $M_{acc} \sim 2 \times 10^{10} h^{-2} M_\odot$ over a Hubble time. This mass is similar to that of the Large Magellanic Cloud and is roughly one–third the mass of our own disk (e.g., Binney and Tremaine 1987).

Note, however, that this estimate of M_{acc} is subject to considerable uncertainty. Halos are not infinite in extent, as equation (1) assumes. The rate of mass accretion would be greatly reduced if galactic halos are relatively compact and do not extend much further than the luminous matter. Even if M_{acc} were as large as the estimate above, it does not necessarily follow that all this mass would be added to disks. Loosely–bound satellites would be tidally destroyed before sinking deep into the potential well of the primary; a rigorous treatment of the response of disks to mergers must ultimately account for the structure of the companion and mass stripping. Equation (1) was derived for a scale–free two–point correlation function, $\xi(r) \propto r^{-1.8}$, a Schechter luminosity function, and a constant mass–to–light ratio, all of which were taken to be independent of epoch. It is difficult to see how any of these assumptions could be strictly correct.

Based on these caveats, qualitative estimates for M_{acc} could easily be in error by factors $\gtrsim 10$. Although observations of high–redshift objects may help by providing plausible choices for $\xi(r)$, it is likely that these various complications will not be completely resolved independently of large–scale modeling. Definitive answers will probably not be forthcoming until our theories of galaxy formation have crystallized and cosmological simulations become significantly more detailed than those currently available. In the meantime, some progress can be made by inverting the problem: the consequences of individual events can, in principle, be used to constrain cosmological parameters. Owing to the complexity of the merger process, quantitative predictions of the effects of merging are not possible without appealing to numerical simulation.

2. Numerical models

The simulations performed by Quinn and Goodman (1986) using a restricted method demonstrate that the accretion of satellites with as little as a few percent the primary mass can severely damage disks. However, as emphasized recently by Ostriker (1989) a proper treatment of self–gravity is critical to our interpretation of this phenomenon; the self–consistent response of disks can, in fact, enhance dynamical heating. On the other hand, tidal stripping can significantly reduce the damage sustained by disks if most of the companion's mass is shed before merging is completed.

An example of the self–consistent response of a disk galaxy to a decaying satellite is shown in Figure 1. In this calculation, the primary consists of a disk, a bulge, and a halo. The density profile of the disk is based on the Bahcall–Soneira model of our own galaxy (Bahcall and Soneira 1980) and the halo is a truncated version of an isothermal sphere (Hernquist 1990b). Its properties are such that that the rotation curve of the composite system resembles that of our own galaxy. The bulge and satellite are both derived from the density distribution

$$\rho(r) = \frac{M}{2\pi} \frac{a}{r} \frac{1}{(r+a)^3}, \qquad (2)$$

where M is the mass and a is a scale–length. As noted by Hernquist (1990c) the projected light distribution of this model well-approximates the de Vaucouleurs $R^{1/4}$ law.

Initially, the satellite had a mass 10% that of the victim disk and was on a circular orbit in the disk plane at $6h$, where h is the exponential scale-length of the disk. The disk-to-bulge mass ratio is $D : B = 10 : 1$. The full self-gravity of all components is included using a hierarchical tree algorithm (Barnes and Hut 1986; Hernquist 1987).

Note, in particular, the spiral structure induced by the tidal field of the decaying satellite. An analogous response has been seen in similar models with rigid halos (Hernquist 1989a,b) and also in simulations of transient encounters resembling M 51 (Hernquist 1990a). At later times when the merger is essentially complete, a weak bar bar is excited near the center of the disk. Owing to the combined friction of the disk and halo the orbit decays more quickly than in models which ignore the self-consistent response of the halo. In fact, the satellite in Figure 1 barely finishes two revolutions before it is absorbed and destroyed by the disk.

3. Physical effects

3.1. Warps

Numerical experiments like that pictured in Figure 1 demonstrate that warps can be excited in the outer parts of disks by decaying satellites (Quinn et al. 1990). Figure 2 shows the structure resulting in a disk following the accretion of a companion with 10% its mass; in this merger the orbit was initially inclined by 30° with respect to the disk plane. A warp is clearly visible in both the projected particle data and its contoured image.

While the model in Figure 2 shows that warps can be excited by tidal effects, it is not clear how long they can survive. As is well-known, radial features in disks wind-up owing to differential precession. A similar fate awaits the warp in Figure 2 and it will eventually fade away. However, the time-scale for this to occur can be long, especially if the potential is nearly spherical in the warped regions. Figure 3 shows the ratio of precession to rotation frequency for the disk used in the simulation which generated Figure 2. For radii $4h \lesssim r \lesssim 6h$, where the warp is most noticeable, this ratio is $\lesssim 0.05$, implying that the warp will persist for $\gtrsim 10$ rotation peri-

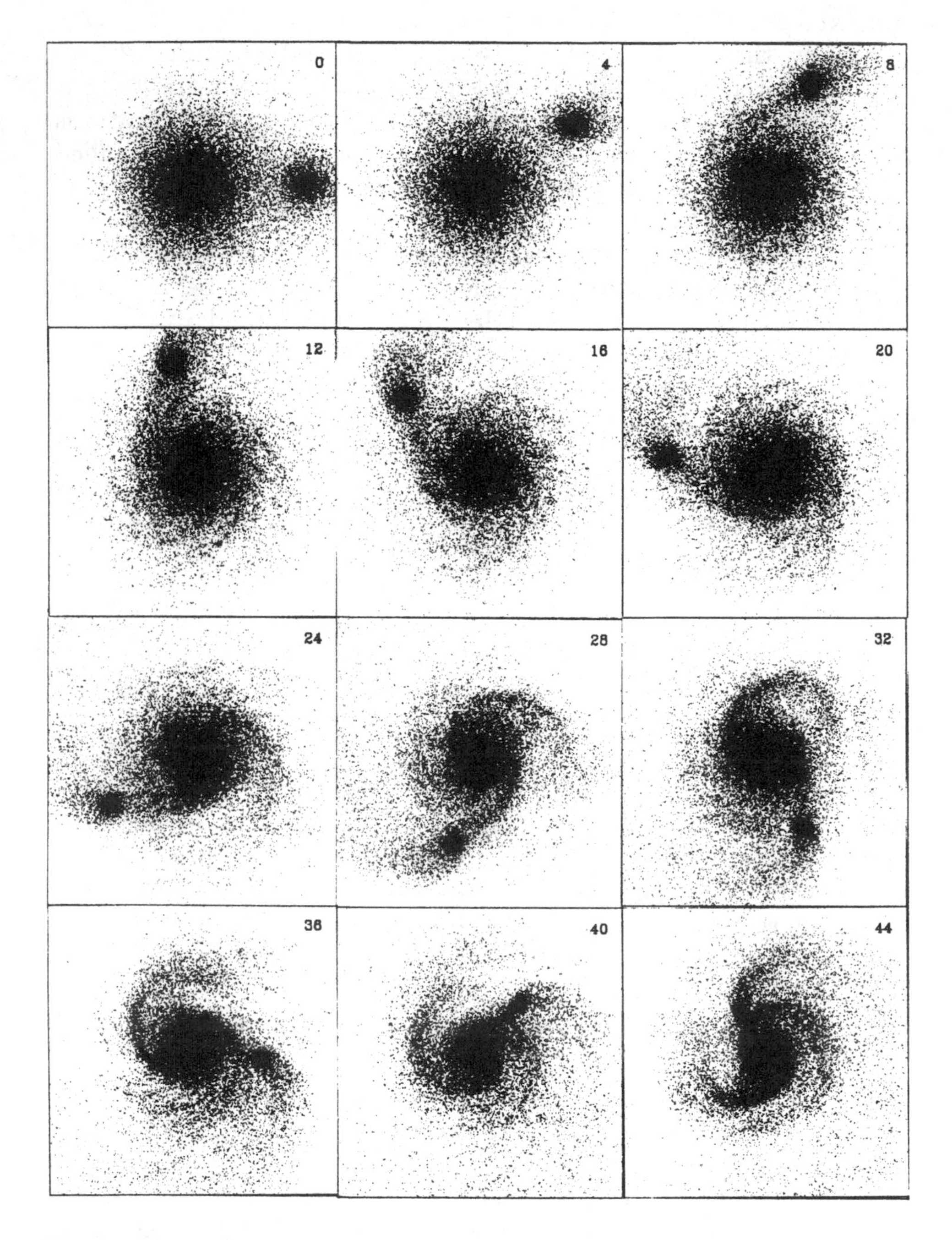

Fig. 1a. Merger between a disk and a satellite of 10% its mass. Time, shown in upper right corner of each frame, is in a dimensionless system of units. Scaled to our own disk, the unit time is roughly 1.3×10^7 years. Each panel measures 15 exponential scale-lengths of the disk per edge.

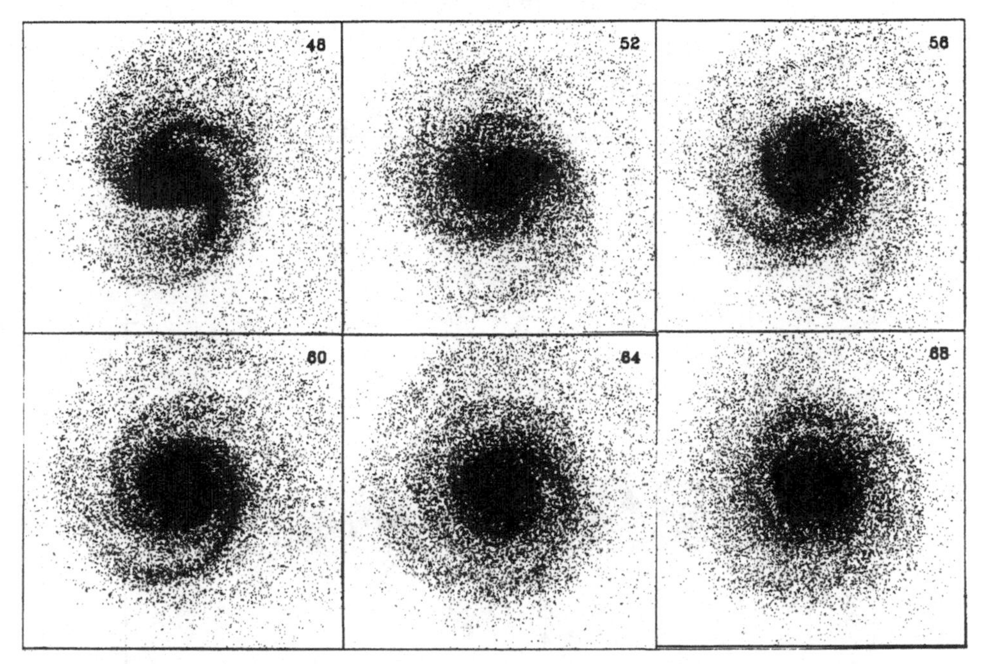

Fig. 1b. Continuation of the run shown in Fig. 1a.

ods. This estimate is, in fact, supported by the long-term behavior of the N-body simulation. Scaling to the properties of a galaxy with a flat rotation curve and an asymptotic circular speed of $220\,\mathrm{km\,s^{-1}}$, the persistence time of the warp is roughly $3-5\times10^9$ years, which is a significant fraction of the age of our own disk.

In principle, then, satellite mergers can excite warps in disks that will live long enough to be of interest to observers. In this regard, it is interesting to note that most galaxies with warps do not appear to have nearby companions. However, this does not imply that mergers are an attractive or even plausible scenario for explaining all warps. The inner parts of the disk in Figure 2 were dynamically heating during the final stages of the merger, more than doubling its scale-height. Unless satellites are destroyed before sinking to the centers of galaxies one would expect most disks with warps to show other evidence for tidal damage; this does not appear to be the case. The precession rate will be higher and so warps will fade more quickly if the halo is non-spherical. This latter objection is not necessarily fatal, however, since a misalignment between the symmetry axes of the disk and a non-spherical halo induced by accreted matter can sustain warps (Ostriker 1991).

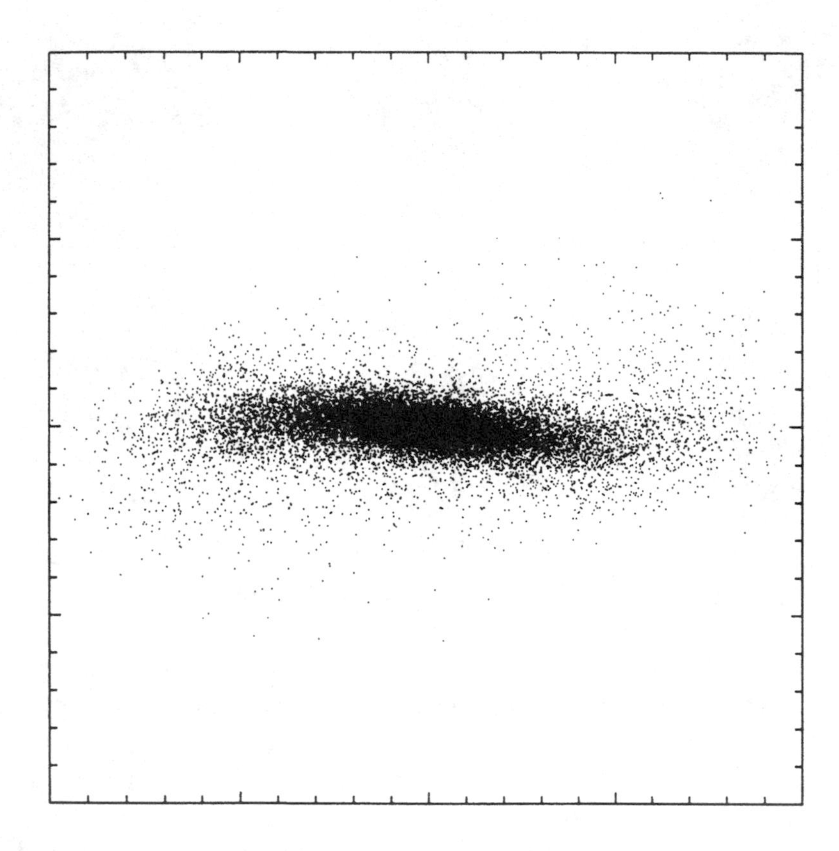

Fig. 2a. Edge–on projection of structure in a disk following the accretion of a satellite of 10% its mass. Particle data from N–body simulation. The frame measures roughly 17 exponential scale-lengths of the disk per edge.

3.2. Mass stripping

Like the satellite shown in Figure 1, companions of real galaxies continuously lose material through tidal stripping. Debris from satellites may explain a number of galactic phenomena. It is generally believed that the Magellanic Stream consists of material liberated from the Magellanic Clouds. This inference has been used to argue that the Magellanic Clouds are on nearly polar orbits around the Milky Way (see Binney and Tremaine 1987). Dynamical models of the Stream have also been used to constrain the galactic potential (e.g., Lin and Lynden–Bell 1982).

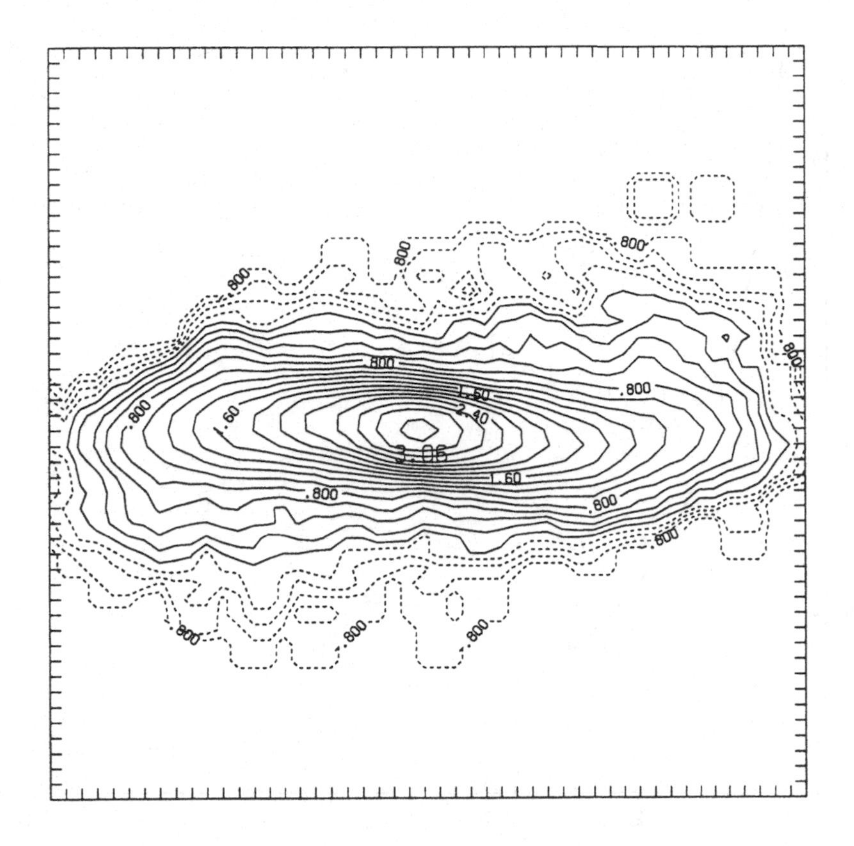

Fig. 2b. As in Fig. 2a, but density shown as contour plot.

A significant fraction, perhaps as many as half of stars in our own halo are on retrograde orbits. It is difficult to account for this observational fact in simple models for the formation of the galaxy through the monolithic collapse of nearly homogeneous gas clouds in rotation (e.g., Eggen et al. 1962). Models in which disk galaxies are formed by the agglomeration of sub–clumps may produce retrograde halo stars, provided that random velocities are sufficiently large (e.g., Larson 1990). Alternatively, some fraction of these stars may be debris from satellites which were tidally destroyed. A few moderately large companions on retrograde orbits would contain enough stars to significantly modify the halo population. It remains to be determined, however, if such a scenario could account for the spatial distribution of halo stars and their metallicities.

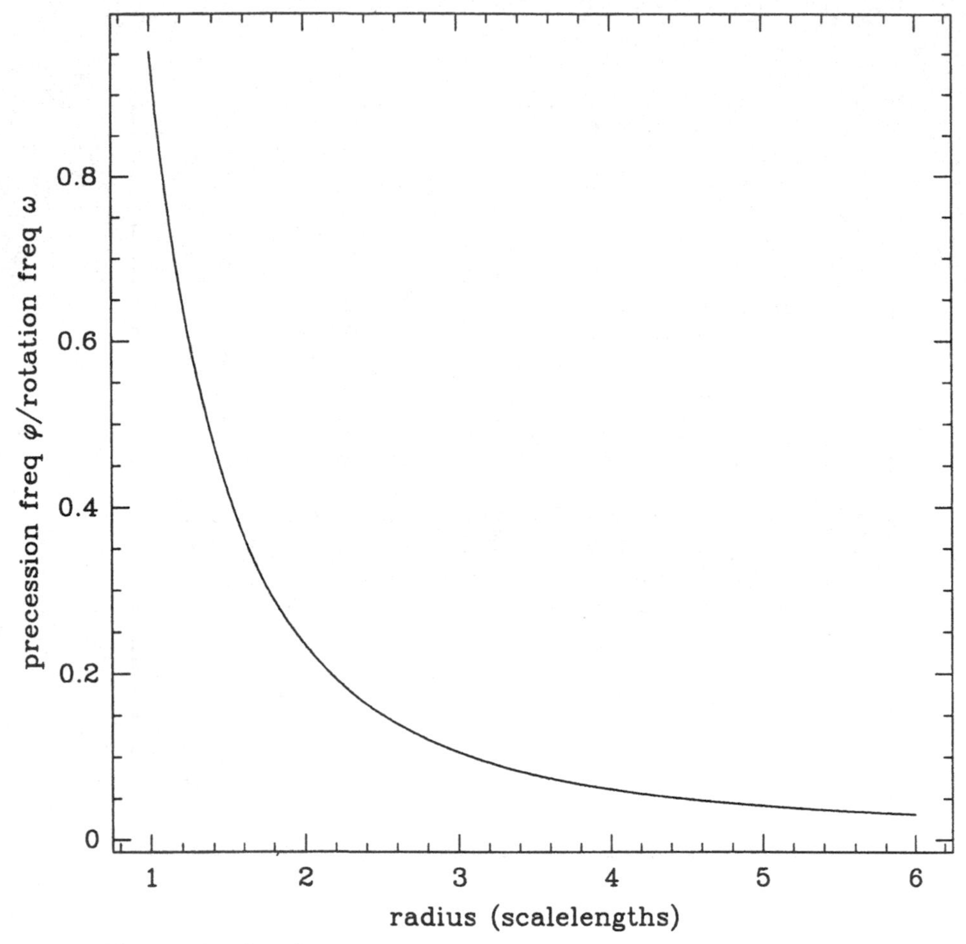

Fig. 3. Ratio of precession frequency to rotation frequency as a function of radius, for the disk used in Fig. 2.

Similar confusion exists over the origin of globular clusters in our own galaxy. It appears that globulars around the Milky Way are separated into two components: an older population in the inner halo with little age spread, and the other in the outer halo which possesses a larger age spread and which is relatively younger (Zinn 1980, 1985). Searle and Zinn (1978) proposed that globular clusters form in satellite galaxies accreted by the Milky Way, a picture that is especially appealing for the outer halo globulars, owing to their large age spread. Recently, Freeman (1990) has noted that nucleated dwarf ellipticals (dEn) have properties well–suited for building globular clusters. The capture and subsequent tidal disruption of ~ 100 dEn could account for the halo globulars as well as a significant fraction of the field population.

Shells are a common property of elliptical galaxies (e.g., Malin and Carter 1983). It is generally believed that these features result from the accretion of material from less–massive companions (Quinn 1984; Hernquist and Quinn 1988). Faint, sharp–edged features like shells are more difficult to detect in disk galaxies owing to confusion with spiral arms. However, deep CCD observations by Schweizer and Seitzer (1988) show that shells exist in spiral galaxies, in accord with theoretical expectations (e.g., Hernquist and Quinn 1989a). If these features also represent the aftermath of accretions of less–massive companions, then those disks with shells should contain other signatures of merger events.

3.3. Gas–dynamics

Though gas typically constitutes only a small fraction of the luminous mass in present–day galaxies its dynamics are critical to the understanding of many galactic phenomena. The notion that the accretion of small satellites might somehow be responsible for strange behavior seen in the nuclei of some galaxies was suggested already in 1979 by Gunn for ellipticals and then later by Gaskell (1985) for disks. Only within the past few years have computational software and hardware evolved to the point where processes involving galactic gas can be modeled with some confidence.

Detailed simulations of collisions between gas–free satellites and disks consisting of both gas and stars show that such events are capable of triggering unusual behavior in galactic nuclei (Hernquist 1989a,b). Unlike the mechanism proposed by Gunn (1979) and Gaskell (1985) the gas which accumulates in the nucleus is not carried there by the companion, but is from the disk itself. Owing to dissipation, the gas behaves differently in response to the tidal field than do the disk stars giving rise to a strong collective interaction between the two. This interaction is such that the stars exert a strong gravitational torque on the gas, removing its angular momentum. This process appears to be quite efficient and the time–scales and inflow rates are similar to those inferred for starburst nuclei.

The relevance of these results to active galactic nuclei is unclear. The numerical models lack sufficient dynamic range to predict whether or not the gas which accumulates on scales $\sim 100\,\mathrm{pc}$ will continue to shed sufficient angular momentum to be accreted by a central black hole. Moreover, since the models have thus far ignored star formation, it remains to be seen whether or not energy input by supernovae could significantly modify the rate of gas inflow. It is possible that a sufficiently violent release of thermal and mechanical energy could disperse the gas before it reaches the nucleus.

The remnants of mergers between satellites and disks containing both gas and stars resemble distorted spiral galaxies which lack gas in their inner regions. Processes similar to those described above can also occur in mergers between comparable–mass disks consisting of both gas and stars (Barnes and Hernquist 1990; Hernquist 1991).

Additional effects may occur if the satellite itself contains gas. Since many dwarfs are gas–rich, the addition of their mass to a disk like ours would significantly modify its gas content. Ram–pressure stripping is likely responsible for observed properties of some disks. For example, it has been suggested that high velocity gas and disk disturbances in some spiral galaxies, such as M101, may have resulted from the accretion of gas–rich companions (van der Hulst and Sancisi 1988).

3.4. Dynamical heating

Since disks are supported mainly by rotation, they are dynamically cold and the accretion of even a small amount of material can damage them considerably. If all kinetic energy of a sinking satellite goes into random velocities of disk stars then the change in the velocity dispersion of the disk is roughly (Ostriker 1989)

$$\Delta\sigma^2 \approx \frac{M_{sat}}{M_{disk}} v_{orb}^2, \tag{3}$$

where M_{sat}/M_{disk} is the satellite–to–disk mass ratio and v_{orb} is the orbital velocity of the satellite. If $M_{sat}/M_{disk} = 0.1$ and $v_{orb} = 220\,\mathrm{km\,s^{-1}}$, then $\Delta\sigma \approx 70\,\mathrm{km\,s^{-1}}$. The corresponding one–dimensional velocity dispersion is similar to that of our own disk locally. The effect of this dynamical heating is to increase both the vertical scale–height and the Toomre Q–value of disks.

More generally, not all the energy of a decaying satellite will be available for increasing random motions of disk stars. Some will be deposited into the halo, also through dynamical friction. A fraction of the energy actually absorbed by the disk is stored as potential energy, rather than kinetic, as the disk expands vertically (Toth and Ostriker 1990). Gas in the disk can also reduce the damage somewhat by radiating; however, this effect is certainly small unless the disk contains much more gas than those in present–day galaxies. Other effects can enhance the response of the disk. For example, the self–consistent response of disks to tidal perturbations can scatter orbits, resulting in additional heating.

A more detailed analysis, taking into account these various factors, shows that if the vertical thickness of our own disk resulted entirely from dynamical heating due to satellite mergers then the accreted mass is limited to (Toth and Ostriker 1990)

$$M_{acc} \lesssim 0.05 \frac{M_{disk}}{\eta}, \tag{4}$$

where η is an efficiency factor. Typically, $\eta \sim 1.7$ and roughly equal amounts of energy are deposited into vertical and planar motions. This implies that less than 3% the mass of our disk could have been added in the form of discrete entities. If we make the reasonable assumption that part of the disk thickness is intrinsic, then this value is reduced even further. As emphasized by Ostriker and Toth, this is a severe constraint since most cosmological models with $\Omega = 1$ predict a significantly larger infall rate over the lifetimes of disks.

In principle, the dynamical "coldness" of disks we see today can limit cosmological parameters. However, a number of issues must be resolved before these arguments can be made rigorous. Mass stripping can reduce the damage done to disks by decaying satellites: loosely bound companions may be tidally destroyed before causing significant damage. Numerical models (Quinn and Goodman 1986; Hernquist and Quinn 1989b) show that the heating is highly non–uniform. The vertical damage to disks is most noticeable at large radii since satellites sink into the disk plane before decaying radially inwards. This suggests that Q–values may be a more sensitive diagnostic than disk thicknesses. The halos of at least some galaxies may be compact, not extending much beyond the luminous matter. In that event satellite mergers would be rare.

In addition to constraining cosmological models, the dynamical heating of disks by satellite mergers likely has observational implications. Moderate amounts of heating may explain the "thick disk" component thought to exist in some spiral galaxies (e.g., Gilmore and Reid 1983).

Dynamical heating can disrupt spiral structure in disks (Hernquist 1989a,b). Swing–amplification theory predicts that the fastest growing wave–numbers are given roughly by $m = (M_{disk} + M_{halo})/M_{disk}$, where m is the number of arms (Toomre 1981). If halo masses within the region occupied by the luminous matter are a few times that of the disk, them one expects $m \sim 4$. Through dynamical heating some fraction of the disk will, crudely speaking, no longer participate in spiral–making and behave more like a background contribution to the potential. In that case, $m = (M_{disk} + M_{halo})/M_{active}$, where M_{active} is the mass which can still produce noticeable spiral structure. Simulations of mergers between satellites and disks with both gas and stars show that in extreme cases $M_{active} \sim M_{gas}$, where M_{gas} is the mass in gas (Hernquist 1989a,b). If $M_{gas}/M_{disk} \sim 0.1$, then $m \sim 40$ for the same halo parameters as above. It seems unlikely that any realistic disk could support such high–wavenumber structure for any reasonable length of time. Remnants of disk–satellite mergers resemble amorphous galaxies (Sandage and Brucato 1979): objects with distorted disks exhibiting recent or ongoing star formation, but which have little or no spiral structure. Models showing that tidal effects can also drive gas into the nuclei of disks support the view that amorphous Seyfert galaxies result from mergers (MacKenty 1989, 1990).

3.5. Formation of bulges

Some mass from sufficiently dense sinking satellites is likely to penetrate to the nucleus of the victim galaxy. It has been suggested by various authors that such events may be responsible for forming bulges in spiral galaxies (e.g., Schweizer 1990). In fact, accreted material may also explain boxy isophotes and X–structures seen in some bulges (Hernquist and Quinn 1989a).

These hypotheses suffer from a number of potential problems. It may be difficult to accumulate significant quantities of mass in a nucleus without destroying the disk. This would not be an issue, of course, if disks form long after most satellite mergers have been completed. Observations indicate that bulges generally rotate and that the sense of the rotation is always identical to that of the disk (Kormendy 1990). If bulges form from the accretion of satellites then it is not obvious that the rotation axes of bulges and disks should always be aligned. Many stars in our own bulge appear to be old, yet metal–rich (Frogel 1988). Larson (1990) has argued that these stars are too metal–rich to have been born in dwarfs and were instead formed *in situ* from gas deposited into the nucleus. As noted by Schweizer (1990) some disk galaxies have unusually blue bulges, suggesting that they resulted from a recent merger. However, it is not clear that this interpretation is unique. Pfenniger and Norman (1990) have proposed that most bulges form from gas driven to the centers of galaxies by bars. Bulges produced in this way would appear unusually blue, at least initially.

3.6. Evolution along the Hubble sequence

Proceeding along the Hubble sequence from late–type to early–type spirals through ellipticals, it is generally the case that: the hot component increases in dynamical importance, disks become thicker and eventually vanish, the gas content drops, and spiral structure becomes less dominant. Based on discussion similar to that above, it has been suggested that the Hubble sequence represents a diary of merger events rather than being determined by conditions early in the formation of galaxies (e.g., Schweizer 1990). Thus, mergers of pure disks with less–massive companions may simultaneously yield bulges and thickened disks having little spiral structure, while mergers involving comparable–mass disks may produce ellipticals (Toomre 1977). Though appealing for its simplicity, such a grand–scheme is likely too simple–minded to account for all aspects of galaxy formation and evolution. Nevertheless, mergers between disks and less–massive satellites do occur and the remnants of these events will probably lie earlier along the Hubble sequence than their progenitors.

4. Discussion

Observations indicate that mergers between disks and less–massive satellites are an ongoing phenomenon. Simple estimates imply that the impulsive addition of even a small amount of mass can have a significant influence on the structure and evolution of spiral galaxies. Such events may play a role in the origin of warps, halo stars and globulars, nuclear starbursts and activity, thick disks, amorphous galaxies, and bulges. Though suggestive, the modeling to date is certainly not conclusive; many important issues remain to be addressed. It is hoped that advancements in computational technology will accelerate our understanding of these fundamental processes.

Acknowledgements

This work was supported in part by a grant from the Pittsburgh Supercomputing Center.

References

Bahcall, J.N. & Soneira, R.M. (1980). Astrophys. J., Suppl. Ser. **44**, 73.
Barnes, J. (1988). Astrophys. J. **331**, 699.
Barnes, J. & Hernquist, L. (1990). Astrophys. J., Lett., in press.
Barnes, J. & Hut, P. (1986). Nature **324**, 446.
Binney, J. & Tremaine, S. (1987) *Galactic Dynamics.* Princeton: Princeton University Press.
Dahari, O. (1984). Astron. J. **89**, 966.
Eggen, O.J., Lynden-Bell, D., & Sandage, A.R. (1962). Astrophys. J. **136**, 748.
Freeman, K.C. (1990). In *Dynamics and Interactions of Galaxies* (ed. R. Wielen), p. 36. Berlin: Springer.
Frogel, J.A. (1988). Annu. Rev. Astron. Astrophys. **26**, 51.
Fuentes-Willams, T. & Stocke, J.T. (1988). Astron. J. **96**, 1235.
Gaskell, C.M. (1985). Nature **315**, 386.
Gilmore, G. & Reid, N. (1983). Mon. Not. R. Astron. Soc. **202**, 1025.
Gunn, J.E. (1979). In *Active Galactic Nuclei* (eds. C. Hazard and S. Mitton), p. 213. Cambridge: Cambridge University Press.
Gunn, J.E. (1982). In *Astrophysical Cosmology* (eds. H.A. Brück et al.), p. 233. Vatican City: Specola Vaticana.
Heckman, T.M., Smith, E.P., Baum, S.A., van Breugal, W.J.M., Miley, G.K., Illingworth, G.D., Bothun, G.D. & Balick, B. (1986). Astrophys. J. **311**, 526.
Hernquist, L. (1987). Astrophys. J., Suppl. Ser. **64**, 715.
Hernquist, L. (1989a). Nature **340**, 687.
Hernquist, L. (1989b). Ann. N. Y. Acad. Sci. **571**, 190.
Hernquist, L. (1990a). In *Dynamics and Interactions of Galaxies* (ed. R. Wielen), p. 108. Berlin: Springer.
Hernquist, L. (1990b). In preparation.
Hernquist, L. (1990c). Astrophys. J. **356**, 359.
Hernquist, L. (1991). This volume.
Hernquist, L. & Quinn, P.J. (1988). Astrophys. J. **331**, 682.
Hernquist, L. & Quinn, P.J. (1989a). Astrophys. J. **342**, 1.
Hernquist, L. & Quinn, P.J. (1989b). In *The Epoch of Galaxy Formation*, (ed. C.S. Frenk et al.), p. 427. Dordrecht: Kluwer.
Kennicutt, R.C. & Keel, W.C. (1984). Astrophys. J., Lett. **279**, L5.
Kormendy, J. (1990). Private communication.
Larson, R.B. (1990). Publ. Astron. Soc. Pac. **102**, 709.
Larson, R.B. & Tinsley, B.M. (1978). Astrophys. J. **219**, 46.
Lin, D.N.C. & Lynden-Bell, D. (1982). Mon. Not. R. Astron. Soc. **198**, 707.
MacKenty, J.W. (1989). Astrophys. J. **343**, 125 .
MacKenty, J.W. (1990). Astrophys. J., Suppl. Ser. **72**, 231 .

Malin, D.F. (1979). Nature **277**, 279.
Malin, D.F. & Carter, D. (1983). Astrophys. J. **274**, 534.
Ostriker, E. (1991). This volume.
Ostriker, J.P. (1989). In *The Evolution of the Universe of Galaxies*, in press.
Ostriker, J.P. & Tremaine, S. (1975). Astrophys. J., Lett. **202**, L113.
Pfenniger, D. & Norman, C. (1990). In *Dynamics and Interactions of Galaxies* (ed. R. Wielen), p. 485. Berlin: Springer.
Quinn, P.J. (1984). Astrophys. J. **279**, 596.
Quinn, P.J. & Goodman, J. (1986). Astrophys. J. **309**, 472.
Quinn, P.J., Hernquist, L., & Fullager, D. (1990). In preparation.
Sandage, A. & Brucato, R. (1979). Astron. J. **84**, 472.
Sanders, D.B., Scoville, N.Z., Sargent, A.I. & Soifer, B.T. (1988). Astrophys. J., Lett. **324**, L55.
Schweizer, F. (1990). In *Dynamics and Interactions of Galaxies* (ed. R. Wielen), p. 60. Berlin: Springer.
Schweizer, F. & Seitzer, P. (1988). Astrophys. J. **328**, 88.
Searle, L. & Zinn, R. (1978). Astrophys. J. **225**, 357.
Toomre, A. (1977). In *The Evolution of Galaxies and Normal Stellar Populations* (eds. B.M. Tinsley and R.B. Larson), p. 401. New Haven: Yale University Observatory.
Toomre, A. (1981). In *The Structure and Evolution of Normal Galaxies* (eds. S.M. Fall and D. Lynden–Bell), p. 111. Cambridge: Cambridge University Press.
Toomre, A. & Toomre, J. (1972). Astrophys. J. **178**, 623.
Toth, G. & Ostriker, J.P. (1990). Preprint.
Tremaine, S. (1981). In *The Structure and Evolution of Normal Galaxies* (eds. S.M. Fall and D. Lynden–Bell), p. 67. Cambridge: Cambridge University Press.
van der Hulst, J.M. & Sancisi, R. (1988). Astron. J. **95**, 1354.
Zinn, R. (1980). Astrophys. J. **241**, 602.
Zinn, R. (1985). Astrophys. J. **293**, 424.

Formation of polar rings

HANS-WALTER RIX & NEAL KATZ
Steward Observatory, University of Arizona

I. Introduction

One of the few points regarding polar rings over which there seems to be general agreement is that the material constituting the rings must have been acquired from outside the host galaxy into its present location after the formation epoch of the host galaxy (e.g., Schechter and Gunn, 1978; Schweizer et al. 1983; van Gorkom et al. 1987). Two possible scenarios for such an acquisition process are a) the orbital decay of a gas rich satellite and its ensuing tidal disruption, and b) the transfer of gas into the potential well of the host galaxy during an encounter with a neighbouring galaxy.

Subsequently, the gas will "smear out" to a more or less axisymmetric configuration, a "ring" or a "disk". Almost all existing theoretical work in the field (e.g., Tohline et al. 1982, Steiman-Cameron and Durisen 1988, 1990) assumes such an axisymmetric configuration as the starting point for its analysis, i.e., differential precession and settling of a ring or disk towards a preferred plane is calculated assuming axisymmetric initial conditions. The rationale behind this is simple: With the additional symmetry, an analytic treatment of precession and settling (e.g., Steiman-Cameron and Durisen 1988, 1990) is possible and the fixed geometry facilitates the use of grid based numerical methods (e.g., Christodoulou and Tohline, this volume). However, it is not clear *a priori* whether the separation of the azimuthal smearing and the settling is justified. In any situation where the precession time scale is comparable to the smearing time scale, the two processes will proceed simultaneously. Furthermore, if the gas is smearing out along a strongly precessing orbit, it is conceivable that it may never reach the configuration that the above mentioned settling calculations assume as their starting point.

In investigating the formation of polar rings, or gas rings and disks in early type galaxies in general, one would like to address the following questions:

- What is the typical timescale for "smearing out" and what role does self-gravity play in this process?
- How long does it take the gas to "find" the closed, non-intersecting orbits on which it should eventually end up?
- Does the gas smear out into a ring or disk-like structure for all (or most) plausible orbital parameters and potential shapes?
- What radial profiles result from the azimuthal smearing and how do they depend on the initial orbital eccentricity?

This paper reports on preliminary steps towards answering these questions. Our approach to studying the formation of polar rings is to investigate the tidal disruption and subsequent long-term evolution of initially spherical, gaseous satellites inside a rigid potential well of a specified shape. As we will illustrate below, it is practical to conduct at least a very coarse survey of the relevant parameter space, which is essentially spanned by the shape of the potential (the radial profile and the axis ratios) and the initial orbit of the gas (the size of the orbit and its eccentricity). Obviously, numerical means are necessary to study this process since no simplifying assumptions, based on symmetry, can be made.

In the remainder of this paper, we outline some numerical experiments, designed to address the above questions: In Section II we briefly discuss the numerical method, in Section III we outline in general the initial conditions for the simulations and in Section IV we describe the simplest possible scenario, the disruption of a purely gaseous satellite on a circular orbit in a spherical potential. We summarize in Section V.

II. The numerical method

A realistic simulation of the scenario outlined above must take into account a number of phenomena:

1) The numerical method must be flexible regarding the geometry of the gas distribution, which changes from a compact, spherical distribution to a much larger, thin ring. This change of geometry and the subsequent flattening of the ring excludes the use of grid-based techniques.
2) To study the smearing of the gas into a ring and the settling

toward a preferred orientation, an appropriate treatment of the gas hydrodynamics is needed.

3) Radiative losses from the gas (cooling) are an important sink of energy. This point will be justified *a posteriori* by the simulations described below.
4) If we envision the initial condition to be a gas sphere which is (at least partly) confined by self-gravity, we must include self-gravity in our calculations, since it will to some extent retard the disruption and smearing process.

An efficient numerical method which fulfills all these requirements is the TREESPH algorithm developed by Hernquist and Katz (1989). An extensive description is given in the mentioned reference. Here we only briefly mention a few details pertaining to our particular application. The TREESPH code can advance 1000 particles (which should be thought of as tracers of a continuous gas, rather than point masses) for one Hubble time using only 1.5 hours of CPU time on the Cray YMP at the Pittsburgh Supercomputing Center. Thus a number of runs large enough to explore parameter space is feasible. The time steps are chosen small enough to conserve the total energy to within a few tenths of one percent over the entire length of each run.

Is is worth emphasizing that this algorithm treats the gas as a continuous medium. This is in contrast to the "sticky particle" method used by T. Quinn (this volume) for similar simulations. In the latter algorithm the particles represent individual compact gas clouds which do not interact with each other (except gravitationally) unless they undergo an inelastic collision. In the "sticky particle" method forces due to pressure gradients are not included.

III. Setting up the numerical experiment

In this section we discuss general constraints on the initial conditions for simulations of polar ring formation arising from both the observed properties of polar rings and their host galaxies and from our preconceptions about plausible formation scenarios.

a) The satellite

As a first step it seems sensible to assume that the material forming the polar ring is purely (or predominantly) gas. As a simple model for a self-gravitating spherical gas configuration we choose an isothermal sphere. Since gas can cool very efficiently at temperatures above $10^4 K$, we assume an isothermal sphere of $10^4 K$, which we truncate to obtain the desired total mass for the satellite. Observations suggest that several $10^8 M_\odot$ is a typical ring mass (Schweizer et al. 1983; van Gorkom et al. 1987). For an isothermal structure, this leads to a satellite radius of $2.5\,\mathrm{kpc}(10^4 K/T)(m/10^8 M_\odot)$. Note that lowering the assumed gas temperature will make the satellite more extended for a given mass.

b) The galaxy potential well

The structure and mass of dark haloes in early type galaxies appears to be, at present, an open question. As extremes we consider both the "null hypothesis" of mass tracing light as well as the case of a massive isothermal halo.

At a radius of 30–40 kpc, the ring material lies well outside the luminous material of the host galaxy. Therefore, insensitive to assumptions about the shape of the luminous mass distribution, the potential would be very close to spherical at the radius of the polar ring if mass traces light. Specifically, for particular Staeckel models (de Zeeuw and Pfenniger 1988) which are scaled appropriately to match the luminous parts of galaxies, the orbital precession times at 30 kpc are far in excess of the Hubble time. Assuming such a potential, polar rings should form at all angles to the major axis of the host galaxy with equal probability.

Since there is some observational evidence (Schweizer et al. 1983), which manifests itself in the terminology "Polar Ring", that long-lived gas rings are preferentially found at high inclinations, it seems more plausible that the potential deviates substantially from sphericity even at radii of 30 kpc. A potential well with constant axis ratios can be conveniently represented by a triaxial, scale-free, logarithmic potential (Richstone 1980) which has been used (e.g., Steiman-Cameron and Durisen 1988, 1990) to study the evolution of settling gas disks and rings. We have chosen this particular potential, with a circular velocity of $230\,\mathrm{km\,s^{-1}}$, to allow an eventual comparison of our results with the above works and the SPH results of Habe and Ikeuchi (1985, 1988). It should be noted, however, that our SPH simulations are not scale free since the inclusion of cooling introduces constants which are fixed in physical units.

c) The initial orbit

To minimize the arbitrariness of the initial conditions we would like to pick an orbit whose pre-history is consistent with the subsequent dynamical evolution in the simulation, i.e., the gaseous satellite should have had a physically sensible way of getting to the state where we start our simulation. Two types of simple orbits satisfy this condition.

First, a circular, or nearly circular, orbit which keeps the satellite outside its tidal disruption limit. Dynamical friction would then cause the orbit to decay, leading to the disruption and azimuthal smearing of the satellite material. However, although only poorly constrained by observations, mass ratios $M_{\text{galaxy}}/M_{\text{satellite}}$ seem to be typically about 100. For such large mass ratios, the orbital decay time (from a radius at which the satellite would not be tidally disrupted) due to dynamical friction is long, for most cases in excess of the Hubble time. Only if the gas were confined in the potential well of a more massive, collisionless component would such a scenario appear plausible.

As a different means of enforcing orbital decay one could consider the ram pressure from the hot gaseous component present in the halo of the host galaxy. However, if the gas density estimates for ellipticals from Fabian and Thomas (1986) are appropriate, this mechanism is also too inefficient.

Second, one could consider a plunging orbit from infinity as the initial orbit, leading to the efficient disruption of the satellite. However, the debris would escape to large radii, causing the timescale for azimuthal smearing, and thus the formation time scale for polar rings, to exceed the Hubble time.

The arguments just outlined seem to make it necessary to invoke an encounter with a third body, such as a nearby galaxy, if one would like to consider purely gaseous satellites. Such an encounter could either move a previously existing satellite onto an orbit closer to the host galaxy or could cause mass transfer from the neighboring galaxy into the host potential well. This later process is directly observed in NGC 3808 (Whitmore et al. 1990).

We adopt this assumption that the gas was brought to its present orbit rather abruptly during the passage of a perturber. We mimic this situation by starting our simulation with the satellite on a roughly circular orbit at a radius such that half of the gas should be stripped instantly by tidal forces.

IV. Formation of gas rings in a spherical potential

As the simplest possible case we consider the disruption and the subsequent azimuthal smearing of gas on a circular orbit in a spherically symmetric potential. The potential is a logarithmic potential with $v_{circ} = 230\,\mathrm{km\,s^{-1}}$, the gas has a total mass of $2 \cdot 10^8 M_\odot$ and a temperature of $10^4 K$. The initial radius of the gas sphere is 5 kpc and the radius of the orbit is 38 kpc.

The time evolution of this model is shown in Figure 1. At least initially, the azimuthal smearing (as shown in the second column of Figure 1) is dominated by the tidal disruption and subsequent phase wrapping of the gas. For the geometry at hand, the orbital time at the outer edge of the initial configuration is about 30% longer than at the inner edge, thus the gas catches up to itself after about three orbital periods. The center manages to retain its identity for about five orbital periods, but after stripping the outer layers of the satellite the now unbalanced gas pressure at the center aids the disruption of the core. Note that even after 3.6 Gyrs the density varies with azimuth by over a factor of three. It takes over ten orbital periods (7 Gyrs) for the gas to become azimuthally homogeneous.

The right column of Figure 1 shows the vertical flattening of the ring, as the gas changes from being pressure supported to being predominantly rotationally supported. The resulting ring has a scale height of $\approx$500 pc. During this flattening and the circularization of the orbits (see below), thermal energy is lost from the cooling gas. Even in this "non-violent" case of smearing and settling the energy loss amounts to more than twice the initial thermal energy.

The left column of Figure 1 illustrates the evolution of the radial profile of the ring. Not surprisingly, the final ring forms at the radius of the initial orbit. Its radial extent is not much larger than the size of the initial configuration. Simulations with non-circular orbits show that the radial extent is determined by either the size of the initial gas sphere or by the radial variation of the initial orbit, whichever is larger. The "flatness" of the final *radial* profile is remarkable, most of the mass is contained in a region with a density contrast of less than a factor of two! This result also holds for orbits of moderate eccentricity and does not seem to be sensitive to the density profile of the initial configuration (which was much steeper).

One of the questions we raised in the introduction was how long it would take the gas to "find" closed, non-intersecting orbits, which constitute the equilibrium orbits for the gas. In spherical potentials,

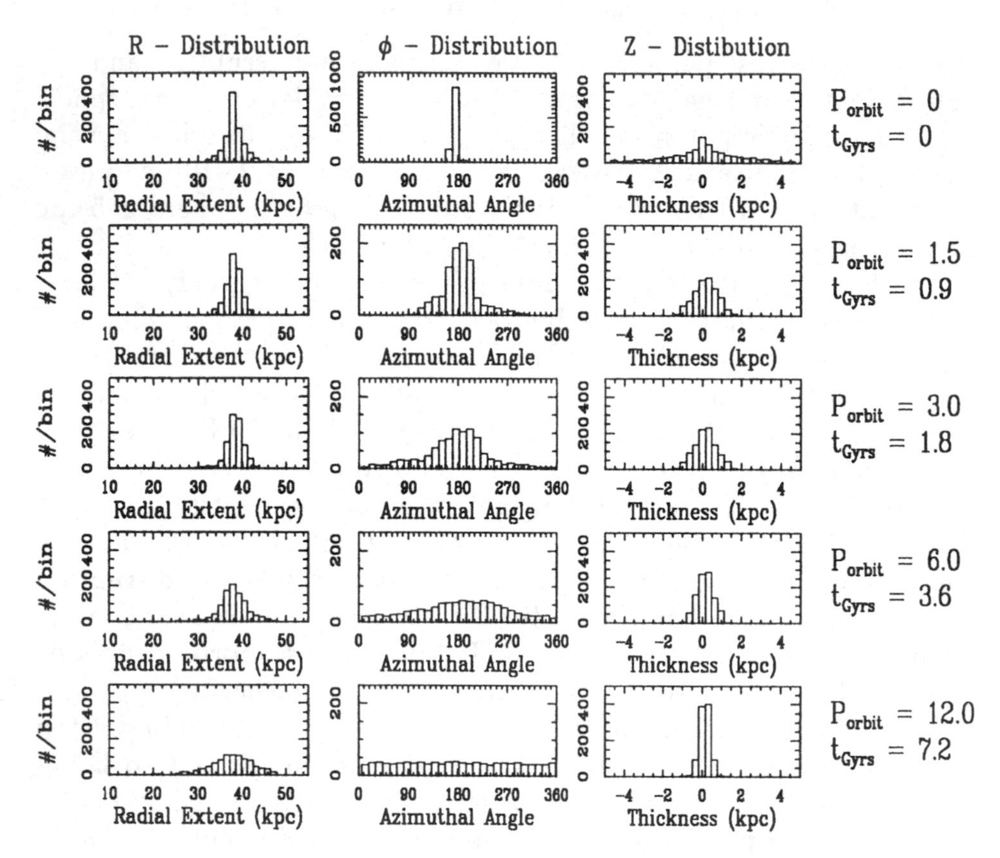

Fig. 1. Disruption and subsequent smearing of a gas sphere on a circular orbit (R_{orbit} = 38 kpc) in a spherically symmetric logarithmic potential of $v_{circular}$ = 230 km s^{-1}. Each panel shows a binned distribution of the particles, representing the gas distribution. The three colums show the evolution of the R, ϕ and z distribution, respectively. The elapsed time is indicated on the right hand side in both orbital periods and Gyrs.

except the Keplerian potential, the only orbits satisfying these conditions are circular orbits. Thus the deviations of the gas motion from the equilibrium orbits can be trivially assessed by measuring the radial component of the velocity field. As a global measure of the radial motions in the gas we have chosen the median of the distribution of radial velocities, normalized by the tangential velocities. Figure 2 shows the time evolution of this quantity in a simulation where the initial ellipticity of the orbit is 0.2. The initial fluctuations ($P < 5$) reflect the orbital phase of the undisrupted material. The decay of the radial velocity component sets in, as expected, only

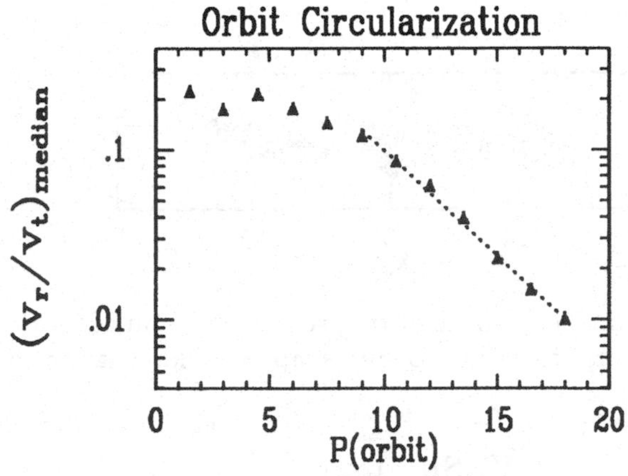

Fig. 2. Decay of the radial velocity component as the gas settles into the circular orbits that constitute its equilibrium configuration. The initial orbit had an ellipticity of 0.2. Note that the radial component declines only after the gas ring has closed on itself ($\approx 6P$). The time scale for the subsequent exponential decay is $3.7P$.

after the ring closes on itself ($P \approx 6$) and the impossibility of orbit intersections drives the gas towards circular motion. This latter phase is well described by an exponential decay of the median radial velocity. The decay time scale, τ, of the radial velocity component in this model is found to be $\tau \approx 3.7\ P$. In general, τ is expected to depend to some extent on the rotation curve of the potential and in more general potentials on the structure of the equilibrium orbits and the initial orbit of the material. However, if the time scale found in our particular model is any indication for such behavior in more general situations, then τ is substantially shorter than the settling times in all realistic potentials. Thus, the limiting factor in reaching the equilibrium would always be the settling of the gas ring.

In the beginning of this section we showed that the "smearing time" for gas at 30–40 kpc is a significant fraction of the Hubble time ($\approx$ 4 Gyrs). Therefore, we expect a significant fraction of observed gas rings to have strong azimuthal variations. Realizing this helps understand the observed properties of some polar rings. The "banana-shaped" polar rings (e.g., UGC 7576) could be explained naturally by an inhomogeneous ring that is seen nearly edge-on. If in such an object the density were only high enough to be observed in part of the ring, then the appearance of a nearly edge-on ring would resemble the observed shape quite well, as illustrated in Figure 3.

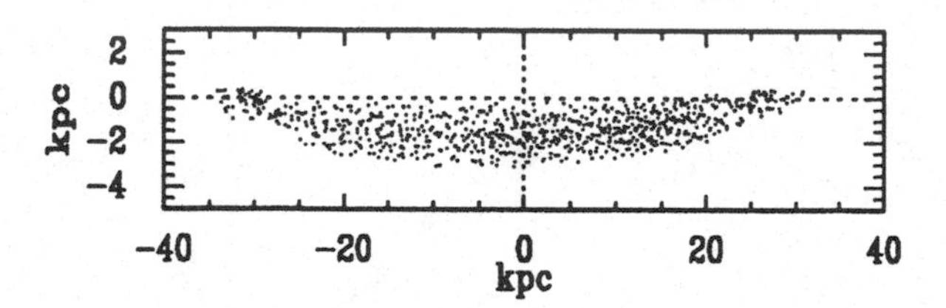

Fig. 3. An incompletely smeared ring seen at 5° from edge-on. It resembles the morphology of "banana-shaped" polar rings, such as UGC 7576.

V. Summary

We have started to investigate the formation of gaseous rings at large radii in early type galaxies by studying the disruption and the subsequent smearing of gaseous satellites in rigid potentials using an SPH method.

For the simple case of a spherically symmetric potential and a purely gaseous satellite in a nearly circular orbit we find the following results:

1) The initial azimuthal smearing is dominated by phase wrapping. Only after rough azimuthal symmetry has been established does dissipation play an important role.
2) For a typical gas ring at 30 – 40 kpc radius the smearing process takes up a significant fraction of the Hubble time. Even after 4 Gyrs (seven orbital periods), our particular simulation showed an azimuthal density variation of a factor of three.
3) For the simple case of a spherically symmetric potential we were able to estimate numerically the time scale for the gas to "find" the circular orbits, which are the only permissible orbits in equilibrium. Measuring the decay of the radial velocity component we find an exponential time scale of $\approx 3.7P$.
4) A number of observed properties of gas rings in early type galaxies may be understood in terms of their non-axisymmetric nature. In particular, "banana-shaped" rings may represent only partly smeared rings, seen nearly edge-on.

In future work we will generalize these results to axisymmetric and triaxial potential shapes and examine, in addition, the settling of rings toward a preferred orientation.

References

de Zeeuw, T. & Pfenniger, D. (1988). Mon. Not. R. Astron. Soc. **235**, 949, Appendix C.

Fabian, A. & Thomas, P. (1986). In *Structure and Dynamics of Elliptical Galaxies*, IAU Symp. 127 (ed. T. de Zeeuw), p. 155. Dordrecht: Reidel.

Habe, A. & Ikeuchi, S. (1985). Astrophys. J. **289**, 540.

Habe, A. & Ikeuchi, S. (1988). Astrophys. J. **326**, 84.

Hernquist, L. & Katz, N. (1989). Astrophys. J., Suppl. Ser. **70**, 419.

Richstone, D.O. (1980). Astrophys. J. **238**, 103.

Schechter, P. & Gunn, J. (1978). Astron. J.**83**, 1360.

Schweizer, F., Whitmore, B.C. & Rubin, V.C. (1983). Astron. J. **88**, 909.

Steiman-Cameron, T.Y. & Durisen, R.H. (1988). Astrophys. J. **325**, 26.

Steiman-Cameron, T.Y. & Durisen, R.H. (1990). Astrophys. J. **357**, 62.

Tohline, J.E., Simonson, G.F., & Caldwell, N. (1982). Astrophys. J. **252**, 92.

van Gorkom, J.H., Schechter, P. & Kristian, J. (1987). Astrophys. J. **357**, 62.

Whitmore, B.C., McElroy, D.B. & Schweizer, F. (1987). Astrophys. J. **314**, 439.

Whitmore, B.C., Lucas, R.A., McElroy, D.B., Steiman-Cameron, T.Y., Sackett, P.D. & Olling, R.B. (1990). Astron. J. **100**, 1489.

Gas-dynamical models of settling disks

DIMITRIS M. CHRISTODOULOU
Steward Observatory, University of Arizona

AND

JOEL E. TOHLINE
Louisiana State University

General remarks

The formation, long-term stability, and evolution of warped galaxy disks and polar rings is an active area of galactic dynamics that has drawn the attention of many theorists and observers. The interested reader may consult the reviews by Athanassoula and Bosma (1985), Schweizer (1986), and Tohline (1990), the dissertations of Bosma (1978), Simonson (1982), Steiman-Cameron (1984), Varnas (1986a) and Christodoulou (1989), and the contributions of Bosma, Briggs, Casertano, Galletta, Rix and Katz, Sackett, Sparke, Steiman-Cameron and Whitmore in this volume.

The entire subject can be subdivided into five broad categories: (a) Existence and kinematics of warped galaxy disks (Bosma 1981a, b; Christodoulou, Tohline and Steiman-Cameron 1990 and references therein). (b) Preferred orientation theory (Tohline and Durisen 1982; Tohline, Simonson, and Caldwell 1982; Durisen et al. 1983; Simonson and Tohline 1983; Steiman-Cameron and Durisen 1984; David, Durisen, and Steiman-Cameron 1984). (c) Influence of self-gravity of luminous matter (Sparke 1984, 1986; Sparke and Casertano 1988). (d) Polar rings (Schweizer, Whitmore, and Rubin 1983; Whitmore 1984; Whitmore, McElroy, and Schweizer 1987). (e) Numerical fluid/hydrodynamical models of evolving disks (Habe and Ikeuchi 1985, 1988; Varnas 1986a,b; Steiman-Cameron and Durisen 1988, 1990; Christodoulou 1989, 1990).

The results that we describe below are related to categories (b), (d), and (e). We have used a 3-D, explicit, first-order accurate, Eulerian hydro code to study the dynamical evolution of disks that are inclined out of the equatorial plane of a spheroidal halo potential of the form:

$$\Phi = -\frac{1}{r}\left\{1 + \frac{1}{2}J_{20}\left(\frac{a}{r}\right)^2 \frac{R^2 - 2z^2}{r^2}\right\}, \tag{1}$$

where (R, z) denote cylindrical coordinates, $r = \sqrt{R^2 + z^2}$, a is the equatorial radius of the spheroid, and the quadrupole coefficient J_{20} is a free parameter.

The initial equilibrium model is a Papaloizou-Pringle (1984) torus of zero mass and constant specific angular momentum. The fluid in the torus is constrained to be polytropic throughout its entire evolution with a polytropic index of $n = 3/2$. Its radial extent is defined by the locations of the inner (R_-) and outer (R_+) edges on the equatorial plane of the ring itself, or, equivalently, by the Mach number at pressure maximum m_o. If the radius of the pressure maximum is defined to be $R_o = 1$, then:

$$R_\pm = \left(1 \mp \frac{\sqrt{3}}{m_0}\right)^{-1}. \tag{2}$$

Although the initial models do not simulate realistic conditions in observed galaxies, they are useful because they are obtained analytically with a high degree of accuracy and do incorporate differential precession of gas-particle orbits and viscous dissipation, the necessary physical effects which can effectively drive the settling of a disk toward a preferred orientation. In the present analysis, the preferred orientation is planar and coincides with the equator of the spheroidal potential.

Model Evolutions

The characteristic model evolutions presented in this section are part of a detailed study of settling gaseous disks in spheroidal potentials (Christodoulou 1989, 1990 and Christodoulou and Tohline 1990). The relevant parameters are $J_{20}a^2 = +0.02$, $m_o = 4.2$, and $R_+ - R_- = 1.0$ corresponding to a fairly *extended* ring under the influence of an *oblate* spheroidal halo potential that deviates only *slightly* from spherical symmetry. Models produced from variation of these parameters will be briefly discussed in the following section.

a) Low inclinations

Figure 1 shows vertical cross-sections—90° away from the intersection of the ring plane with the halo plane—through a ring originally inclined by $i_0 = 10°$ from the preferred orientation. In our simulations, the ring always starts horizontal in the computational grid and the preferred orientation is indicated by the straight line at an angle from the horizontal line. Both isodensity contours and velocity vectors are displayed on each cross-section and the elapsed time is indicated in initial rotation periods at the pressure maximum of the ring.

As was expected from the smoothed-particle hydrodynamical simulations of Habe and Ikeuchi (1985) and from the cloud-fluid simulations of Steiman-Cameron and Durisen (1988), the ring settles to the preferred plane without showing signs of inflow toward the nuclear region of the potential. In the process, a transient warp appears and gradually gets weaker as the outer regions approach the preferred plane. The location of that plane is clearly emphasized at later times by the strong shock front that forms on it.

Finally, notice how thin the ring appears to be at the end of the evolution. The Papaloizou-Pringle (1984) instability is by no means responsible because it has been damped out by the numerical viscosity of the code earlier on. Furthermore, in a similar evolution with a prolate potential ($J_{20}a^2 = -0.02$) the ring flattens out 90° away from the position in Figure 1. It appears, then, that this $m = 2$ distortion is a remnant from the apparent nonaxisymmetry of the potential that the ring was feeling throughout its entire evolution. The fact that the $m = 2$ distortion shifts its phase by 90° between oblate and prolate potentials may be used to identify the gross geometric shape of dark halos in real galaxies if such asymmetries can be observed (see also Tohline and Christodoulou 1990).

b) Moderate inclinations

Figure 2 shows cross-sections from the evolution of a ring initially inclined by $i_0 = 40°$. As the evolution proceeds, a prominent warp forms and disappears when the ring approaches the preferred plane. But, as was accurately predicted by Steiman-Cameron and Durisen (1988), inflow plays now a major role; the ring is simultaneously settling and flooding the central hole with gas. (Mass and angular momentum density that flow into the inner few zones of the grid are taken out in order to speed up the computation.) At this moderate

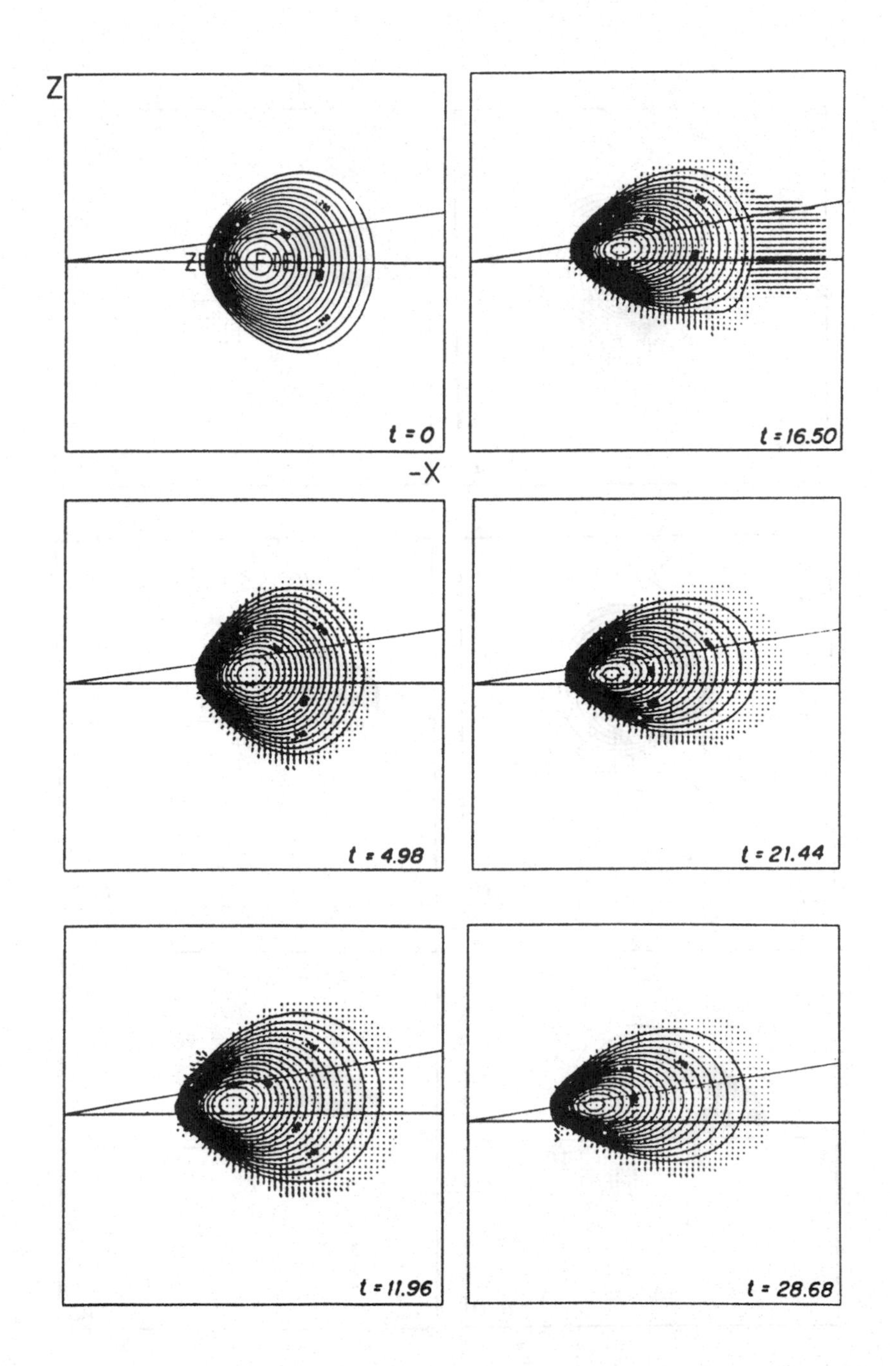

Fig. 1. Vertical cross-sections at different times through a ring originally inclined by $i_0 = 10°$ from the preferred orientation. The preferred orientation is indicated by the line at an angle from the horizontal line. The external potential is oblate spheroidal with $J_{20}a^2 = +0.02$.

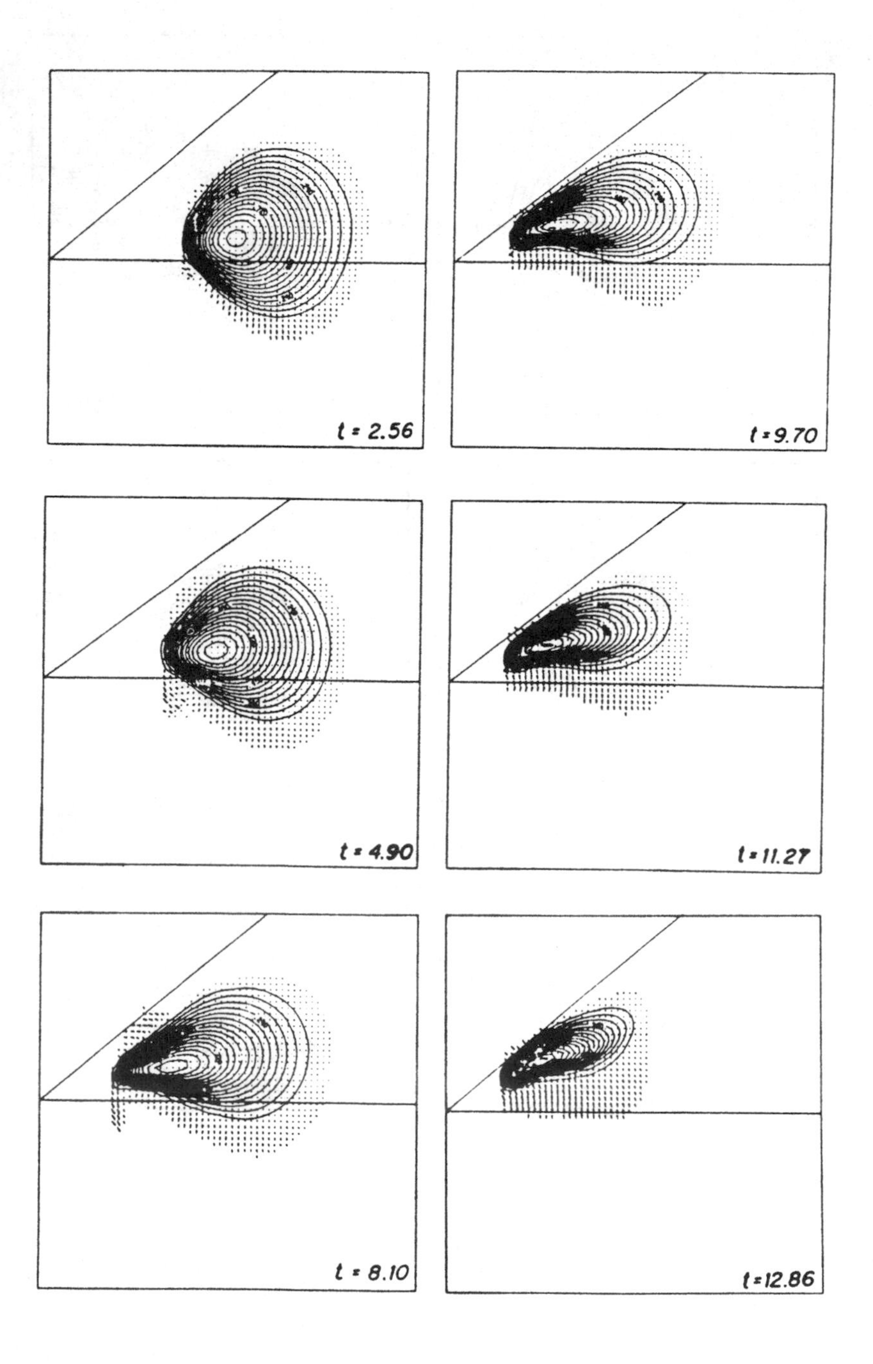

Fig. 2. As in Fig. 1, but for a ring inclined by $i_0 = 40°$.

inclination, the inflow time-scale appears to be comparable to the settling time-scale.

c) High inclinations

Figure 3 shows cross-sections from the evolution of a ring initially at $i_0 = 80°$. At this inclination, the differential precession is very small resulting in extremely long settling times. Naturally, then, no significant inflow or settling have yet occurred by the end-point of our simulation (see also Christodoulou 1989 for more details).

Conclusions

1) Gaseous disks settling from small or moderate inclinations toward the preferred planar orientation of an external, weakly spheroidal potential develop smooth but transient warps in their outer regions. Matter inflow toward the nucleus of the potential is absent at low inclinations but becomes important at moderate inclinations. Results obtained in this region of the parameter space from different numerical techniques tend to agree remarkably well (Habe and Ikeuchi 1985; Steiman-Cameron and Durisen 1988; Christodoulou and Tohline 1990).
2) Our hydrodynamical simulations have indicated that two important features dominate in settled disks: a prominent shock front forms exactly on the preferred plane and a strong $m = 2$ distortion appears in the disk and switches its phase by 90° between oblate and prolate spheroidal potentials (Tohline and Christodoulou 1990).
3) Rings placed at high inclinations from the preferred plane of a weakly spheroidal potential survive comfortably for more than 30 rotations. This result provides an explanation for the existence of polar rings around galaxies—polar rings survive for a Hubble time because the dark halos are only slightly distorted from spherical symmetry and the differential precession rates are extremely small at high inclinations—but it is in disagreement with the results of Habe and Ikeuchi (1985). This disagreement is about to be resolved, however, with a new SPH code (Hernquist and Katz 1989). Preliminary results show that rings forming at high inclinations do not collapse to the nucleus but, instead, are orbiting for at least a Hubble time. The dif-

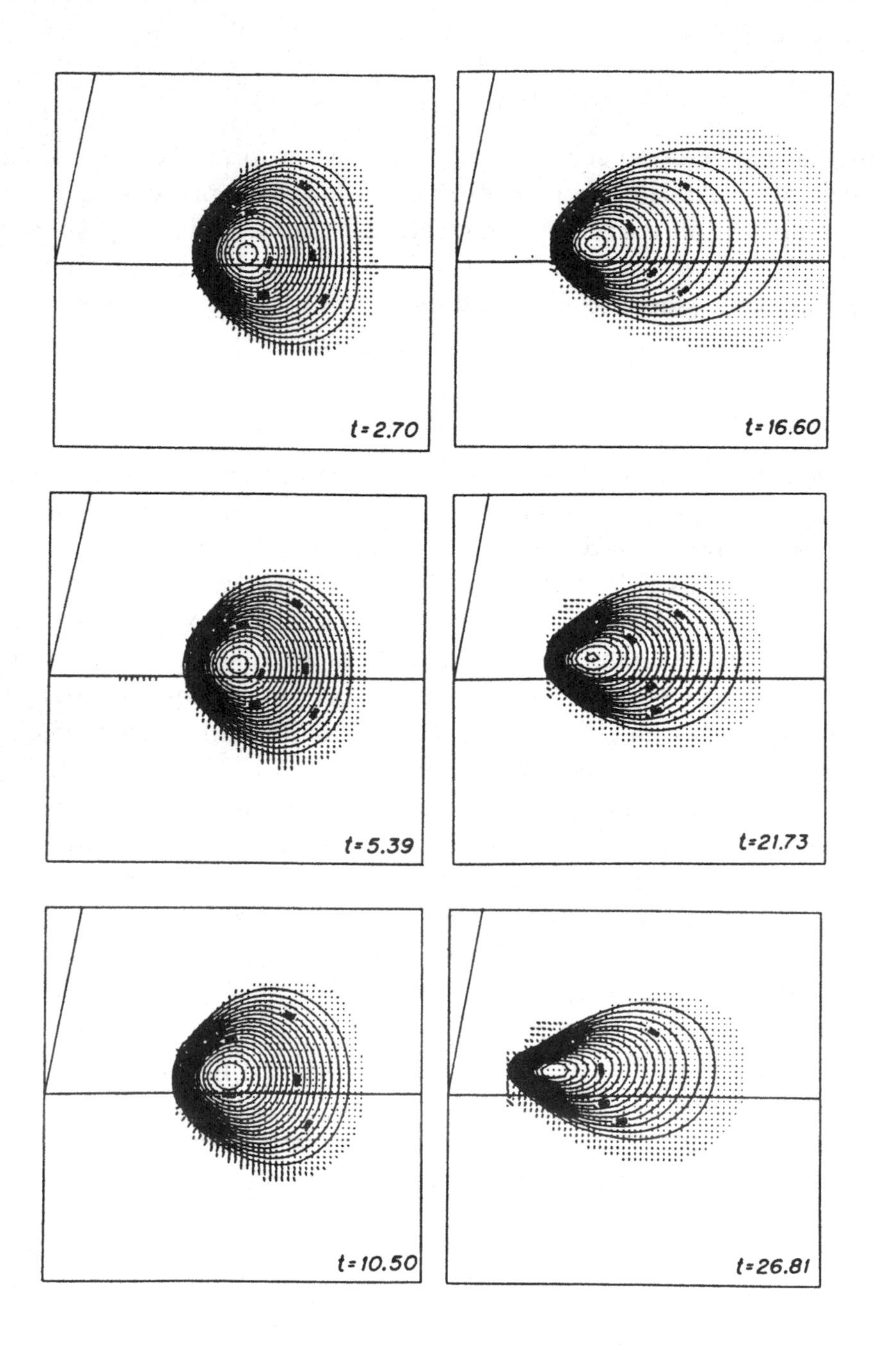

Fig. 3. As in Fig. 1, but for a ring inclined by $i_0 = 80°$.

ference between SPH results is probably due to different initial conditions (see Rix, Katz, and Christodoulou 1991 for details).

4) In the course of our investigation, we have also performed simulations with thinner rings ($m_o = 7.2$, $R_+ - R_- = 0.5$) and stronger quadrupole distortions ($J_{20}a^2 = 0.09$, 0.19). Dynamical evolutions that have utilized thin rings have produced qualitatively similar results. On the other hand, stronger distortions in the potential have resulted in dramatic changes: rings starting at low or moderate inclinations still recognize the preferred plane, but end up in the nuclear region before they have time to settle smoothly; rings at high inclinations undergo violent readjustments and end up with irregular distortions and surrounded by debris. Strongly distorted dark halos may, therefore, be responsible for the appearance of some galaxies which show extreme irregularities, pronounced asymmetries, or chaotic dynamics.

Acknowledgements

This work was supported in part by grant n. AST 87–01503 to J.E.T., by a Young Presidential Investigator award (AST 89–57107) to R. Narayan, both from the National Science Foundation, and by a grant from the IBM to D.M.C.

References

Athanassoula, E. & Bosma, A. (1985). Annu. Rev. Astron. Astrophys.23, 147.

Bosma, A. (1978). Ph.D. thesis, University of Groningen.

Bosma, A. (1981a). Astron. J. **86**, 1791.

Bosma, A. (1981b). Astron. J. **86**, 1825.

Christodoulou, D.M. (1989). PhD Thesis, Louisiana State University.

Christodoulou, D.M. (1990). In *Galactic Models* (eds. J.R. Buchler, S.T. Gottesman & J.H. Hunter, Jr.), Annals of the New York Academy of Sciences, vol. 596, p. 207.

Christodoulou, D.M. & Tohline, J.E. (1990). In preparation.

Christodoulou, D.M., Tohline, J.E. & Steiman-Cameron, T.Y. (1990). Astrophys. J., Suppl. Ser., submitted.

David, L.P., Durisen, R.H. & Steiman-Cameron, T.Y. (1984). Astrophys. J. **286**, 53.

Durisen, R.H., Tohline, J.E., Burns, J.A. & Dobrovolskis, A.R. (1983). Astrophys. J. **264**, 392.

Habe, A. & Ikeuchi, S. (1985). Astrophys. J. **289**, 540.

Habe, A. & Ikeuchi, S. (1988). Astrophys. J. **326**, 84.

Hernquist, L. & Katz, N. (1989). Astrophys. J., Suppl. Ser. **70**, 419.

Papaloizou, J.C.B. & Pringle, J.E. (1984). Mon. Not. R. Astron. Soc. **208**, 721.
Rix, H.-W., Katz, N. & Christodoulou, D.M. (1991). In preparation.
Schweizer, F. (1986). In *Structure and Dynamics of Elliptical Galaxies*, IAU Symp. 127 (ed. T. de Zeeuw), p. 109. Dordrecht: Reidel.
Schweizer, F., Whitmore, B.C. & Rubin, V.C. (1983). Astron. J. **88**, 909.
Simonson, G.F. (1982). PhD Thesis, Yale University.
Simonson, G.F. & Tohline, J.E. (1983). Astrophys. J. **268**, 638.
Sparke, L.S. (1984). Astrophys. J. **280**, 117.
Sparke, L.S. (1986). Mon. Not. R. Astron. Soc. **219**, 657.
Sparke, L.S. & Casertano, S. (1988). Mon. Not. R. Astron. Soc. **234**, 873.
Steiman-Cameron, T.Y. (1984). PhD Thesis, Indiana University.
Steiman-Cameron, T.Y. & Durisen, R.H. (1984). Astrophys. J. **276**, 101.
Steiman-Cameron, T.Y. & Durisen, R.H. (1988). Astrophys. J. **325**, 26.
Steiman-Cameron, T.Y. & Durisen, R.H. (1990). Astrophys. J. **357**, 62.
Tohline, J.E. (1990). In *Galactic Models* (eds. J.R. Buchler, S.T. Gottesman & J.H. Hunter, Jr.), Annals of the New York Academy of Sciences, vol. 596, p. 198.
Tohline, J.E. & Christodoulou, D. M. (1990). In preparation.
Tohline, J.E. & Durisen, R.H. (1982). Astrophys. J. **257**, 94.
Tohline, J.E., Simonson, G.F. & Caldwell, N. (1982). Astrophys. J. **252**, 92.
Varnas, S.R. (1986a). PhD Thesis, Monash University.
Varnas, S.R. (1986b). Proc. Astron. Soc. Australia **6**, 458.
Whitmore, B.C. (1984). Astron. J. **89**, 618.
Whitmore, B.C., McElroy, D.B. & Schweizer, F. (1987). Astrophys. J. **314**, 439.

Evolutionary processes affecting galactic accretion disks

THOMAS STEIMAN-CAMERON
NASA Ames Research Center

Compelling evidence exists that a significant fraction of all galaxies have experienced accretion events. Disks seen in many early type galaxies, including dust lane ellipticals and polar-ring galaxies, typically display warps and twists or have angular momentum vectors that are misaligned with the kinematic axes of the galaxy. These disks are generally believed to be the result of accretion. When an interaction involves the accretion of material by a larger galaxy, dissipative processes can be expected to smear the captured material into a disk or annulus in a few orbital periods. Newly formed disks are unstable and are driven towards an equilibrium configuration by the interplay between gravitational and dissipational processes. The manner in which a captured disk evolves and the steady-state orientation towards which it settles depends upon the detailed structure of the gravitational potential into which it is captured. Thus, accreted disks serve as powerful probes of a galaxy's three-dimensional mass distribution, providing information not available from any other source. In addition, insight into the interaction history of galaxies, including merger/interaction rates, can be gained from a detailed understanding of the dynamical properties of galactic accretion disks.

To interpret the structure of nonsteady-state disks in the context of the geometry of the host galaxy and the history of the disk, we need to understand how a thin disk with arbitrary initial orientation evolves in the gravitational potential of a nonspherical galaxy. The principal forces involved are: 1) precessional forces arising from the nonsphericity of the galaxy into which the material is accreted; 2) dissipative forces arising from interactions between mass elements of the newly formed disk; 3) torques arising from the self-gravity of the accreted disk; and 4) effects arising from rotation of the surface figure or potential of the galaxy. While they have not been adequately explored, nonradial instabilities and magnetic torques (see Battaner in these proceedings) may also play roles in the evolution of some accreted disks.

It is convenient to describe a disk as a series of concentric but not necessarily coplanar rings representing annular Lagrangian mass elements of the disk. Forces acting on these rings correspond to orbit-averaged forces. The configuration of a disk at any time t is provided by the sizes and orientations of these rings. If a ring's size is specified by its circular semimajor axis a, and its orientation by the inclination i and longitude of ascending node Ω, then the dynamical evolution of a disk is described by three coupled differential equations, i.e.,

$$\begin{aligned}(\mathrm{d}a/\mathrm{d}t) &= (\mathrm{d}a/\mathrm{d}t)_D,\\ (\mathrm{d}i/\mathrm{d}t) &= (\mathrm{d}i/\mathrm{d}t)_G + (\mathrm{d}i/\mathrm{d}t)_D + (\mathrm{d}i/\mathrm{d}t)_S + (\mathrm{d}i/\mathrm{d}t)_R, \qquad (1)\\ (\mathrm{d}\Omega/\mathrm{d}t) &= (\mathrm{d}\Omega/\mathrm{d}t)_G + (\mathrm{d}\Omega/\mathrm{d}t)_D + (\mathrm{d}\Omega/\mathrm{d}t)_S + (\mathrm{d}\Omega/\mathrm{d}t)_R,\end{aligned}$$

where subscripts G, D, S, and R refer to changes arising from galactic precessional forces, dissipation, disk self-gravity, and rotation of the galaxy's surface figure, respectively. While precessional forces are always important, the relative importance of the latter terms in eqs. (1) will vary from case to case.

Precessional forces

A galaxy's gravitational potential is the primary driver of disk evolution. Because of its fundamental importance, a long-standing goal of galactic dynamics has been the determination of potentials which accurately portray real systems. The scale-free logarithmic potential has been frequently used to model galaxies with flat rotation curves (cf. Richstone 1980; Monet, Richstone and Schechter 1981; Miller 1982; Levison 1987). While this potential clearly provides a reasonable model in the "equatorial plane" of galaxies over the radial range where rotational velocities are approximately constant, the appropriateness of this potential outside the equatorial plane has not been as apparent. As will be discussed later, it now appears that in at least some cases this potential provides an extremely good three-dimensional representation of real galaxies over regions where circular velocities are approximately constant (Steiman-Cameron, Kormendy and Durisen 1990).

The scale-free potential has the properties that the rotation curve is flat, isopotential surfaces are similar and concentric ellipsoids, and isodensity surfaces of the galactic mass distribution are also similar and concentric. The potential is given by

$$\Phi(s) = v_c^2 \ln(ks), \tag{2}$$

where v_c is the constant circular velocity, k is a scaling factor, $s^2 = x^2 + y^2/p^2 + z^2/q^2$ represents isopotential surfaces, and p and q are the axis ratios of these surfaces.

If the galactic mass distribution is not highly flattened, then the precession rate for circular orbits in this potential is given by (Steiman-Cameron 1990)

$$\begin{aligned} (\mathrm{d}i/\mathrm{d}t)_G &= (3v_c/r)\,\beta \sin i \sin 2\Omega, \\ (\mathrm{d}\Omega/\mathrm{d}t)_G &= (3v_c/r)\left[\beta \cos i \cos 2\Omega - (\eta/2)\cos i\right], \end{aligned} \tag{3}$$

where i is the angle between the angular momentum vector of the orbit and the z axis, Ω is measured from the x axis in the (x, y) plane, and η and β are functions of the axis ratios (b/a) and (c/a) of the galactic mass distribution. If a, b, and c are semiaxes of an isodensity surface along the x, y, and z axes, respectively, and b is the intermediate length axis, then

$$\begin{aligned} \eta &= \frac{[1 + (b/a)^2 - 2(c/a)^2]}{6[1 + (b/a)^2 + (c/a)^2]}, \\ \beta &= \frac{[(b/a)^2 - 1]}{12[1 + (b/a)^2 + (c/a)^2]}. \end{aligned} \tag{4}$$

The coefficient β is zero for axisymmetric galaxies. As long as $0.6 \leq (b/a) \leq 1.6$ and $0.6 \leq (c/a) \leq 1.6$, the density distribution is approximately ellipsoidal. Systems in which these ratios have been reasonably estimated fall in this regime (Whitmore 1984; Steiman-Cameron, Kormendy and Durisen 1990). For (b/a) or (c/a) outside this range, the mass distribution shows significant "dimpling" along a semiaxis.

It is useful to introduce the concept of *preferred* orientations. As will be discussed later, dissipation causes a differentially precessing disk to settle towards a preferred orientation. The orbit of a disk element lying in one of these orientations: 1) does not precess, and 2) orbits slightly inclined to this orientation precess about it. The number and nature of preferred orientations depends on the geometry of the galaxy and on any rotation of the surface figure or gravitational potential of the galaxy. Preferred orientations in static axisymmetric galaxies correspond with the equatorial plane, while these orientations in static triaxial systems are planes normal to the long axis and to the short axis of the galaxy. Rotation of the surface figure of a galaxy or the "slewing" of its potential alters the nature of some of the preferred orientations (see below). Characteristics of preferred orientations in both axisymmetric and triaxial galaxies have been investigated by a number of authors (Tohline and Durisen 1982; Tohline, Simonson and Caldwell 1982; Steiman-Cameron and Durisen 1982, 1984; Durisen et al. 1983; David, Durisen and Steiman-Cameron 1984; Steiman-Cameron 1984).

Determination of the preferred orientations is equivalent to locating the 1:1:1 resonance periodic orbits (cf. Heisler, Merritt and Schwarzschild 1982; de Zeeuw and Merritt 1983; David, Steiman-Cameron and Durisen 1985). With the exception of orbits in spherical galaxies and in the equatorial plane of axisymmetric galaxies, these periodic orbits are noncircular. However, deviations from circularity generally will be small. Orbit eccentricity is a second order effect in precession (cf. Danby 1962). Therefore, while eqs. (3) are derived for circular orbits, they provide reasonable approximations for moderately noncircular orbits.

Precession is faster at smaller radii, thus accreted disks become twisted. Numerical simulations suggest that accreted material will smear into a disk in a few orbital periods (Varnas 1986a; Habe, private communication). Since this timescale is much shorter than the differential precession time, except for highly flattened galaxies, newly accreted disks will be approximately planar. Thus the assumption of planar geometry is an appropriate initial condition for studies of accreted disks.

The sense of the twist arising from differential precession, i.e., leading or trailing with respect to rotation, unambiguously determines whether the mass distribution is prolate-like or oblate-like. The twisted accretion disk in NGC 4753 clearly demonstrates the oblate-like nature of the halo of this S0 galaxy (Steiman-Cameron, Kormendy and Durisen 1990). The warped HI disks in many spiral galaxies also display distinct twists (Christodoulou, Tohline and Steiman-Cameron 1988, 1990; Briggs 1990). Christodoulou et al., using the assumption that these twists arise from differential precession, found that the twisted disks of NGC 5033 and NGC 5055 are consistent with precession in a prolate-like galaxy, with the long-axis of the mass distribution normal to the equatorial plane of the spiral galaxy. If this interpretation is correct, then this result has important repercussions for the formation of galaxies and also appears to rule out theories of non-Newtonian gravity proposed to account for flat rotation curves.

Figure rotation

Rotation of the surface figure of a galaxy, or, equivalently, slewing of its potential, causes some of the preferred orientations to change with radius (see preferred orientation references cited above; see also the contribution of Ostriker in this volume). This is readily seen by the definition of a preferred orientation as that orientation in which $(\mathrm{d}i/\mathrm{d}t) = (\mathrm{d}\Omega/\mathrm{d}t) = 0$. The introduction of rotational terms in eq. (1) can change the values of i and Ω which satisfy the definition and, in addition, these may be functions of radius. Thus in the presence of rotation of the gravitational potential some preferred orientations may lie outside of a principal plane of the galaxy. In addition, disks residing in one of these orientations will experience steady-state warps.

Attempts to explain the warps seen in some disks, particularly in early-type galaxies, as manifestations of the rotation of a galaxy's surface figure have been made by a number of authors (cf. Tohline, Simonson and Caldwell 1982; de Zeeuw and Merritt 1983; Steiman-Cameron and Durisen 1984, and references therein). Many of these attempts have in common the assumption that the surface figure rotates about its short axis. Indeed, numerical n-body investigations of large stellar systems generally result in tumbling about this axis. Warps arising as a result of rotation about the short axis lead to disks which rotate in a retrograde sense relative to the sense of surface rotation. If stellar streaming is in the same sense as rotation, then observations do not support this mechanism as an explanation for observed warps.

While surface figures are commonly assumed to rotate about their short axis, Miller and Smith (1982) in their numerical test of the triaxial model of Schwarzschild (1979) were able to show apparent stability for figure rotation about *both* the short and long axes of the mass distribution. More recently, n-body collapse models of Duncan and Levison (1989) produced stable or quasi-stable figure rotation about the intermediate axis. Therefore caution may be in order in the assumption of rotation axis. For figure rotation about either the long or intermediate axis, steady-state warped disks which rotate in a prograde sense are possible. Warps resulting from rotation of the surface figure about a *principal* axis have have a common line of nodes. Rotation about a *nonprincipal* axis results in a steady-state warp and twist, i.e., the line of nodes changes with radius (David et al. 1984). Such a condition could arise if an interaction excited a transverse shear mode in the host galaxy.

Two additional notes of caution may also be in order concerning the sense of figure rotation. It is commonly assumed that figure rotation is in the same sense as stellar streaming. There exists some evidence, however, that this may not always be true (Freeman 1966a, b, c; Vandervoort and Welty 1982; Schwarzschild 1983). In addition, it is the shape of the mass distribution which determines dynamical effects. While one would hope that the distribution of matter roughly mimics the distribution of light, i.e., the short axis of the light distribution corresponds with the short axis of the mass distribution, our experience with halos of disk galaxies suggests that this is not necessarily a safe assumption.

Self-gravity

The effects of disk self-gravity in the limit where dissipation is unimportant have been examined by Hunter and Toomre (1969), Sparke (1984, 1986), and Sparke and Casertano (1988) (see also Sackett and Sparke 1990; and the contributions of Casertano, Sackett, Sparke, Pitesky, and Hofner in this volume). Normal modes which produce steady-state warps can arise in self-gravitating disks. These modes yield simple warps, i.e., the line of nodes is constant with radius. While observations of warped disks suggest that simple warps are not common (Christodoulou, Tohline and Steiman-Cameron 1988, 1990; Briggs 1990), self-gravity can nonetheless be an important effect in the evolution of accreted disks. The term due to disk self-gravity in eqs. (1) can be found by determining the gravitational torque on ring j due to ring k, evaluating how this torque changes i_j and Ω_j, and summing over all rings $k \neq j$. If the inclination between rings j and k is given by i_{jk}, $\Delta\Omega = (\Omega_k - \Omega_j)$, and $x = a_j/a_k$, then

$$\begin{aligned} \left(\frac{\mathrm{d}i_j}{\mathrm{d}t}\right)_S &= \sum_{k \neq j} \frac{Gm_k}{v_c a_j a_k} L(x, i_{jk}) \sin i_k \sin \Delta\Omega, \\ \left(\frac{\mathrm{d}\Omega_j}{\mathrm{d}t}\right)_S &= \sum_{k \neq j} \frac{Gm_k}{v_c a_j a_k} L(x, i_{jk}) \left[\cos i_k - \cot i_j \sin i_k \cos \Delta\Omega\right], \end{aligned} \tag{5}$$

where m_j and m_k are the ring masses and L is a function given by

$$L(x, i_{jk}) = \frac{x}{(2\pi)^2} \int_0^{2\pi} \int_0^{2\pi} \frac{\sin\alpha \sin\beta \, \mathrm{d}\alpha \mathrm{d}\beta}{\left[1 + x^2 - 2x\cos\psi\right]^{-3/2}}. \tag{6}$$

Here $\cos\psi = \cos\alpha\cos\beta + \sin\alpha\sin\beta\cos i_{jk}$. Note that certain symmetries exist for L. In particular, $L(x, \pi - i_{jk}) = -L(x, i_{jk})$ and $L(x, i_{jk}) = x^{-1}L(x^{-1}, i_{jk})$.

Dissipation

Three approaches have been used to examine the effects of dissipation on a differentially precessing galactic disk. These include a cloud-fluid approach (Steiman-Cameron 1984; Steiman-Cameron and Durisen 1988, 1990, hereafter referred to as SCD1 and SCD2 respectively), smooth particle hydrodynamic simulations (SPH) (Habe and Ikeuchi 1985, 1988; Varnas 1986a, b), and finite-difference hydrodynamic simulations (Christodoulou 1989, 1990; see also Christodoulou and Tohline in these proceedings). Although differing in their methodologies and the form of the dissipative forces, all three methods give results which are in very good agreement.

In the absence of dissipation, a differentially precessing disk will eventually twist to such an extent that it loses its identity as a disk. The presence of dissipation causes the disk to settle towards the preferred orientation about which it precesses and, at the same time, causes material to inflow. Settling is essentially a local process, depending primarily on local properties of the gravitational potential and the disk. In differentially rotating planar galactic disks subject to dissipation, Keplerian shear causes inflow (cf. Pringle 1981). This inflow is accelerated in disks which are settling. For disks which are captured at small or moderate inclinations relative to a preferred orientation, the timescale for settling is much less than the inflow time. On the other hand, the inflow time for disks captured at large relative inclinations becomes comparable to or shorter than the settling time (SCD1).

The equations describing the time-dependent structure of a settling disk are somewhat complex and must, in general, be solved numerically. In the approach of SCD1 and SCD2, the disk is treated as a two-dimensional, continuous, smoothly warping and twisting cloud fluid which is assumed to exhibit zero-order circular bulk motion. (An excellent discussion of the validity of a "cloud-fluid" approach to modeling the interstellar medium is given by Scalo and Struck-Marcell 1984). Equations were derived which describe changes in the disk resulting from precession and viscosity. These equations represent the viscous transport of angular momentum by the stress terms of the Navier-Stokes equations and are time-averaged over one orbit to remove all periodic effects with time scales of an orbit period. The result is a set of three coupled differential equations [eqs. (30a-c) of SCD1] which contain all the terms of eqs. (1a-c) except for the self-gravity term. (Note that the signs of Γ_{n3} and Γ_{n4} in eq. [30c] of SCD1 are wrong and should be reversed. The corresponding equation in the Appendix of SCD1 possesses the correct signs).

Conditions which are plausibly met in real galaxies allow for analytic solutions of these equations (SCD1). The requisite conditions are that: 1) settling occurs in an axisymmetric gravitational potential; 2) the orbital precession rate is dominated by gravitational rather than viscous forces; 3) the viscous time scale for inflow is much longer than the time scale for disk settling; and 4) the initial inclination of the disk relative to a preferred orientation is small. The applicability of the first assumption to real systems is an open question. However, the analytic solution is approximately valid in triaxial systems if the other conditions are fulfilled. There are indications that analytic solutions can be found and used in galaxies possessing a more complex geometry (Dobrovolskis et al. 1989), but these solutions have not yet been determined. The second condition is generally met for parameters thought to be typical of real galaxies. The third condition is satisfied if condition four is met. The latter is less restrictive than it might at first seem. Comparisons between the approximate analytic solution and numerical computations in SCD1 demonstrated very good agreement for initial disk inclinations as large as 40° in oblate model galaxies.

Subject to the above conditions, the time-dependent structure of an accreted disk can be described by the two dimensionless parameters $\aleph = (\tau_p/\tau_e)$ and $T = t/\tau_p$, where $\tau_p = 2\pi/\dot{\Omega}_p$ is the precession period, $\tau_e = [(\nu/6)(d\dot{\Omega}_p/dr)^2]^{-1/3}$ is the effective settling time, and ν is the effective coefficient of kinematic viscosity for the cloud fluid. If the disk is initially planar with orientation $(i, \Omega) = (i_o, 0)$ at $t = 0$, then the time-dependent configuration of the disk is given by (see SCD2)

$$\begin{aligned} i/i_o &= \exp[-(\aleph T)^3], \\ \Omega &= 2\pi T. \end{aligned} \tag{7}$$

These equations provide a very simple description of a process that is, in detail, quite complicated.

In principle, observations of settling disks allow for the determination of $\aleph$ and T. These, in turn, provide insight into the shape of the host galaxy and the age of the disk. Unfortunately, our limited understanding of dissipative processes in galactic disks does not permit a precise description of ν, thus inhibiting the interpretation of $\aleph$. However, the weak dependence of τ_e on ν, going only as $\nu^{-1/3}$, allows reasonable approximations for ν to serve for the determination of τ_e and hence the time-dependent disk structure. Assuming cloud-cloud interactions to be the dominant dissipative mechanism for settling in galactic disks, SCD1 developed a simple analytic estimate for the maximum permissible value of ν. For a gas disk possessing a Maxwellian velocity distribution and residing in a flat rotation curve galaxy, $\nu_{max} = 0.184\, r v_{rms}^2 / v_c$, where v_{rms} is the velocity dispersion and v_c is the circular velocity. This expression has the advantage that ν_{max} is expressed entirely in terms of observable quantities. The use of ν_{max} produces minimum settling times. In a scale-free galaxy, the minimum settling times, normalized to the precessional period τ_p is

$$\begin{aligned}(\tau_e/\tau_p)_{min} &= 0.46\,(v_c/v_{rms})^{2/3}\,|\eta \cos i_o|^{1/3}, \\ &= \aleph^{-1},\end{aligned} \tag{8}$$

where η is given by eq. (4) with $(b/a) = 1$. The dimensionless parameter $\aleph$ is thus constant for constant v_{rms}.

The timescale for settling into a preferred orientation is typically on the order of 0.5 to 3 precession periods. This result holds for a wide range of galactic geometries. Settling times in this range have been determined with the cloud-fluid approach, the SPH simulations of Habe and Ikeuchi (1985, 1988) and the 3-d hydrodynamical simulations of Christodoulou (1989, 1990). This in spite of differences in approach, both in terms of methods and the manner in which dissipation is treated. This agreement suggests that the time scale for settling is not strongly dependent on the dissipative mechanism, consistent with the weak dependence of τ_e on ν. The differential precession rate is the dominant determinant of settling times, as argued earlier by Simonson (1982) and by Tohline, Simonson and Caldwell (1982).

Equations (3) and (7) have been used to investigate the structure of the S0 galaxy NGC 4753. This galaxy, shown in *The Hubble Atlas* (Sandage 1961), displays prominent and somewhat chaotic appearing dust lanes. Steiman-Cameron, Kormendy and Durisen (1990) have found that the dust lanes can be reproduced extremely well as optical depth effects in a highly twisted disk. The disk was accreted at an inclination of 18° in an oblate galaxy. Differential precession has subsequently twisted the disk by $\sim 4\pi$ over a factor of seven in radius. The inner edge of the disk is still very near its original inclination, implying a value of ν somewhat smaller than ν_{max}. The extremely good fit implies that a massive, unseen halo exists and that its isodensity surfaces all have the same shape; the galaxy does not become rounder or flatter with radius. Unfortunately the flattening of the halo and the age of the disk cannot be determined independently. However, the halo can be no flatter than an E3 and no "rounder" than an E0.5. In either case, the flattening of the halo is *constant* to within $\sim$ 5% over a factor of seven in radius! If an independent means could be found to date the captured disk, then the exact flattening of the halo could be obtained (or vice versa).

The model for NGC 4753 places limits on the degree of triaxiality of the halo. Unfortunately, because of the low capture inclination these limits are not particularly stringent. Small amounts of triaxiality can have profound effects on disks accreted at high inclinations. Dissipation will lead disks captured at high inclinations to settle towards a polar orientation in triaxial galaxies. This mechanism has been proposed as the origin of polar ring galaxies (Steiman-Cameron and Durisen 1982; Habe and Ikeuchi 1985). Assume the short axis of the galactic mass distribution is normal to the stellar disk in S0s and the long and intermediate axes lie in the equatorial plane. Then disks accreted at low inclinations precess about the short axis and ultimately settle into the equatorial plane. Such systems might be observationally classified as anemic spirals. Disks accreted at high inclinations precess about the long axis and settle into the polar orientation. A slight amount of figure rotation about the short axis will lead the polar orientation to differ somewhat from 90°. Steiman-Cameron and Durisen (1982) presented a simple analytic expression for the fraction of all capture orientations which lead to precession about the long axis, as a function of axis ratios. Their expression was derived for an "external" potential. For a scale-free galaxy with axis ratios (b/a) and (c/a), where $(b/a) > (c/a)$, this fraction is given by

$$f = 1 - (2/\pi)\arcsin\left[\frac{1 + 2(b/a)^2 - 3(c/a)^2}{2 + (b/a)^2 - 3(c/a)^2}\right]^{1/2}. \qquad (9)$$

While this differs from the original expression of Steiman-Cameron and Durisen, it retains the character that even small deviations from axisymmetry lead to a significant fraction of all captured disks settling into the polar orientation.

Equation (9) was derived in the limit where disk self-gravity is unimportant. The situation where *both* dissipation and self-gravity are important is more complex and currently under investigation

(Steiman-Cameron 1990). In the case of NGC 4753, disk self-gravity appears to have had a negligible effect on the disk's evolution.

References

Briggs, F.H. (1990). Astrophys. J. **352**, 15.

Christodoulou, D.M. (1989). PhD Thesis, Louisiana State University.

Christodoulou, D.M. (1990). In *Galactic Models* (eds. J.R. Buchler, S.T. Gottesman & J.H. Hunter, Jr.), Annals of the New York Academy of Sciences, vol. 596, p. 207.

Christodoulou, D. M., Tohline, J. E. & Steiman-Cameron, T.Y. (1988). Astron. J. **96**, 1307.

Christodoulou, D. M., Tohline, J. E. & Steiman-Cameron, T.Y. (1990). Astrophys. J., Suppl. Ser., submitted.

Danby, J.M.A. (1962). *Fundamentals of Celestial Mechanics*, New York: Macmillan.

David, L.P., Durisen, R.H. & Steiman-Cameron, T.Y. (1984). Astrophys. J. **286**, 53.

David, L.P., Steiman-Cameron, T.Y. & Durisen, R.H. (1985). Astrophys. J. **295**, 65.

de Zeeuw, T. & Merritt, D. (1983). Astrophys. J. **267**, 571.

Dobrovolskis, A., Steiman-Cameron, T.Y. & Borderies, N. (1989). Geophys. Res. Letters **16**, 949.

Durisen, R.H., Tohline, J.E., Burns, J.A. & Dobrovolskis, A.R. (1983). Astrophys. J. **264**, 392.

Duncan, M.J. & Levison, H.F. (1989). Astrophys. J., Lett. **339**, L17.

Freeman, K.C. (1966a). Mon. Not. R. Astron. Soc. **133**, 47.

Freeman, K.C. (1966b). Mon. Not. R. Astron. Soc. **134**, 1.

Freeman, K.C. (1966c). Mon. Not. R. Astron. Soc. **134**, 15.

Habe, A. & Ikeuchi, S. (1985). Astrophys. J. **289**, 540.

Habe, A. & Ikeuchi, S. (1988). Astrophys. J. **326**, 84.

Heisler, J., Merritt, D. & Schwarzschild, M. (1982). Astrophys. J. **258**, 490.

Hunter, C. & Toomre, A. (1969). Astrophys. J. **155**, 747.

Levison, H. (1987). Astrophys. J., Lett. **320**, L93.

Miller, R.H. (1982). Astrophys. J. **254**, 75.

Miller, R.H. & Smith, B.F. (1982). Astrophys. J. **257**, 103.

Monet, D., Richstone, D. & Schechter, P. (1981). Astrophys. J. , 245, 454.

Pringle, J.E. (1981). Annu. Rev. Astron. Astrophys. **19**, 137.

Richstone, D. (1980). Astrophys. J. **238**, 103.

Sackett, P. D. & Sparke, L.S. (1990). Astrophys. J. **361**, 408.

Sandage, A. (1961). *The Hubble Atlas.* Washington D.C.: Carnegie Institution.

Scalo, J.M. & Struck-Marcell, C. (1984). Astrophys. J. **276**, 60.

Schwarzschild, M. (1979). Astrophys. J. **232**, 236.

Schwarzschild, M. (1983). Astrophys. J. **263**, 599.

Schweizer, F., Whitmore, B.C. & Rubin, V.C. (1983). Astron. J. **88**, 909.

Simonson, G.S. (1982). PhD Thesis, Yale University.

Sparke, L.S. (1984). Mon. Not. R. Astron. Soc. **211**, 911.

Sparke, L.S. (1986). Mon. Not. R. Astron. Soc. **219**, 657.

Sparke, L.S. & Casertano, S. (1988). Mon. Not. R. Astron. Soc. **234**, 873.

Steiman-Cameron, T.Y. (1984). Ph. D. thesis, Indiana University.
Steiman-Cameron, T.Y. (1990). In preparation.
Steiman-Cameron, T.Y. & Durisen, R.H. (1982). Astrophys. J., Lett. **263**, L51.
Steiman-Cameron, T.Y. & Durisen, R.H. (1984). Astrophys. J. **276**, 101.
Steiman-Cameron, T.Y. & Durisen, R.H. (1988). Astrophys. J. **325**, 26.
Steiman-Cameron, T.Y. & Durisen, R.H. (1990). Astrophys. J. **357**, 62.
Steiman-Cameron, T.Y., Kormendy, J. & Durisen, R.H. (1990). In preparation.
Tohline, J.E. & Durisen, R.H. (1982). Astrophys. J. **257**, 94.
Tohline, J.E., Simonson, G.F. & Caldwell, N. (1982). Astrophys. J. **252**, 92.
Toomre, A. (1982). Astrophys. J. **259**, 535.
Vandervoort, P.O. & Welty, D.E. (1982). Astrophys. J. **248**, 504.
Varnas, S.R. (1986a). PhD Thesis, Monash University.
Varnas, S.R. (1986b). Proc. Astron. Soc. Australia **6**, 458.
Whitmore, B.C. (1984). Astron. J. **89**, 618.
Whitmore, B.C., Lucas, R.A., McElroy, D.B., Steiman-Cameron, T.Y., Sackett, P.D., Olling, R.B. (1990). Astron. J. **100**, 1489.

Particle simulations of polar rings

THOMAS QUINN
University of Oxford

The modeling of rings as a cloud system

Most modeling of the settling of galactic rings has used the cloud fluid approximation. That is, either the material of the ring is continuous, or, if it is clumped, the parameters of the clouds would be such that they could be approximated by a continuous medium. In order to see if this approximation is valid, let us review the characteristic scales of a galactic cloud system using parameters for our own galaxy, following Scalo and Struck-Marcell (1984).

These dimensions can be given in terms of the observationally constrained average number of clouds per unit length, ν, and mean cloud radius a. In terms of these parameters, the mean separation is

$$\bar{s} = (6a^2/\nu)^{1/3} = 18 a_{\rm pc}^{2/3} \nu_{\rm kpc}^{-1/3}\ {\rm pc}.$$

For a fluid simulation to be valid, the volume element has to have linear dimensions much larger than $\bar{s}$, or about 18 pc with the galactic parameters.

The collision time is

$$t_c = 1/4p\nu\langle V_{rel}\rangle = 2\times 10^7/p\nu_{\rm kpc}c_{10}\ {\rm yr}.$$

where V_{rel} is the relative velocity, p is the collision cross section divided by the geometrical cross section, and c_{10} is the three dimensional rms random velocity in km s^{-1}. A related quantity is the mean free path in the rotating frame:

$$\lambda = \frac{160}{p\nu_{\rm kpc}}\ {\rm pc}.$$

These two quantities determine whether the system will be influenced by non-equilibrium effects, and whether these effects will be spatially resolved. For the larger clouds, the collision time is typically 10^8 yr which is comparable to the dynamical time, and it is not that small compared to the age of a galaxy. Therefore, a single cloud does not suffer a great number of collisions (on the order of a few hundred) in a Hubble time. The formation of polar rings in galaxies is thought to be due to the cannibalization of a smaller body. In this situation, the number of collisions will be even smaller since phase mixing will reduce the coarse grain phase space density. Because of this situation, to the extent that a polar ring galaxy can be considered as a collection of clouds, it is appropriate to use a *particle* simulation rather than a gas-dynamical calculation (smoothed-particle hydrodynamics included) to model the evolution of a polar ring.

Computational method

The evolution of a system of non-self-gravitating clouds was simulated in a fixed logarithmic triaxial potential (Richstone 1980). The potential is

$$\Phi(x,y,z) = \Phi_c \ln\left[\left(\frac{x}{a}\right)^2 + \left(\frac{y}{b}\right)^2 + \left(\frac{z}{c}\right)^2 + d\right],$$

where

$$\Phi_c = 4\pi G\rho_c\left[2\left(\frac{1}{a^2}+\frac{1}{b^2}+\frac{1}{c^2}\right)\right]^{-1} d.$$

The case $a = 150$ pc, $b = 180$ pc, $c = 200$ pc and $d = 2.7183$ was studied in order to make comparisons with the smoothed-particle hydrodynamics simulations of Habe and Ikeuchi (1985, 1988).

The cloud-cloud interactions are governed by two forces: a pressure or velocity independent force, and a viscous or velocity dependent force. For the pressure force, we assume the clouds are composed of infinitely compressible, isothermal gas. The pressure acceleration is then of the form $a_p \approx c_s^2/r$, where r is the interparticle separation, and c_s is the isothermal sound speed. To calculate the viscous force we assume that with zero impact parameter, the particles will not pass through each other, that is, given any initial approach velocity, their line-of-centres velocity will drop to zero before a critical separation $r > 0$ is reached. This implies a deceleration of $a_v \approx u^2/r$, where u is the line-of-centres velocity. This viscosity is of the same form as the artificial viscosity used in hydrodynamic simulations in order to follow shocks (Whitehurst, 1988). Such a viscous force law corresponds to the limit of highly super-sonic cloud collisions.

The total acceleration law is then:

$$a_r = \begin{cases} C^2/r + Qu^2/r, & \text{if } u < 0; \\ C^2/r & \text{if } u \geq 0, \end{cases}$$

for $r < \epsilon$, the cloud diameter. The interparticle acceleration is zero otherwise. Q is a dimensionless viscosity parameter of order 1. No self-gravity is included and the collisions were followed through by the integration scheme, rather than simulated using a Monte-Carlo method.

Since the intercloud forces are zero except for clouds that are touching, a fast method for locating neighboring clouds was required for efficient integration of the system. The method used for these calculations is based on a data structure called a *k-d tree* as described by Bently (1975). This is a binary tree, using the particle coordinates as discriminators. The tree construction can be done in $N \log(N)$ time, and a search for near neighbors of a single particle in this structure is done in $\log(N)$ time. The tree is rebuilt every timestep since the time to build the tree is about 1/4 the time to search for the neighbors of every particle and this search time is reduced by having an "optimum" tree. Storage overhead for the structure is 4 words per particle in this implementation, but can be reduced to less than 2 words per particle.

The leapfrog method was used as the integrator. Its advantages are the low storage overhead and that it is simplectic. Being simplectic implies that the integrals of motion are approximately conserved, so that particles are prevented from spiraling in or out due to inexact energy conservation. About 260 steps per orbit were needed to follow the collisions correctly.

Evolution of the cloud system

One thousand particles are placed in a ring segment of inner radius 5 kpc, outer radius 7 kpc, length 2 kpc and thickness 0.4 kpc. Within an order of magnitude, this number of particles is about the correct number of large molecular clouds for a small galaxy; there should be no problems about scalings with N. All particles are given the same velocity: the circular velocity at the center of the arc. The direction of the angular momentum of all the particles starts at 18° from the long axis towards the intermediate axis. This a crude mimic of an infalling satellite whose orbit is circularized by dynamical friction.

The pressure constant, C, corresponded to $T = 10^4$ K. The viscos-

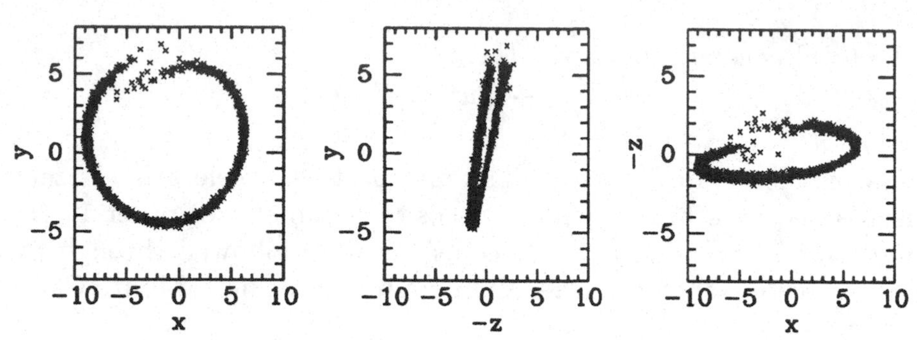

Fig. 1a. Three orthogonal projections of the cloud system at a time of 0.36 Gyr after the start of the simulation. The axes are labeled in kiloparsecs.

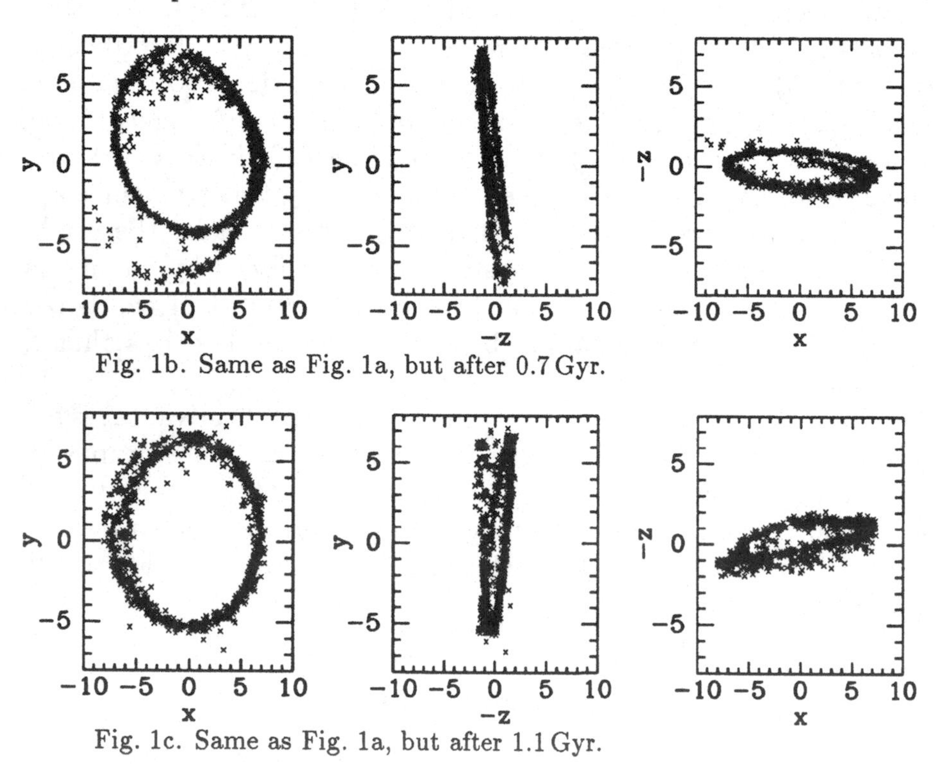

Fig. 1b. Same as Fig. 1a, but after 0.7 Gyr.

Fig. 1c. Same as Fig. 1a, but after 1.1 Gyr.

ity parameter was 1, that is, clouds will turn around well before their centers meet. The cloud diameter was chosen to be 160 pc. These initial conditions were chosen for comparison with the simulations of Habe and Ikeuchi.

Figures 1a through 1d show several snapshots of a simulation run over the period of 2.25 Gyr. Three orthogonal views are displayed. The long axis of the potential is the z axis, and the short axis is the x axis. Figure 1a shows the phase wrapping of the simulation after

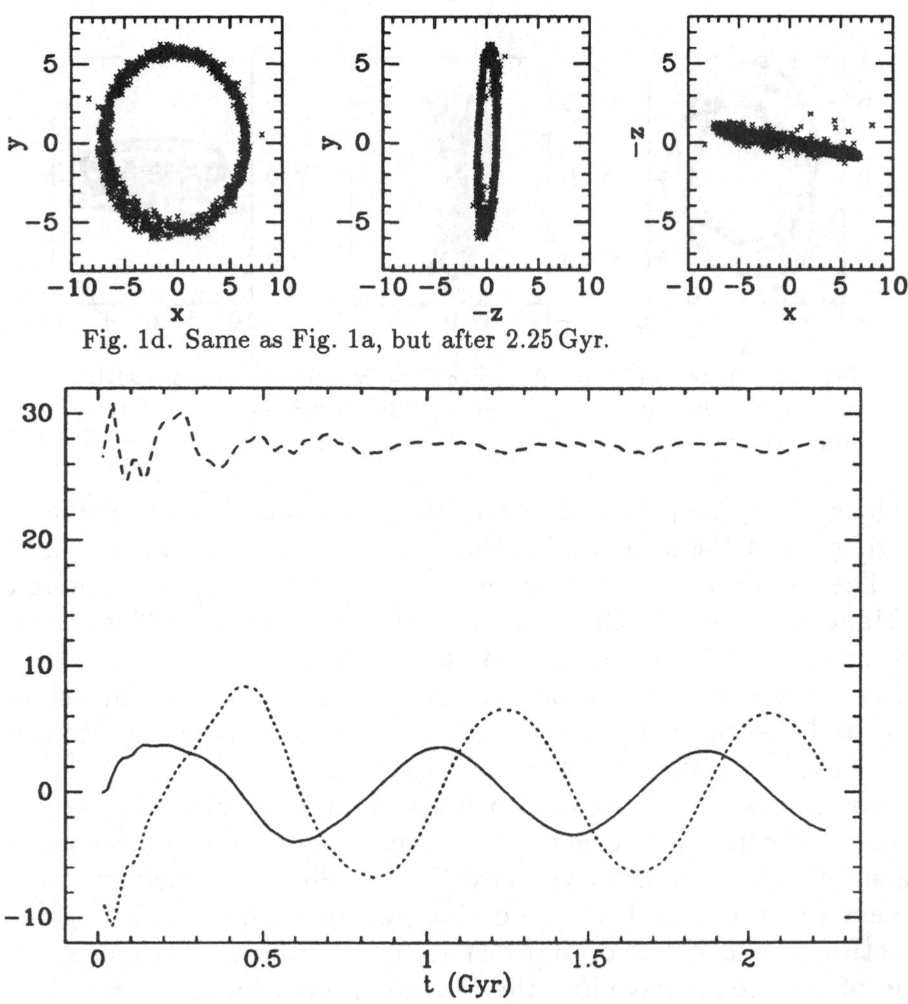

Fig. 1d. Same as Fig. 1a, but after 2.25 Gyr.

Fig. 2. The evolution of the x, y, and z components of the total angular momentum of the clouds in arbitrary units, as a function of time in Gyr. The solid line is l_x, the dotted line is l_y, and the dashed line is l_z.

0.36 Gyr. After 0.7 Gyr a helical structure appears, which persists for over 0.5 Gyr. At the end of the simulation (2.25 Gyr) the helical structure is gone, but the ring remains inclined to the axes of the potential. Another feature of the ring is that it is thinner in the radial direction than when it started out. The initial ring segment was 2 kpc thick while the final ring is 1 kpc thick in places. This is in contrast to what is expected from fluid dynamical simulations where a ring will spread into a disk by the transport of angular momentum outward.

Figure 2 shows the evolution of the total angular momentum of the

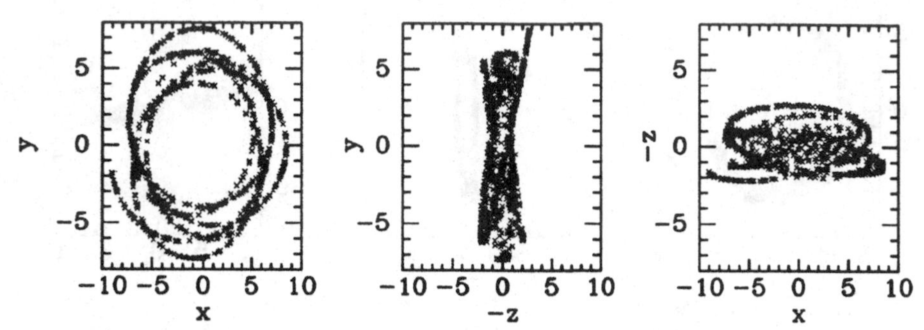

Fig. 3. Three orthogonal projections of the simulation with no cloud collisions at a time of 2.25 Gyr after the start of the simulation.

particles along the principal axes of the potential. The precession of the ring about the long axis is clearly seen in the variation of l_x and l_y. The period of this precession matches well the periods found by Habe and Ikeuchi (1985) in their SPH simulations. This is not surprising since the potential parameters here are the same as theirs, and the ring that forms is the same size as their ring. In contrast to Habe and Ikeuchi, the precession shows no sign of damping into the plane of the potential over these timescales.

Several runs have been done with several cloud sizes to see the effect of the collision rate on the evolution. Figure 3 shows a snapshot of a simulation after 2.25 Gyr run with the collisions turned off. This is effectively accomplished by choosing a cloud size of zero. This gives a picture of the initial orbital distribution of the simulation, since none of the clouds have had their orbits altered by collisions. The evolution of the total angular momentum in this simulation is shown in Figure 4. Here the angular momentum becomes aligned with the z axis, but it is clear that this is simply due to phase wrapping rather than any viscous damping effects.

A simulation was done with a cloud diameter of 320 pc instead of the original 160 pc. Figure 5 shows a snapshot of this simulation after 2.25 Gyr. Comparison of this figure with 1d shows that this ring is thicker, both in radial width and vertical height. This could be evidence that the higher collision rate is beginning to transfer angular momentum outward, and spread the ring. However, Figure 6 shows the evolution of the total angular momentum, and this indicates no signs of the ring settling into the preferred plane.

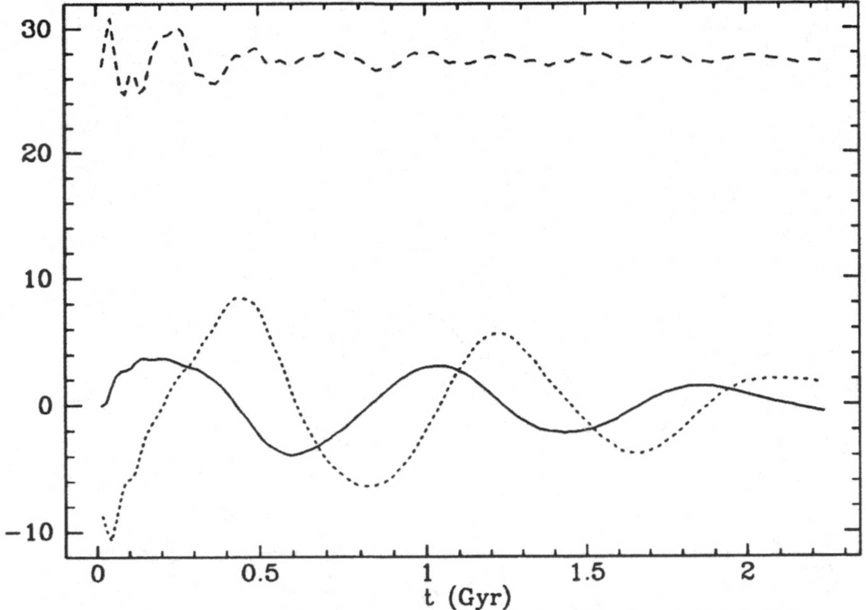

Fig. 4. The evolution of l_x, l_y, and l_z for the simulation with no cloud collisions. The lines are the same as in Fig. 2.

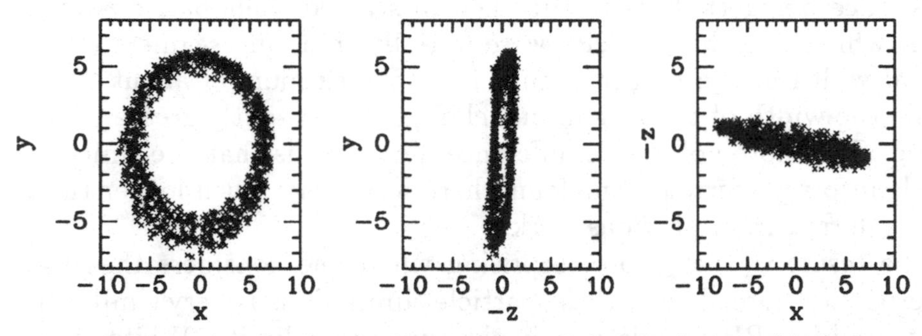

Fig. 5. Three orthogonal projections of the simulation with large clouds at a time of 2.25 Gyr after the start of the simulation.

Conclusions

Perhaps the most surprising aspect of these simulations is the disparity of the results with the hydrodynamical calculations of previous works. (see Habe and Ikeuchi 1985, Steiman-Cameron and Durisen 1988 and Christodoulou and Tohline, this volume) The first indication of differences can be seen by comparing the radial width of the ring in Figure 1d with a similar figure of Habe and Ikeuchi's (1985) model TA. This is a model run for a similar length of time, and with similar initial conditions as the simulations in this paper, the major

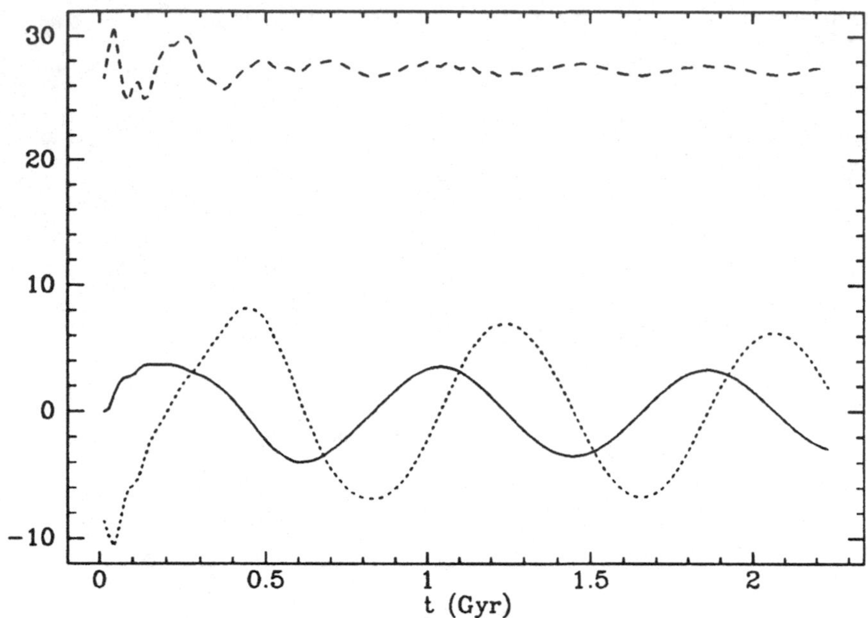

Fig. 6. The evolutions of l_x, l_y, and l_z for the simulation with large clouds. The lines are the same as in Fig. 2.

difference being that Habe and Ikeuchi started their particles in a ring, while here the particles were initially in a ring segment. The radial width of the ring in Figure 1d is a maximum of about 2 kpc while the width of the ring in model TA is consistently greater than 4 kpc. The more noticeable difference, however, is that the times to settle into a preferred plane found here were very much longer than that inferred from previous work.

The result is even more surprising when one considers that the interparticle force law in these particle simulations is very similar to that used in SPH calculations in the supersonic limit. (Whitehurst, 1988) In fact, the main difference is that here a typical cloud will only interact with another cloud less than once per dynamical time, whereas the parameters of an SPH calculation are usually adjusted so that each particle interacts with several others every timestep. Some insight into this problem may be gleaned from studies of systems of colliding particles in a Kepler potential done the context of solar system dynamics (Brahic, 1976, 1977). In these papers, it has been shown that the rate of evolution of the system as measured by the dispersion in radius or flattening scales with $N(a/R)^2$, where R is a characteristic size of the system. From this we can infer that if we either increase the particle size, or integrate for a longer period of time, the system will begin to behave like a fluid.

These simulations only crudely model a cloud system. In reality, cloud collisions involve fragmentation, coalescence, star formation, and supernova energy injection. If clouds on very similar orbits collide, the timescale for the collision can be long enough that self-gravity can become important. The big caveat, however, is that the ring was modeled as being entirely clouds. The addition of an intercloud medium would introduce a drag force and evaporation. Given enough intercloud gas, the clouds would cease to be important for the evolution at all, and the relevant calculations would be the gasdynamical calculations. However, if polar rings can be modeled as a cloud system, they will survive and remain out of the preferred planes much longer than the hydrodynamic calculations indicate.

References

Bently, J. (1975). C. A. C. M. **18**, 509.
Brahic, A. (1976). J. Comp. Phys. **22**, 171.
Brahic, A. (1977). Astron. Astrophys. **54**, 895.
Habe, A. & Ikeuchi, S. (1985). Astrophys. J. **289**, 540.
Habe, A. & Ikeuchi, S. (1988). Astrophys. J. **326**, 84.
Richstone, D.O. (1980). Astrophys. J. **238**, 103.
Scalo, J.M. & Struck-Marcell, C. (1984). Astrophys. J. **276**, 60.
Steiman-Cameron, T.Y. & Durisen, R.H. (1988). Astrophys. J. **325**, 26.
Whitehurst, R. (1988). Mon. Not. R. Astron. Soc. **233**, 529.

A bending instability in prolate stellar systems

DAVID MERRITT
Rutgers University

Abstract

Prolate stellar systems are unstable to bending modes when sufficiently elongated. N-body experiments suggest that, in one family of nonrotating prolate models, the critical axis ratio for instability is about 2:5. This result may explain in part the observed absence of elliptical galaxies flatter than E6.

Edge-on spiral galaxies often exhibit warps, and this fact has motivated a large number of theoretical investigations into the origin and persistence of warps in rapidly rotating disks (as reviewed by Toomre 1983). Less is known about the response of *hot* stellar systems to bending perturbations. The purpose of this contribution is to report the existence of a bending instability in a family of nonrotating prolate models. The instability appears to persist in models as round as 1:3, i.e. with "Hubble types" of E6 or E7. To the extent that this instability, or similar ones, are characteristic of pressure-supported models, it is possible that the absence of elliptical galaxies flatter than about E6 may be attributable entirely to dynamical instabilities.

The basic features of the instability in prolate models may be reproduced by a simple linear model. Imagine a stellar-dynamical "needle" with linear mass density $\eta(x)$ and thickness ϵ. If ϵ is sufficiently small, the gravitational force will constrain all stars at a given point to move together in their perpendicular displacements. Let $X(t)$ be the unperturbed motion of a star along the axis of the needle, and $h(x,t)$ the perpendicular displacement. Following Toomre (1966), the perpendicular acceleration of a star with this equilibrium trajectory is

$$a_\perp = \left(\frac{\partial}{\partial t} + \frac{dX}{dt}\frac{\partial}{\partial x}\right)^2 h(x,t) \tag{1a}$$

$$= \frac{\partial^2 h}{\partial t^2} + 2\frac{\partial^2 h}{\partial x \partial t}\frac{dX}{dt} + \frac{\partial^2 h}{\partial x^2}\left(\frac{dX}{dt}\right)^2 + \frac{\partial h}{\partial x}\frac{d^2X}{dt^2}. \tag{1b}$$

Averaged over all particles at x, this acceleration must equal the gravitational restoring force per unit mass $F_\perp$:

$$\frac{\partial^2 h}{\partial t^2} + \sigma^2(x)\frac{\partial^2 h}{\partial x^2} + a(x)\frac{\partial h}{\partial x} - F_\perp(x,t) = 0, \tag{2}$$

where $\sigma(x)$ is the equilibrium velocity dispersion of stars at x and $a(x)$ is their unperturbed acceleration. In the small-slope approximation, the perturbed force is

$$F_\perp(x,t) = -G\int_{-\infty}^{\infty} \eta(x')\frac{[h(x,t) - h(x',t)]}{[(x-x')^2 + \epsilon^2]^{3/2}}dx'. \tag{3}$$

Setting $h(x,t) \propto \exp\left[i\left(kx - \omega t\right)\right]$ gives, for the *homogeneous* needle,

$$F_\perp(x,t) = -\frac{2G\eta}{\epsilon^2}\left[1 - k\epsilon K_1(k\epsilon)\right] h(x,t), \tag{4}$$

where K_1 is the modified Bessel function of the second kind. The corresponding dispersion relation is

$$\omega^2 = \frac{2G\eta}{\epsilon^2}\left[1 - k\epsilon K_1(k\epsilon)\right] - \sigma^2 k^2 \tag{5}$$

or, in the long-wavelength ($k\epsilon \ll 1$) limit,

$$\omega^2 = -G\eta k^2 \log(k\epsilon/2) - \sigma^2 k^2. \tag{6}$$

Note that short-wavelength perturbations are the most destabilizing; the maximum unstable wavelength is $\pi\epsilon\exp(\sigma^2/G\eta)$.

This simple model may be carried one step further by computing the normal modes of an infinite, *inhomogeneous* needle, again under the simplifying assumption that all stars remain confined to the axis of the needle as it bends. Let

$$\eta(x) = \eta_{KZ}(x) = \frac{M}{\pi a}\frac{1}{1 + x^2/a^2}, \tag{7}$$

the linear mass density of the Kuzmin-de Zeeuw prolate spheroid in the limit of large elongation. Expressing the z displacement as

$H(x)e^{st}$ then yields the linearized bending equation

$$s^2H(x)+\sigma_{KZ}^2(x)\frac{d^2H}{dx^2}+a_{KZ}(x)\frac{dH}{dx}=$$
$$\frac{GM}{\pi a}\int_{-\infty}^{\infty}\frac{[H(x')-H(x)]}{(1+x'^2/a^2)\left[(x-x')^2+\epsilon^2\right]^{3/2}}dx', \tag{8}$$

where $a_{KZ}(x)=-d\Phi_{KZ}/dx$, $\eta_{KZ}(x)\sigma_{KZ}^2(x)=\int_{\Phi_{KZ}}^0\eta_{KZ}(\Phi')d\Phi'$, and the equilibrium potential is

$$\Phi_{KZ}(x)=-G\int_{-\infty}^{\infty}\frac{\eta_{KZ}(x')dx'}{[(x-x')^2+\epsilon^2]^{1/2}}. \tag{9}$$

The eigenvalue equation (8) may be solved by rewriting it in finite-difference form at a set of N points and casting the resulting N equations into matrix-eigenvalue form (e.g. Erickson 1974). The spectrum of eigenvalues turns out to be continuous. Figure 1 shows a selection of eigenmodes and their growth rates for $\epsilon/a=0.1$. Modes are strictly even or odd in their dependence on x. The growth rate increases with decreasing wavelength, in a manner consistent with the dispersion relation (5) for the homogeneous needle.

The stars in a real prolate galaxy are not constrained to move together as the galaxy deforms; the response to any bending perturbation will therefore be weaker than in the simple model just analyzed, and it is reasonable to expect that sufficiently round models will not be unstable at all. To investigate the response of realistic prolate models to bending perturbations, Lars Hernquist and I carried out 40,000-particle integrations using a tree N-body code of a set of equilibrium models generated from the "regular" distribution functions discussed by Hunter et al. (1991). The "regular" models are composed entirely of thin-walled tube orbits that circulate around the long axis; the model density is $\rho_{KZ}=\rho_0\left[1+x^2+(y^2+z^2)/c^2\right]^{-2}$, with $c<1$ the axis ratio. Initial conditions were chosen such that there was no streaming. Models with initial ellipticities of E9, E8 and E7 were found to be unstable; the E6 model showed no sign of unstable evolution. Figure 2 illustrates even and odd bending modes extracted from the E8 model, at a time corresponding to slightly more than one full period of an axial orbit of unit amplitude. The modes were computed from the expressions

$$\Phi^{even}(x,R,\phi)=-\cos\phi\int_0^{k_{max}}dkS_k(R)\cos(kx), \tag{10a}$$

$$\Phi^{odd}(x,R,\phi)=-\cos\phi\int_0^{k_{max}}dkA_k(R)\sin(kx), \tag{10b}$$

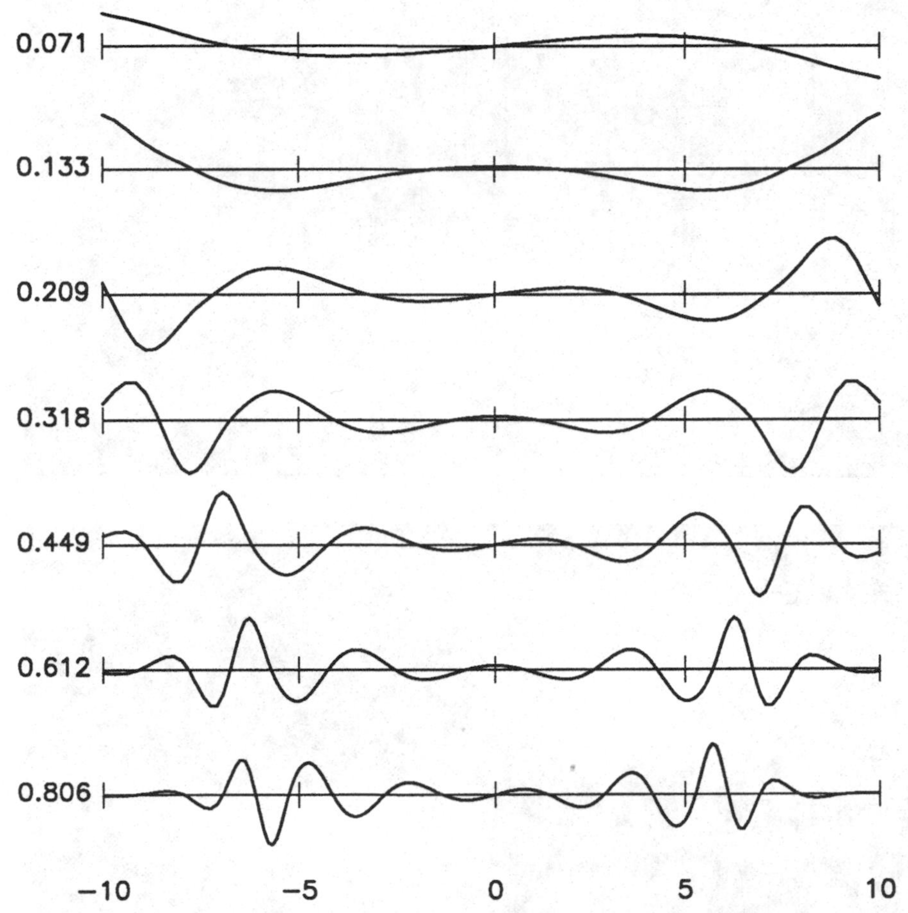

Fig. 1. Selected bending eigenfunctions of the inhomogeneous needle and their growth rates, as derived from equation (8) with $\epsilon = 0.1$. Units are $G = M = a = 1$.

where $\{x, R, \phi\}$ are the usual cylindrical coordinates, and $k_{max} = 5$; the coefficients $S_k(R)$ and $A_k(R)$ are obtained by expanding the potential Green function in cylindrical coordinates, e.g.

$$S_k(R) = \frac{4Gm}{\pi}[K_1(kR) \sum_{R'<R} I_1(kR') \cos(kx') \cos\phi' + I_1(kR) \sum_{R'>R} K_1(kR') \cos(kx') \cos\phi']. \tag{11}$$

The even and odd contributions to the bending (corresponding to banana- and S-shaped deformations, respectively) were found to lie in distinct planes and to grow at different rates, as expected for dis-

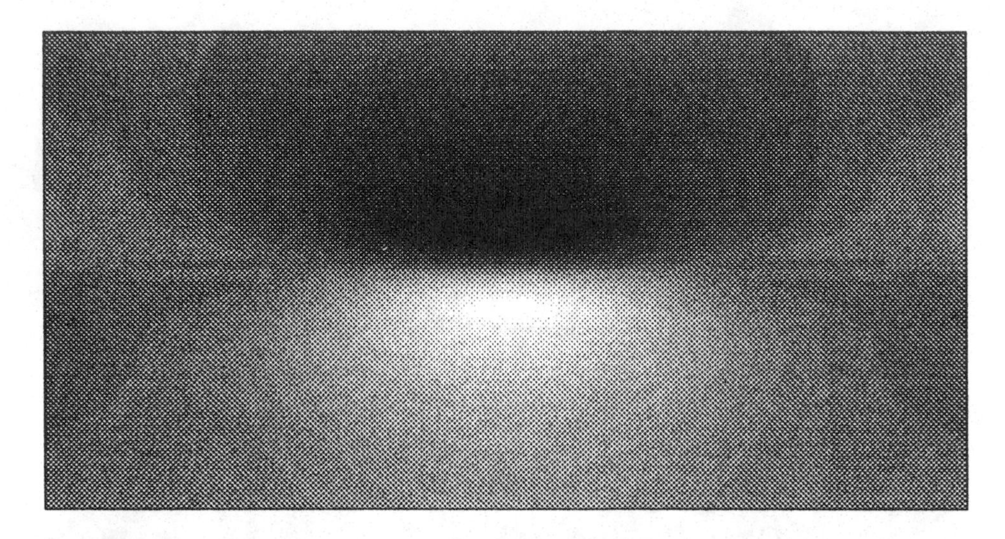

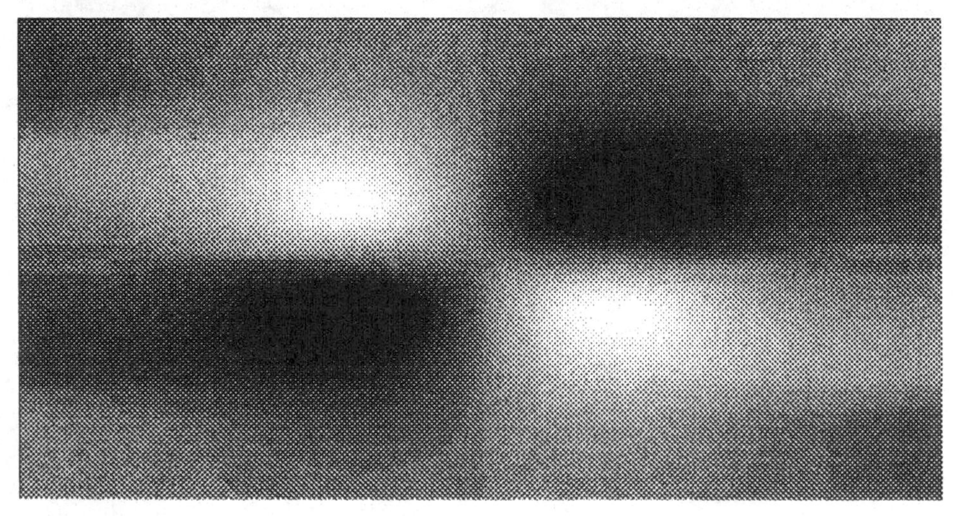

Fig. 2. Gray-scale plots of the even and odd components of the perturbed potential in the integration of the E8 prolate model. The scale is -2 to +2 on the horizontal (x) axis, and -1 to +1 on the vertical (R) axis.

tinct modes. The growth rates were consistent with those predicted by the simple linear model just described.

After a few crossing times, these models look distinctly "boxy" (Figure 3). We do not know whether this boxiness would persist.

Fridman and Polyachenko (1984) proposed that instabilities similar to the one reported here might be responsible for the absence of elliptical galaxies flatter than about E6. A convincing proof of

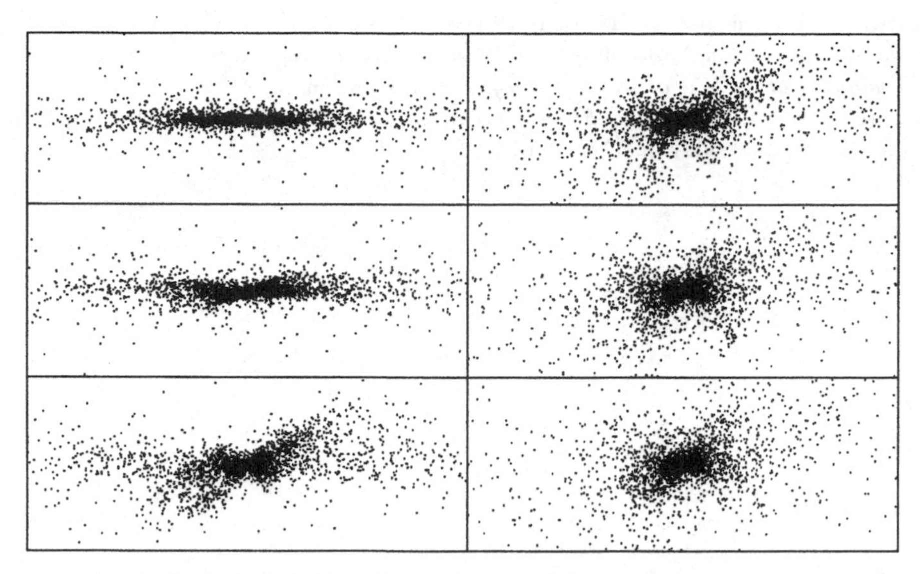

Fig. 3. Snapshots of the E9 prolate model at six times.

this hypothesis will require substantially more work. Perhaps the most important question to be answered is whether bending instabilities are important in slowly-rotating *oblate* models. At least one nonrotating disk model is known to be unstable to bending (Hunter and Toomre 1969, p. 752), but most work on disks has focussed on the case of circular orbits. The response to bending perturbations of thin disks with isotropic or radially-biased velocity distributions has apparently never been investigated; such a study would constitute an important first step toward understanding the effect of model kinematics on the stability of oblate models of elliptical galaxies.

This contribution is based on the more complete discussion of prolate shell-orbit models in Merritt and Hernquist (1991). The work described here was supported in part by a Fullam/Dudley Award from the Dudley Observatory, and by a grant from the Pittsburgh Supercomputing Center.

References

Erickson, S. A. (1974). PhD Thesis, Massachusetts Institute of Technology.

Fridman, A. M. & Polyachenko, V. L (1984). *Physics of Gravitating Systems.* New York: Springer.

Hunter, C. & Toomre, A. (1969). Astrophys. J. **155**, 747.

Hunter, C., de Zeeuw, P.T., Park, Ch. & Schwarzschild, M. (1991). In press.

Merritt, D. & Hernquist, L. (1991). In press.

Toomre, A. (1966). In *Notes on the 1966 Summer Study Program in Geophysical Fluid Dynamics at the Woods Hole Oceanographic Institution*, p. 111.

Toomre, A. (1983). In *Internal Kinematics and Dynamics of Galaxies*, IAU Symp. 100 (ed. E. Athanassoula), p. 177. Dordrecht: Reidel.

The Milky Way: lopsided or barred?

KONRAD KUIJKEN
Canadian Institute for Theoretical Astrophysics

Introduction

Figure 1 shows a map of average brightness temperature T_b of Galactic HI within 10° of the Galactic plane, in the $v_{\rm rad}$-l plane referred to the Local Standard of Rest (LSR), prepared from the surveys of Weaver & Williams (1974) and Kerr et al. (1986). If this gas is assumed to be moving on circular orbits, and a rotation curve is adopted for matter outside the solar circle, then kinematic distances can be derived and the spatial distribution of the gas studied. Gas outside the solar circle has negative velocities in directions $0° < l < 180°$, positive velocities elsewhere; the other emission in the region $-90° < l < 90°$ originates from gas inside the solar circle. For a circularly symmetric Galaxy this map should be symmetric about the point ($l = 0°$, $v_{\rm rad} = 0°$). Broadly speaking this is indeed the case, but as can be seen from a map of $T_b(l, v_{\rm rad}) - T_b(360° - l, -v_{\rm rad})$ (Fig. 1, lower panel) there are systematic deviations from this symmetry. The same is evident from the contours of constant T_b. Both the faint contours of the outer gas, and the "tangent points" (from which the inner rotation curve is usually derived), deviate between the Northern and Southern halves by some tens of $\rm km\,s^{-1}$. The asymmetries in the inner Galaxy are often attributed to spiral arms. In the outer Galaxy, where spiral arms are much weaker, large asymmetries are also in evidence; these are the subject of this paper.

Recently, it has been proposed (Blitz & Spergel 1990, henceforth B&S) that the asymmetries in the observed velocity field of the outer Galaxy can be understood in terms of an outward motion of the LSR of $\sim 14\,\rm km\,s^{-1}$. After considering various possibilities for the origin of this motion, B&S concluded that the observations require that the Galaxy have a non-axisymmetric (in the Galactic plane) bulge,

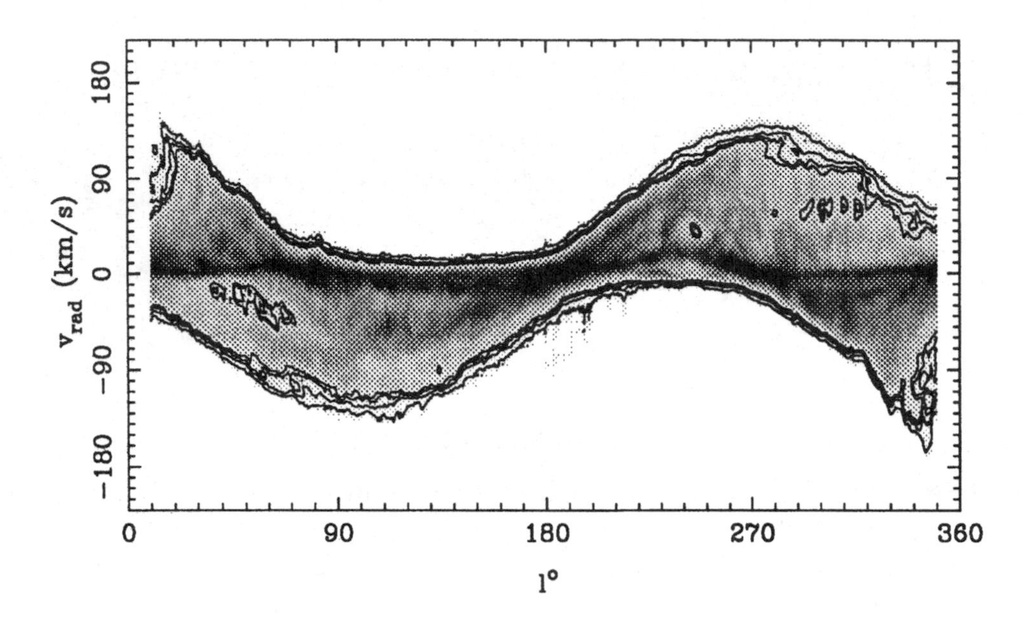

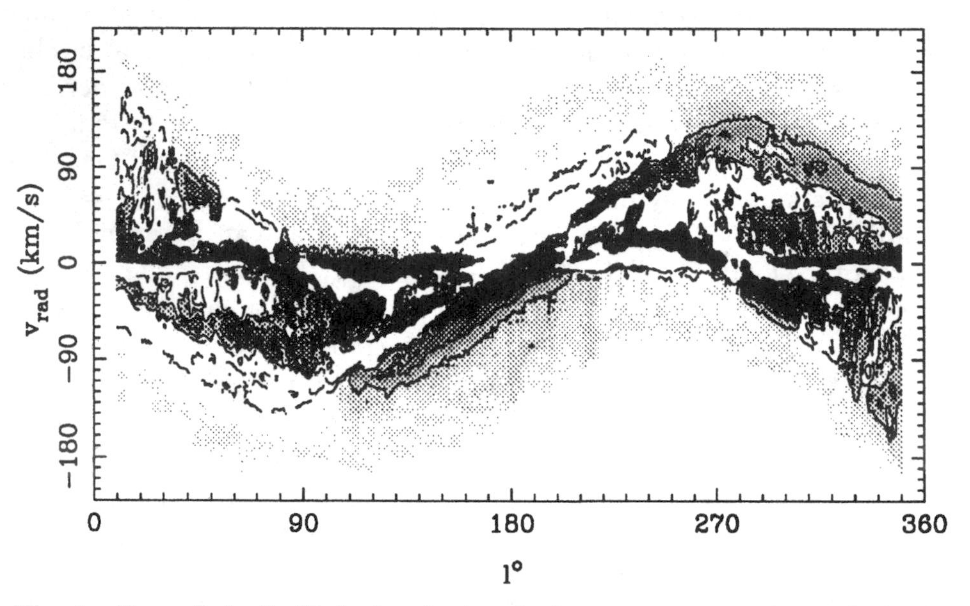

Fig. 1. Top: Galactic HI in longitude-velocity space, averaged over latitudes between 10° and −10°. Contours at 1, 2, 3, and 4K are shown. The gray scale runs from 0.5K (white) to 60K (black). Velocities are referred to the LSR. Bottom: Antisymmetrization of upper panel. Points (l, v) with excess emission over $(-l, -v)$ are shown in gray scale, running from 0 (white) to 10K (black). Contours are shown at ±1 and ±4K (dashed ones for negative excesses). Note the many spiral arm features.

which perturbs the overall potential away from circular symmetry and causes matter near the Sun to move on elliptical orbits. The outer Galaxy orbits are more circular. Given the mass of the bulge, they found it impossible to perturb the LSR's orbit sufficiently without including a slow figure rotation so that the Inner Lindblad Resonance (ILR) of the potential lies just outside the solar circle, at $R \simeq 11$kpc.

The observations B&S advance in favour of their model, apart from the outer Galaxy asymmetry, are mostly also consistent with an inner Galaxy (and LSR) in circular rotation, and so do not provide direct evidence for or against either, although it is interesting that non-axisymmetric models can be constructed which are consistent with such "null-detections". Thus, the asymmetry in the inner Galaxy rotation curve (as measured from the tangent points) is fitted comparably well by a constant zero value as by the prediction from the B&S model, the absorption feature towards the Galactic centre has no net radial motion referred to the LSR, and the independence of mean outward motion on velocity dispersion of stellar populations is predicted by either model. The only data which directly provide evidence for an outward motion of the LSR are the kinematics of the gas in the inner kpc of the Galaxy, which, assuming equilibrium, are broadly consistent with an outward motion, denoted by Π_0, of $14\,\mathrm{km\,s^{-1}}$. On the other hand, the kinematics of 55 OH/IR stars in the bulge within 1° of the Galactic Centre, give a net outward motion of $0.2\pm12\,\mathrm{km\,s^{-1}}$, although unfortunately this sample is rather small. Hopefully further observations will extend this sample (Habing 1987). A diagnostic which was not considered by B&S is the velocity field of the Galactic HII regions up to ca. 5 kpc outside the Sun's orbit (Brand 1986). This shows no irregularities, even though this is where the model of B&S places the ILR. Since in this region a velocity "glitch" of at least $2\Pi_0 \simeq 30\,\mathrm{km\,s^{-1}}$ is predicted (see below), this is a serious problem for the model.

The point was made by B&S that the approximate $\cos l$ asymmetries in the outer contours of the HI map can be explained equally well by an outward motion of the LSR in a circular outer galaxy as by a circular motion of the LSR, surrounded by a "lopsided" (i.e., suffering an $m = 1$ distortion) outer galaxy. This is easy to see with the epicycle approximation (see appendix, equation A10). As discussed by Baldwin et al. (1980), an $m = 1$ distortion can persist for several Gyr before finally winding up. B&S rejected this possibility by considering the asymmetry in the total gas density, which shows

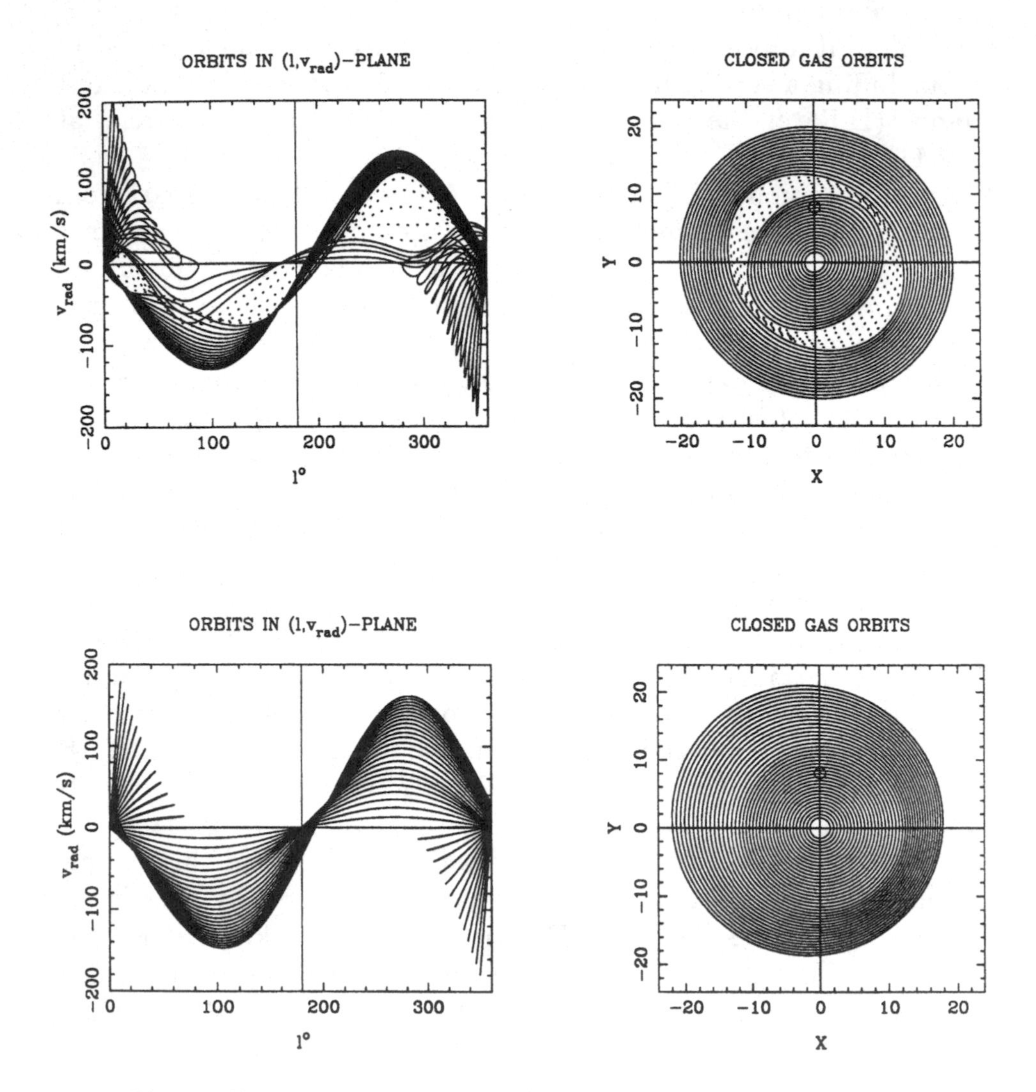

Fig. 2. Closed orbits with epicentric radii between 1.5 and 20 kpc for the two models considered here, projected in the (l, v) plane and in real space. Top: B&S model (dotted lines show the region where closed orbits cross. Here an *ad hoc* prescription has been used to mimic crudely the effects of viscosity). Bottom: a lopsided model. Note that the outer orbits in both models crowd and cross in different places.

evidence for an $m = 2$ distortion, not an $m = 1$ distortion. However, because of the steep radial gradient in the HI density, this quantity is actually rather dominated by fluctuations in the spatial gas distribution just outside the solar circle, which, as can be seen on the antisymmetrized HI map (the region just outside the solar circle lies at small positive velocities for $l > 180°$, and small negative velocities elsewhere) shows several spiral-arm features which give a strong antisymmetric signal while not being central to this discussion. Thus this would appear to be a rather weak argument in favour of the outward motion of the LSR. In fact, as will be shown here, the $m = 1$ model offers explanations of several problems which the B&S model faces.

The closed gas orbits for the B&S model and for a lopsided Galaxy model, projected into the $(l, v_{\rm rad})$ plane and into real space, are shown in Figure 2.

The line profile near the Galactic Anticentre

The B&S model relies crucially on the existence of a nearby ILR to perturb the solar circle. In this way a small perturbation in the potential can lead to large perturbations of the LSR, while not affecting regions far from the Sun, either towards the Galactic centre or away from it.

The exact outward velocity component of the gas near the ILR cannot be determined by studying closed orbits only, since there is a transition between two different orbit families which intersect one another: viscous effects must perforce become important. These are hard to quantify, and also hard to calculate, requiring hydrodynamical simulations; however, the generic behaviour of $v_{\rm rad}(R)$ will be as sketched in Figure 3a, which corresponds to the parameters of the B&S model. The solid curves show $v_{\rm rad}(R)$ calculated with the epicycle approximation (equation A9 in the appendix), with dashed lines indicating the region near the ILR where the closed orbits cross. The dotted curve schematically shows how viscous effects might smooth out the singularity at the resonance. The implication of this picture is that towards the Anticentre the gas with the most negative velocity detected is not the most distant, but rather lies at some intermediate distance, beyond which the deviation from the LSR decreases again. Thus one would expect a rather sharp (governed by the gas velocity dispersion) edge to the line profile, such as is seen at the tangent points of the inner Galaxy gas, which also lies in regions of maximum line-of-sight velocities.

The HI line profile looking out from the Sun towards the Galactic

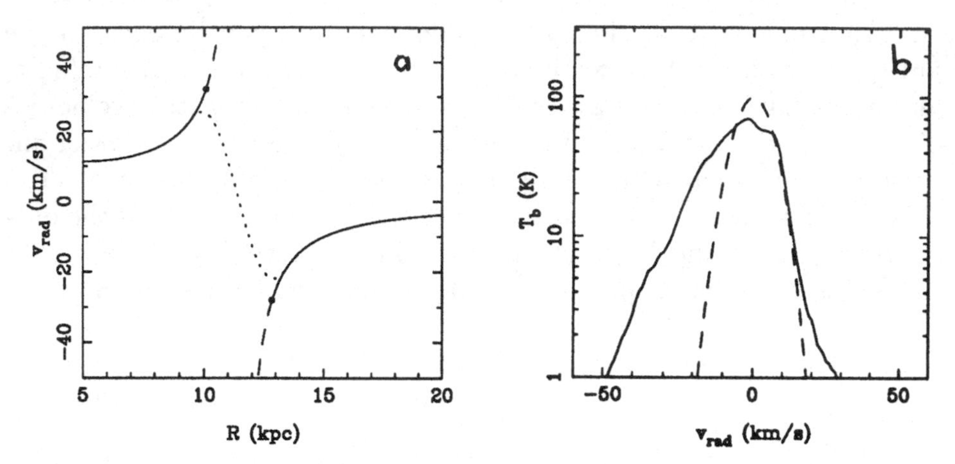

Fig. 3. The Galactic Anticentre. Panel a: outward velocity component of the closed orbits near the Lindblad resonance predicted by the B&S model. The dashed lines indicate regions where these orbits (calculated in the epicycle approximation) cross, and the dotted curve schematically indicates how the singularity will be smoothed out by viscosity. Panel b: Line profile of Galactic HI towards $l = 180°$. The dashed curve is a gaussian with dispersion $6\,\mathrm{km\,s^{-1}}$. Near the peak of this profile optical depth effects become important.

Anticentre is shown in Figure 3b. Except for an amount consistent with the gas velocity dispersion, there is little or no gas which moves outwards more quickly than the LSR ($v_{\rm rad} > 0$), and the bulk of the gas is in fact moving with the LSR. This means that the Sun's radius happens to be close to the positive peak in Figure 3a, at the point of maximum deviation from circular orbits. However, the negative-velocity side of the line profile shows a gentle, exponential intensity decrease as far as can be observed (in the deeper data of Burton 1985, which show brightness temperature contours down to 0.2K, this trend continues), which is inconsistent with the required reversal in $v_{\rm rad}(R)$. As mentioned above, direct measurement of the velocity field from HII regions (Brand 1986) reveals no evidence for a resonance either.

Velocity Crowding in the Outer Galaxy

The spacing of the contours of brightness temperature in the outer Galaxy changes with longitude. In a perfectly symmetrical Galaxy,

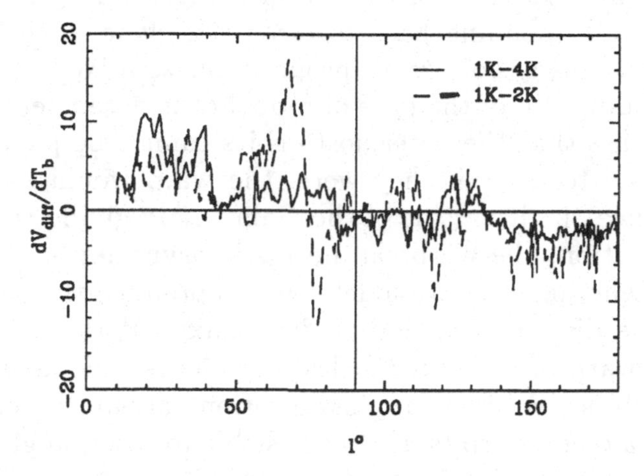

Fig. 4. The asymmetry in the crowding of the brightness temperature contours. The fact that this curve is more odd than even about $l = 90°$ indicates an $m = 1$ disturbance.

all closed (circular) orbits outside the solar circle project into sine curves in the l-$v_{\rm rad}$ plane, and consequently crowd most at longitudes of 0° and 180°. The asymmetry in the crowding, which can be defined as the sum of $dv_{\rm rad}/dr_0$ at l and $-l$, is zero in this case. In the presence of asymmetries, this is no longer so: as shown in the appendix (equation A12), instead this asymmetry has an approximate $\cos ml$ dependence. This is therefore a diagnostic with which the mode of the disturbance can be determined (see also Figure 2). Following B&S, it can be assumed that $dv_{\rm rad}/dT_b$ is a good representation of $dv_{\rm rad}/dr_0$ (as long as there has been no velocity reversal). Figure 4 shows the $dv_{diff}/dT_b(l)$ curve, obtained by subtracting nearby contours in the HI map. Regions within $\sim 40°$ of the Galactic Centre and Anticentre are probably affected by velocity reversals; in these regions $dv_{\rm rad}/dr_0(l)$ has changed sign at large radii, so that measuring dv_{diff}/dr_0 from the T_b-contours underestimates the magnitude of the asymmetry. This curve clearly shows a monotonic behaviour, indicative of $m = 1$.

The Warp of the Outer Galaxy

It is clear from a glance at a $(v_{\rm rad}, b)$ map of Galactic HI (e.g., Burton 1985) near $l = 90°$ and 270° that the outer disk of the Galaxy is warped. Rather than a classic grand design, integral-sign warp, however, the warp is quite asymmetric outside about ~ 2 solar radii from the Galactic centre (Burton 1987), growing in amplitude in the North, but actually bending back towards the plane in the South. Assuming that all the gas orbits are properly phase-mixed and closed (the central assumption of the type of model considered here and by B&S), this implies that the outermost orbits cannot be planar with a pure $m = 2$ distortion: if they were, the warp would extend to the same distance on either side of the Galaxy. By giving the orbits an $m = 1$ distortion, the warp can be made asymmetric, although it is not clear whether such a model could reproduce the return to the plane of the disk on one side of the galaxy. If this cannot be done then the warp of the outer Galaxy challenges the assumption of well-filled orbits. Without any assumptions about the distribution of the gas along its orbits, it is impossible to disentangle purely spatial anomalies in a circular rotation model from a model with kinematic anomalies (Mihalas & Binney 1981).

Conclusions

The explanation of the observed asymmetries in the projected $v_{\rm rad}$–l distribution of Galactic HI in terms of an outwards component to the motion of the LSR of $\sim 14\,{\rm km\,s^{-1}}$, advanced by Blitz & Spergel (1990), is non-unique. A different explanation, involving a lopsidedness of the outer Galaxy, can also account for most of the observations. Neither model is perfect. The B&S model involves a slowly rotating bulge with non-zero quadrupole moment in the Galactic plane, and predicts the presence of an inner Lindblad resonance at galactocentric radius ca. 11 kpc, for which there is no evidence in observations of the velocity field or in the line profile towards the Galactic anticentre. Lopsided models are less tidy, in that they are not in equilibrium but rather wind up on a timescale of some Gyr, and require some explanation for their origin. Nevertheless, the crowding of orbits when projected into the $v_{\rm rad}$-l plane favours an $m = 1$ (lopsided) over an $m = 2$ (quadrupole) asymmetry. On the other hand, the asymmetries in the spatial density of the gas indicate an $m = 2$ distortion, although these are quite likely to be strongly affected by the nearby spiral arm just outside the solar circle. Neither model seems able to satisfactorily explain the asymmetry in the Galactic warp, which may be an indication that the gas in the outer regions of the galaxy is not properly phase-mixed along its orbits.

References

Baldwin, J.E., Lynden-Bell, D.E. & Sancisi, R. (1980). Mon. Not. R. Astron. Soc. **193**, 313.

Binney, J. & Tremaine, S. (1987). *Galactic Dynamics*, Chapter 3.3. Princeton: Princeton University Press.

Blitz, L. & Spergel, D.N. (1990). Astrophys. J., submitted (B&S).

Brand, J. (1986). PhD Thesis, Leiden.

Burton, W.B. (1985). Astron. Astrophys., Suppl. Ser. **62**, 365.

Burton, W.B., (1987). In *The Galaxy* (eds. G. Gilmore and R.C. Carswell), p. 141. Cambridge: Cambridge University Press.

Habing, H.J., (1987). In *The Galaxy* (eds. G. Gilmore and R.C. Carswell), p. 173. Cambridge: Cambridge University Press.

Kerr, F.J., Bowers, P.F., Jackson, P.D. & Kerr, M. (1986). Astron. Astrophys., Suppl. Ser. **66**, 373.

Mihalas, D. & Binney, J. (1980). *Galactic Astronomy*, Chapter 9. San Francisco: Freeman.

Weaver, H. & Williams, D.R.W. (1974). Astron. Astrophys., Suppl. Ser. **17**, 1.

Appendix. Epicyclic Treatment of Non-Axisymmetric Perturbations

In this appendix, expressions are derived for the observed line-of-sight velocities of closed orbits as seen from the Sun. The galactic potential is assumed to be slightly non-axisymmetric, and to figure-rotate. The epicycle approximation is used to calculate the closed orbits, following the treatment of Binney & Tremaine (1987). The warp of the Galactic plane will be ignored in this analysis.

Take the potential to be

$$\Phi(r,\phi) = \Phi_0(r) + \Phi_b(r)\cos m(\phi - \phi_\odot - \Omega_b t) \qquad \text{(A1)}$$

expressed in inertial polar coordinates (r,ϕ). The orbits $(r(t),\phi(t))$ which close in the frame that corotates with the potential are of the form

$$r = r_0 + \alpha\cos m\phi_0, \qquad \phi = \phi_\odot + \Omega_0 t + \beta\sin m\phi_0 \qquad \text{(A2)}$$

with

$$\alpha = -\frac{1}{\Delta}\left[\Phi_b' + \frac{2\Omega_0\Phi_b}{r_0(\Omega_0 - \Omega_b)}\right], \qquad \text{(A3)}$$

$$\beta = \frac{1}{m\Delta r_0(\Omega_0 - \Omega_b)}\left[2\Omega_0\Phi_b' + \frac{4\Omega_0^2 - \Delta}{r_0(\Omega_0 - \Omega_b)}\Phi_b\right]. \qquad \text{(A4)}$$

Here $\Delta = \kappa_0^2 - m^2(\Omega_0 - \Omega_b)^2$, and Ω_0 and κ_0 are the usual circular and epicyclic frequencies of a circular orbit of radius r_0 in the potential

Φ_0 (in an inertial frame). The prime denotes differentiation. The angle $\phi_0 = \phi - \phi_b - \Omega_b t$ denotes the Galactocentric azimuth with respect to the axis of the rotating potential. The velocities along this orbit are easily seen to be

$$v_r = \delta \sin m\phi_0, \qquad v_\phi = \Omega_0 r_0 + \epsilon \cos m\phi_0 \,, \tag{A5}$$

with δ and ϵ given by:

$$\delta = \frac{m(\Omega_0 - \Omega_b)}{\Delta}\left[\Phi_b' + \frac{2\Omega_0\Phi_b}{r_0(\Omega_0 - \Omega_b)}\right], \tag{A6}$$

$$\epsilon = \frac{\Omega_0}{\Delta}\Phi_b' + \frac{2\Omega_0^2 - \Delta}{\Delta r_0(\Omega_0 - \Omega_b)}\Phi_b \,. \tag{A7}$$

Let the LSR move along such an orbit with epicentric radius $r_\odot$, and fix the origins of time and ϕ at the present epoch and solar galactocentric azimuth, respectively. Then, as viewed from the LSR, the line-of-sight velocity at galactic longitude l of the orbit with epicentre at radius r_0 follows from simple trigonometry as

$$v_{\rm rad} = v_{\phi 0}(R_\odot/R_0)\sin l - v_{r0}\cos(\phi + l) - v_{\phi\odot}\sin l + v_{r\odot}\cos l \,. \tag{A8}$$

The radii R_0 and $R_\odot$ are the galactocentric distances of the orbits, not the epicentric radii. ϕ and l are related by $R_\odot \sin l = R_0 \sin(\phi + l)$, hence $\cos\phi = \cos l\sqrt{1 - (R_\odot \sin l/R_0)^2} + (R_\odot/R_0)\sin^2 l$, with $\sin l$ and $\sin\phi$ having the same sign. To first order in Φ_b

$$\begin{aligned} v_{\rm rad} = {} & r_\odot(\Omega_0 - \Omega_\odot)\sin l \\ & + [(r_\odot/r_0)(\epsilon_0 - \alpha_0\Omega_0)\cos m(\phi - \phi_\odot) - (\epsilon_\odot - \alpha_\odot\Omega_\odot)\cos m\phi_\odot]\sin l \\ & + \delta_0 \sin m(\phi - \phi_\odot)\cos(\phi + l) - \delta_\odot \sin m\phi_\odot \cos l \,. \end{aligned} \tag{A9}$$

Subscripts correspond to those of the radii of the guiding centres at which they are evaluated. The quantity v_{diff} constructed by B&S is the difference between $v_{\rm rad}(l)$ and $-v_{\rm rad}(360° - l)$, or

$$\begin{aligned} v_{diff} &= v_{\rm rad}(l) + v_{\rm rad}(360° - l) \\ &= 2(r_\odot/r_0)(\epsilon_0 - \alpha_0\Omega_0)\sin m\phi_\odot \sin l \sin m\phi \\ &\quad - 2\delta_0 \sin m\phi_\odot \cos m\phi \cos(\phi + l) - 2\delta_\odot \sin m\phi_\odot \cos l \,. \end{aligned} \tag{A10}$$

If $(r_\odot/r_0) << 1$, then $\phi \simeq 180° - l$, hence for distant gas

$$v_{diff} \simeq 2(-1)^m \delta_0 \sin m\phi_\odot \cos ml - 2\delta_\odot \sin m\phi_\odot \cos l \,. \tag{A11}$$

If the Sun happens to sit on a symmetry axis of the potential, no North-South asymmetries are seen: hence Φ_b and its azimuth $\phi_\odot$ only enter in the combination $\Phi_b \sin m\phi_\odot$.

The crowding of orbits in the $v_{\rm rad}$-l plane is also of interest. It is described by $dv_{\rm rad}/dr_0(l)$. High values of this gradient correspond to low crowding, and contrariwise. For large orbits, the difference curve of this is readily derived from equation A11 as

$$\frac{dv_{diff}}{dr_0} \simeq 2(-1)^m \delta_0' \sin m\phi_\odot \cos ml\,. \tag{A12}$$

Merger origin of starburst galaxies

LARS HERNQUIST
University of California, Santa Cruz

1. Introduction

Observational evidence supports the view that some starbursts are triggered by collisions between galaxies. Larson and Tinsley (1978) were the first to show that interacting galaxies tend to be bluer than their isolated counterparts; a property most naturally explained by recent bursts of star formation. The IRAS survey revealed a class of galaxies with infrared luminosities L_{IR} exceeding $10^{11}L_{\odot}$ (Soifer et al. 1984a,b). Among the ~ 20 such sources now catalogued, Mrk 231 is the brightest with $L_{IR} \sim 3.5 \times 10^{12} L_{\odot}$ (Scoville 1990). These objects generally possess features typical of galaxies that are interacting strongly or are in the process of merging, such as tidal tails (Joseph and Wright 1985; Sanders et al. 1988a).

It is generally believed that the infrared emission from these objects is reprocessed optical and ultraviolet radiation from newly–formed stars. If so, gas must be converting into stars at an exceptionally high rate, typically $\dot{M}_{gas \rightarrow stars} \sim 10 - 100 M_{\odot}\mathrm{yr}^{-1}$. If these sources live for $\sim 10^{8} - 10^{9}$ years, then the total fuel masses required to sustain the infrared emission are $M_{fuel} \sim 10^{9} - 10^{12} M_{\odot}$. Similar values are inferred from CO observations, assuming a canonical conversion factor between CO and H_2.

Since M_{fuel} for these sources is similar to gas masses throughout present–day galaxies it is natural to suppose that this fuel was initially distributed on galactic scales. However, the emission from these galaxies is highly concentrated within their nuclei. Arp 220 is estimated to contain $\sim 2 \times 10^{10} M_{\odot}$ of gas within $\sim 300\,\mathrm{pc}$ of its center. The mechanism by which galactic gas could accumulate on such small scales is unclear. As first noted by Gunn (1979), this gas must somehow shed $\sim 90 - 99\%$ of its angular momentum. Moreover, it must do so on roughly a dynamical time-scale to account for the lifetimes of starbursts. If starbursts are indeed triggered by mergers, then galactic tides must excite strong dynamical torques on the gas to remove its angular momentum. Owing to the complex nature of these systems, the mysteries of this process have been difficult to unravel.

The evolution of galaxies merging to form starbursts entails both collisionless and gas–dynamics. These two components are coupled since they interact with one another gravitationally. It also appears that the response that ultimately drives gas into galactic nuclei involves non–linear processes, such as shock waves, and dissipation through radiative cooling. Under these circumstances, one is forced to adopt a numerical approach.

2. Numerical models

Josh Barnes and I have begun to model mergers between gas–rich disks using a hybrid code capable of simultaneously evolving gas and collisionless matter. This method is described in detail by Hernquist and Katz (1989). Briefly, collisionless matter, such as stars and dark matter in galaxies, is handled using a conventional N–body approach. Gas–dynamics is treated with a technique known as smoothed particle hydrodynamics (SPH; Lucy 1977; Gingold and Monaghan 1977). In SPH, gas is discretized into fluid elements which are represented computationally as particles. Each particle carries along with it local hydrodynamic and thermodynamic information, such as density and entropy. These properties are updated using Lagrangian forms of the hydrodynamic conservation laws. The particular implementation we use is adaptive in both space and time. This consideration is particularly important for modeling gas in galaxies since dynamic ranges $\sim$ 100–1000 : 1 are often required.

Gravitational forces are computed using a hierarchical tree algorithm (Barnes and Hut 1986; Hernquist 1987). Since the cpu costs of this method scale with the number of particles as $\sim O(N \log N)$, much higher resolution can be attained than in most previous adaptations of SPH. The thermal energy equation includes heating and cooling terms appropriate for interstellar gas, with the proviso that cooling is shut–off below 10^4 K to inhibit local Jeans instability Hernquist (1989a, b). For the relevant physical conditions, cooling times are such that the gas is nearly always at a temperature $T \approx 10^4$ K. Shocks are handled using an artificial viscosity.

The galaxy models employed in the calculations shown and discussed here are similar to those described by Barnes (1988). Each galaxy consists of a bulge, an exponential disk, and a spheroidal halo with mass ratios $B : D : H = 1 : 3 : 16$. The gas constitutes 10% of the disk mass. Results are presented in a dimensionless system of units where the gravitational constant $G = 1$, the mass of each galaxy is 1.25, and the exponential scale–length of the disks is $h = 1/12$. In this system the rotation period at $3h$ is 0.93. Scaled to values appropriate for our own galaxy, the units of length, mass, and time correspond roughly to 40 kpc, $2.2 \times 10^{11} M_{\odot}$, and 2.5×10^{8} yr, respectively.

3. An example

Figure 1 summarizes the outcome of one of our experiments (Barnes and Hernquist 1990). In this model, each galaxy consists of 45056 particles, 8192 of which represent gas. The two galaxies were initially on a parabolic orbit which would have reached a pericenter distance $R_p = 0.2$ at time $t = 1$. The left–most disk in the panel at $t = 0.5$ is seen face–on and suffers an exactly prograde passage. The other is inclined by 71° with respect to the orbital plane.

The second panel in the top sequence of Figure 1 shows a view only 0.25 time units after pericenter. Already the disks are being violently deformed by their mutual interaction and are developing features that quickly evolve into extended tidal tails. At this time the gas and stars in each disk are nearly coextensive. However, by the next time shown, $t = 2$, they are distributed somewhat differently. In particular, a significant fraction of the gas has been driven into the nucleus of each galaxy. Note also the dense knot which develops in the lower tidal tail and the bridge of gas joining the two disks. The final panel shows the two galaxies as they are nearing final merger. By this time more than 50% of all the gas initially distributed throughout both disks has collapsed into a central region too small to resolve in this figure.

The lower panels in Figure 1, enlarged by a factor of 5 relative to those above, illustrate the principal effect leading to this central accumulation of gas. These views show the distribution of gas and stars shortly after pericenter in the face–on disk. As a consequence of the first passage between the two galaxies, this disk suffers bar instability, in a manner similar to that studied by Noguchi (1987). This bar appears in both the stars and gas, but the response of the

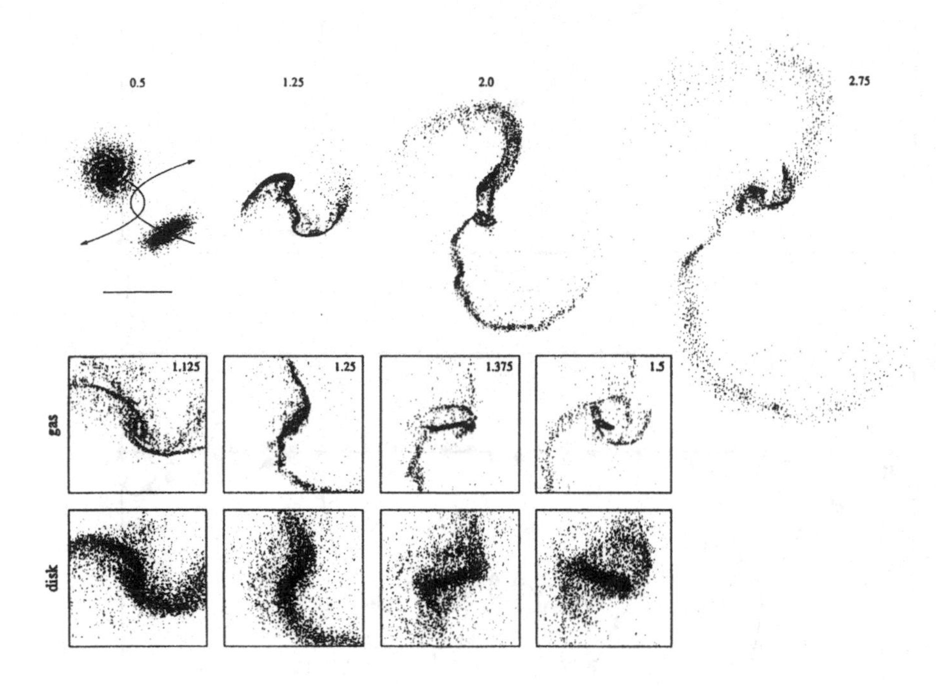

Fig. 1. Merger of two gas-rich disks, projected onto the orbital plane. The unframed plots show the gas only. The first of these frames shows the initial parabolic trajectory and the solid line measures one length unit. Numbers indicate time. Panels below separately show the gas (top) and stars (bottom) in the face-on disk.

gas leads that of the stars. By $t = 1.375$ the gas bar is offset with respect to the stellar bar by an angle of about $\sim 5°$. Owing to this phase difference, the stars exert a strong torque on the gas, removing its angular momentum. By $t = 1.5$ the gas in the inner parts of this disk has contracted into a small ring, containing roughly 50% of the gas initially throughout the disk.

The physical mechanism responsible for removing angular momentum from the gas can be inferred from Figure 2, which shows the torque on the gas ending up in the nucleus of the face-on disk be-

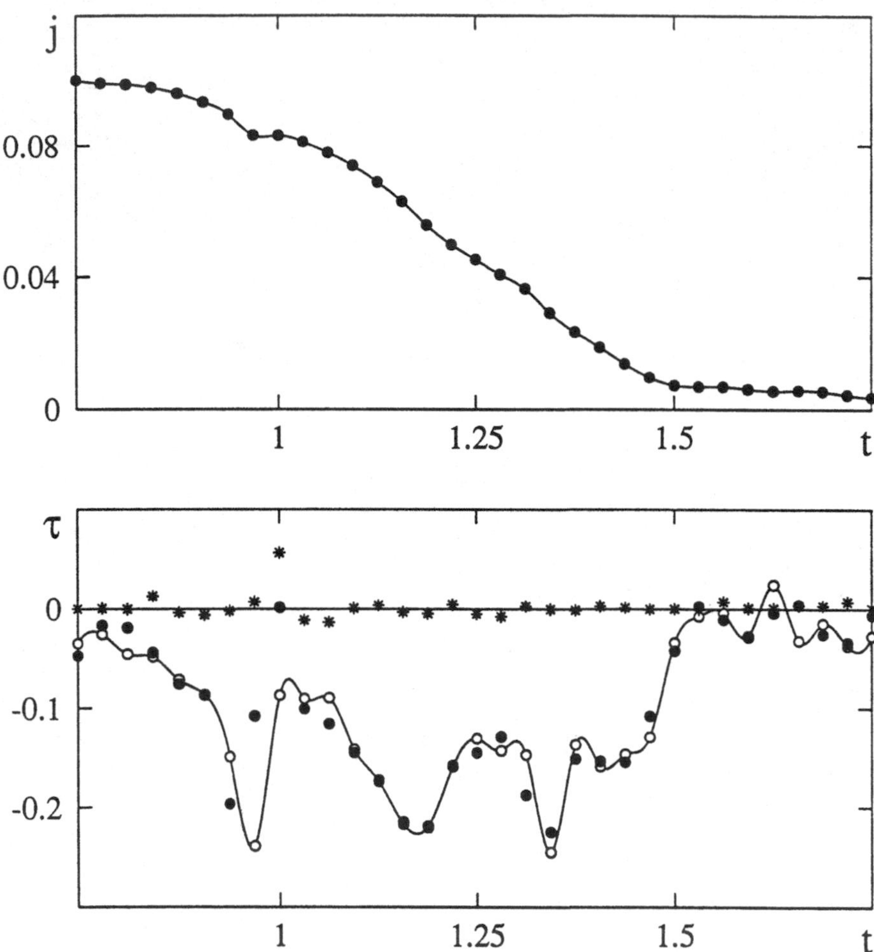

Fig. 2. *Top*: specific angular momentum of the gas which collects at the center of the face-on galaxy following first pericenter. *Bottom*: specific torques acting on this gas. Open circles connected by a smooth curve are gravitational torques exerted on the gas by the rest of the system, while stars are torques due to hydrodynamic forces. Filled circles show the result of numerically differentiating the curve shown in the upper panel.

tween times $t = 0.75$ and $t = 1.75$. The contributions to the torque from gravitational and hydrodynamic forces are separately displayed. Only around pericenter, when the two disks first interpenetrate one another, do hydrodynamic effects have a measurable effect. In fact, the sign of the resulting torque is such as to *add* angular momentum to this gas. By $t = 1.75$ the angular momentum of the gas has declined by a factor of 25, entirely as the result of *gravitational* torques. A more detailed analysis shows that before the bar develops in this disk most of the torque is provided by the other galaxy. After pericenter, the torque is dominated by the gravitational interaction between the disk stars and the gas, owing to the phase difference between their responses.

The gas in the inner parts of the other disk suffers a similar fate. Shortly after pericenter bar instability sets in, the result of which is to drive significant quantities of gas into the nucleus. As the two galaxies finally merge, the central agglomerations of gas within each nucleus coalesce, resulting in some cancellation of angular momentum. Since the resulting mass of gas is quite dense, it quickly sinks to the center of the remnant, transporting additional angular momentum to the surrounding stars and dark matter via dynamical friction. By a time $t = 3$ roughly 60% of all the gas initially distributed throughout both disks ends up within a central region of radius ~ 0.005. Scaling to values appropriate for our own galaxy, this corresponds to a mass $\sim 5 \times 10^9 M_\odot$ and a radius $\sim 200\,\mathrm{pc}$. This linear scale is determined mainly by limited resolution in the computation of the gravitational field; the model here was generated with a fixed softening length $\epsilon = 0.015$ in computing all gravitational forces. If the self-gravity in the gas were treated more precisely on these scales, further contraction would likely result.

The marked segregation between stars and gas in the inner parts of the remnant is shown in Figure 3. The stars are well-fitted by a de Vaucouleurs $R^{1/4}$–law with effective radius $R_e = 0.1$, i.e., roughly 4 kpc in dimensional units. The fact that an $R^{1/4}$–law density profile is established in the stellar distribution so rapidly after merging has been completed is in good agreement with observations of Arp 220 (Wright et al. 1990).

4. Discussion

Models like that described here demonstrate how tidal effects can drive nuclear inflows of gas as galaxies strongly interact or merge. Violently perturbed disks develop bars in both their stellar and gas components. Owing to dissipation, the response of the gas leads that of the stars. Consequently, the stars exert a torque on the gas, removing its angular momentum and causing it to collapse into the center of each disk. In the example here, a further loss of angular momentum occurs as the disks finally merge and the two central gas accumulations coalesce. This additional transport results partly from

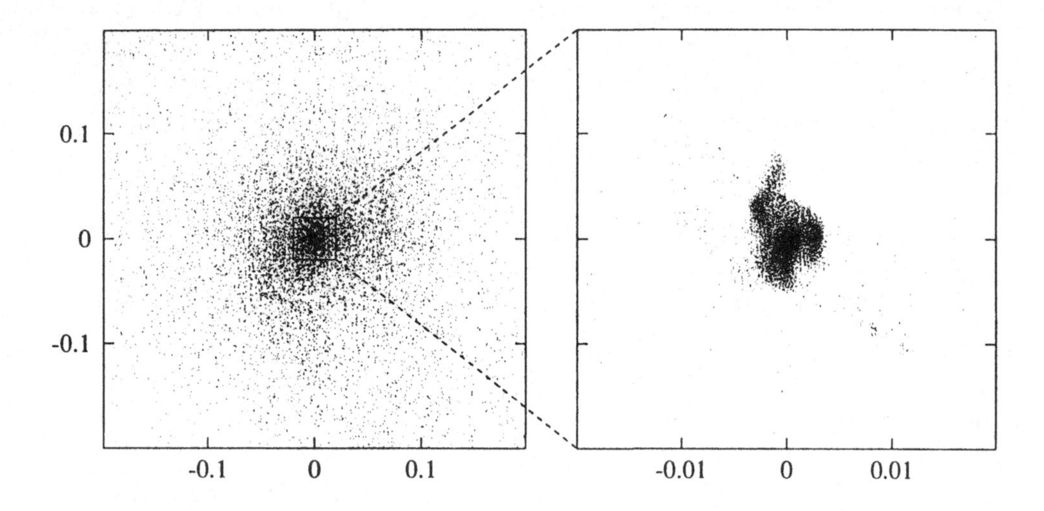

Fig. 3. *Left*: stellar distribution of the remnant resulting from the merger in Figure 1, projected onto the orbital plane. *Right*: 10× enlargement of the central region, showing the distribution of gas there.

direct cancellation between misaligned spins, but mainly through dynamical friction against less–concentrated collisionless matter.

The results reported here are closely related to earlier work by Negroponte and White (1983) and Noguchi (1988) on the fate of gas in interacting galaxies. Unlike the present calculations, which employ continuum gas–dynamics, both of these previous studies modeled the gas using a discrete–cloud method. Significant quantities of gas were driven into the centers of the galaxies in each case by tidal effects. Owing to limited dynamic range, neither investigation isolated the detailed mechanism responsible for removing angular momentum from the gas. Nevertheless, the fact that both found sets of calculations are in good quantitative agreement with those here suggests that the phase structure of the gas is not critical to the onset of nuclear gas inflows in disks.

Additional experiments demonstrate that the effects described in the previous section of the this paper are not sensitive to the magnitude or functional form of the artificial viscosity. This conclusion is not surprising in view of the fact that the gas loses angular momentum by gravitational, rather than viscous transport. It was found, however, that nuclear inflow of gas could be completely suppressed by inhibiting radiative cooling. The inference that cooling may be the primary factor governing differences in the evolution of gas and stars may account for the similarities between the present results and those of Negroponte and White (1983) and Noguchi (1988): continuum gas at a temperature $\sim 10^4$ K dissipates energy at roughly the same rate as an ensemble of discrete clouds with velocity dispersion $\sim 10\,\mathrm{km\,s^{-1}}$.

The final fate of the gas deposited in the nucleus of the merger remnant is problematical since the calculations have not taken into account star formation and heating due to supernovae. Nevertheless, in view of the high gas densities, a burst of star formation seems likely. If so, than the resulting infrared emission might resemble that from the ultraluminous IRAS galaxies. Hence, models like that depicted in Figure 1 lend strong support to the notion that these sources originate from galactic mergers. The models can account for the masses and concentrations of gas observed in the nuclei of these objects. In the particular example presented here, roughly $5 \times 10^9 M_\odot$ of gas is driven into a region with radius $\lesssim 200\,\mathrm{pc}$, similar to the central gas content of NGC 520 (Sanders et al. 1988b).

The stellar distribution of the merger remnant is well–fitted by a de Vaucouleurs profile, as are elliptical galaxies. However, the innermost regions could deviate significantly from an $R^{1/4}$ law, depending on the history of star formation in the gas. Since the gas can dissipate, much higher phase space densities can be achieved than in pure stellar–dynamical mergers. Also, star formation would likely contribute to metallicity gradients in the nucleus. Since the nuclear gas loses nearly all its angular momentum it is largely decoupled from the bulk of the remnant. Merger events are, therefore, a plausible mechanism for generating the kinematically distinct cores observed in many normal ellipticals (Franx and Illingworth 1988).

The fact that these merger remnants have structural properties akin to those of elliptical galaxies implies that such events may be capable of explaining nuclear activity in at least some bright radio galaxies. A particularly strong case can be made for those with on-going star formation and those which possess features commonly seen in reputed merger candidates, such as tails, bridges, and shells (e.g., Heckman et al. 1986; Vader et al. 1989; Baum and Heckman 1989). At present, the simulations cannot address this issue directly, however, owing to their limited dynamic range. Nevertheless, a variety of mechanisms have been proposed that would enable the gas to shed further angular momentum once it becomes self-gravitating (e.g., Begelman et al. 1984).

The results presented here complement those of Hernquist (1989a,b), who showed that mergers between disks containing both gas and stars with less–massive satellites can lead to nuclear inflows of gas. In those experiments, the rotation curve of the disk was such that the stellar component did not suffer dynamical bar instability. Nevertheless, tidal forces acting on the gas compressed it, fragmenting it under its own self–gravity. Efficient transfer of angular momentum from these gas fragments to the stellar disk resulted in rapid accumulation of gas in the nucleus. Additional experiments (Hernquist 1990) demonstrate that the specific mechanism described here can also operate during mergers between disks and satellites provided that the disk is nearly bar–unstable in isolation.

The results here together with those of Hernquist (1989a, b) support the view that unusual activity in at least some galactic nuclei is controlled by events on galactic scales. Mergers between disks and other galaxies can drive gas into the nucleus, resulting in a starburst. Additional evolution may lead to the development of an active nucleus, provided that some of the gas can shed sufficient angular momentum to be accreted by a central black hole. If these speculations are correct, then the type of active galaxy that results is determined in large part by the mass ratio between the two merging galaxies. Mergers involving disks and less–massive companions produce remnants with disturbed disks having little or no spiral structure. Such events are a plausible explanation for the amorphous Seyfert galaxies studied recently by MacKenty (1989, 1990). The remnants of mergers between galaxies of comparable mass resemble elliptical galaxies. If nuclear activity can ultimately be promoted in these remnants then such mergers may account for at least some young radio galaxies (e.g., Baum and Heckman 1989, Vader et al. 1989).

Mergers may also be responsible for at least some quasar activity. Direct observations of low–redshift quasars suggest that many of these objects have features similar to those of local merger candidates (e.g., Stockton 1990). The increased rate of merging predicted in most cosmologies at high redshifts and the likelihood that galaxies at those epochs contained more gas than present–day disks suggest that effects leading to nuclear inflow of gas should have been more common early in the history of the universe. This expectation is in accord with observations implying that active galaxies were more common and more luminous at higher redshifts. However, definitive answers to these various speculations will be forthcoming only with modeling that is significantly more detailed than presented here.

Acknowledgements

I thank my collaborator Josh Barnes for permitting me to present our results in advance of publication. This work was supported in part by a grant from the Pittsburgh Supercomputing Center.

References

Barnes, J. (1988). Astrophys. J. **331**, 699.
Barnes, J. & Hut, P. (1986). Nature **324**, 446.
Barnes, J. & Hernquist, L. (1990). Astrophys. J., Lett., in press.
Baum, S.A. & Heckman, T. (1989). Astrophys. J. **336**, 681.
Begelman, M.C., Blandford, R.D. & Rees, M.J. (1984). Rev. Mod. Phys. **56**, 255.
Franx, M. & Illingworth, G.D. (1988). Astrophys. J., Lett. **327**, L55.
Gingold, R.A. & Monaghan, J.J. (1977). Mon. Not. R. Astron. Soc. **181**, 375.
Gunn, J.E. (1979). In *Active Galactic Nuclei* (eds. C. Hazard and S. Mitton), p. 213. Cambridge: Cambridge University Press.
Heckman, T.M., Smith, E.P., Baum, S.A., van Breugal, W.J.M., Miley, G.K., Illingworth, G.D., Bothun, G.D., & Balick, B. (1986). Astrophys. J. **311**, 526.
Hernquist, L. (1987). Astrophys. J., Suppl. Ser. **64**, 715.
Hernquist, L. (1989a). Nature **340**, 687.
Hernquist, L. (1989b). Ann. N. Y. Acad. Sci. **571**, 190.
Hernquist, L. (1990). In preparation.
Hernquist, L. & Katz, N. (1989). Astrophys. J., Suppl. Ser. **70**, 419.
Joseph, R.D. & Wright, G.S. (1985). Mon. Not. R. Astron. Soc. **214**, 87.
Larson, R.B. & Tinsley, B.M. (1978). Astrophys. J. **219**, 46.
Lucy, L. (1977). Astron. J. **82**, 1013.
MacKenty, J.W. (1989). Astrophys. J. **343**, 125.
MacKenty, J.W. (1990). Astrophys. J., Suppl. Ser. **72**, 231.
Negroponte, J. & White, S.D.M. (1983). Mon. Not. R. Astron. Soc. **205**, 1009.
Noguchi, M. (1987). Mon. Not. R. Astron. Soc. **228**, 635.
Noguchi, M. (1988). Astron. Astrophys. **203**, 259.

Sanders, D.B., Soifer, B.T., Elias, J.H., Madore, B.F., Matthews, K., Neugebauer, G. & Scoville, N.Z. (1988a). Astrophys. J. **325**, 74.
Sanders, D.B., Scoville, N.Z., Sargent, A.I. & Soifer, B.T. (1988b). Astrophys. J., Lett. **324**, L55.
Scoville, N.Z. (1990). Private communication.
Soifer, B.T. et al. (1984a). Astrophys. J. **278**, L71.
Soifer, B.T. et al. (1984b). Astrophys. J. **283**, L1.
Stockton, A. (1990) In *Dynamics and Interactions of Galaxies* (ed. R. Wielen), p. 440. Berlin: Springer.
Vader, J.P., Heisler, C.A. & Frogel, J.A. (1989). Ann. N. Y. Acad. Sci. **571**, 247.
Wright, G.S., James, P.A., Joseph, R.D. & McLean, I.S. (1990). Nature **344**, 417.

Warped and flaring HI disks

A. BOSMA
Observatoire de Marseille

Abstract

We present statistics of warped HI disks, both for edge-on galaxies in which the warp can be seen directly, and for more face-on disks in which the warp is inferred from the kinematics. The fraction of warped HI disks is of order 50% at least. From the face-on sample there is evidence that a concentrated halo inhibits the warp. Circumstantial evidence on low surface brightness giants and very early type spirals seems to reinforce this conclusion.

The problem of flaring HI disks is briefly addressed: some predictions are given, and some difficulties are outlined.

Introduction

Warps in the HI layer of galactic disks present a long standing dynamical problem, for which many solutions have been proposed, none of them very satisfactory. Early observations of our Galaxy demonstrated the presence of a warp in the outer gas layer (Burke 1957, Kerr 1957); also the flaring of the layer, i.e. the increase in its thickness with increasing radius, has been established early-on (see, e.g., Gum et al. 1959).

In external galaxies, warps have been found to be ubiquitous. Large scale deviations from circular motions in the velocity field of M83 led Rogstad et al. (1984) to construct tilted ring models, in which the outer rings are allowed to deviate from the plane of the disk defined by the inner rings. Similar models describe the observations of M33 (Rogstad et al. 1976), NGC 300 (Rogstad et al. 1979), and many of the spirals studied with the Westerbork Synthesis Radio Telescope (Bosma 1978, Wevers 1984, Begeman 1987). For edge-on galaxies the warp can be seen directly, and already the early study of Sancisi (1976) hinted at a high frequency of warping.

Now that many more observations are available, it seems timely to address the question of the statistics of warps, and to see whether there is any correlation between warping or non-warping and other dynamical parameters of the galaxies in question.

Edge-on galaxies

For edge-on galaxies the warp can be seen directly, except in unfavourable circumstances, e.g. if the maximum tilt occurs in the line of sight towards the observer and the change in the line of nodes remains small. In a recent program I observed ten galaxies with the VLA D-array, thereby doubling the sample available for statistics. Out of 20 cases studies so far, 12 are warped, 6 of which show distinct asymmetries. Most of these asymmetries are in position, e.g. either there is a warped gas layer at one side and no gas detected at the other side, or the gas is warped at both sides but the HI extent is different. For one galaxy, NGC 4157, the warp is asymmetric in velocity, in the sense that the position-velocity diagram along the major axis is clearly asymmetric, with at one side a drop in projected velocity of more than $20\,\mathrm{km\,s^{-1}}$. Four more galaxies are slightly warped (i.e., the warp is of relatively low amplitude, like in NGC 891, cf. Sancisi and Allen 1979, Rupen 1990). Finally, 4 galaxies do not show any warping, at least to within the limits set by the angular resolution of the data. In fact, one of these has subsequently been observed at very high resolution, and a slight warp has been found. Note that most of the galaxies in this sample are of type Sbc or later.

Face-on galaxies

For face-on galaxies (understood here as anything but edge-on), warps can be inferred from the kinematics. As discussed by Bosma (1978, 1981b; see also Teuben, this volume), the kinematic signatures in the velocity fields are distinctly different for elliptical orbits in a single plane vs. circular orbits according to tilted ring models. The kinematical major and minor axes stay perpendicular to each other in the case of a warp. This very fact can be used to argue for the dominance of the disk in the central parts of warped galaxies if halos are triaxial.

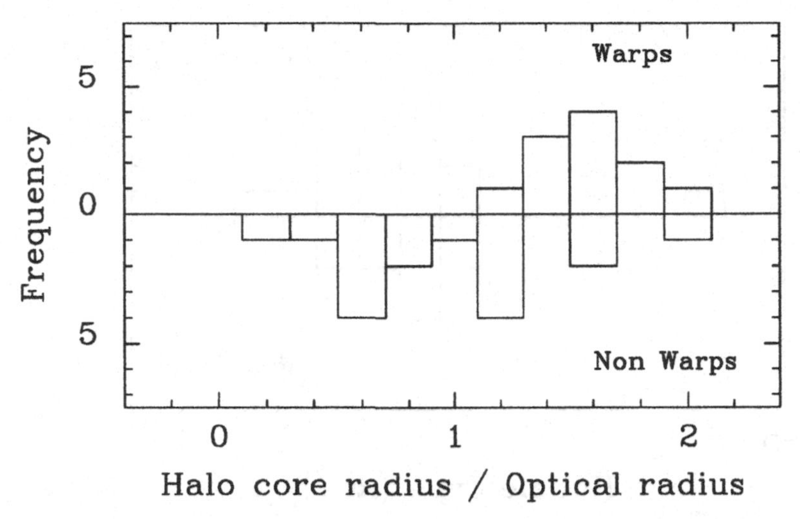

Fig. 1. Histogram of the number of warped and non-warped galaxies as function of the ratio of halo core radius to optical radius.

Athanassoula et al. (1987) used rotation curve and surface photometry data for the construction of multicomponent mass models consisting of a bulge and a disk, each with constant M/L, and a dark halo necessary to account for the high rotation velocities at large radii. These models are constrained using considerations from spiral structure theory. For 27 of the galaxies studied this way, the rotation data comes from two-dimensional HI velocity fields. We will use models in which the disk is not allowed to have $m = 1$ disturbances. These models seem to be the most satisfactory, for reasons described in detail in Athanassoula et al. From the parametrization of the halo rotation curve with an isothermal sphere, we can derive core radii for the halo.

Eleven of the twenty-seven galaxies in this sample show a distinct warp. The frequency of warping vs. non-warping as function of the ratio of halo core radius to optical radius is shown in Figure 1. It can be clearly seen that warps do not occur if the ratio of halo core radius to optical radius is small. In other words, it seems that concentrated halos somehow inhibit a warp.

Corroborating evidence can be found from other data: if halo concentration is a significant quantity in this context, we should look at low surface brightness giants, and at early type spiral galaxies. For the low surface brightness disk galaxy NGC 5963, Bosma et al. (1988) showed that its halo is much more concentrated than for nor-

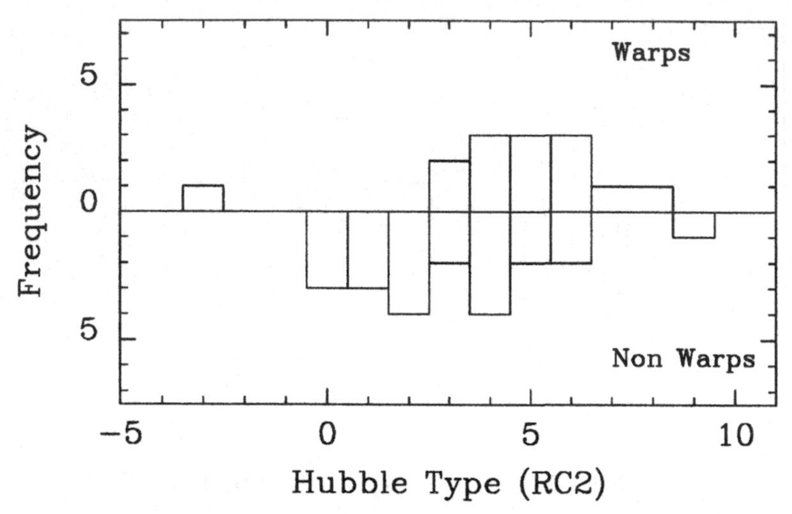

Fig. 2. Histogram of the number of warped and non-warped galaxies as function of Hubble type. Galaxies from the sample of van Driel (1987) have now been included.

mal disk galaxies having a rotation curve with similar amplitude. Yet this galaxy does not show a warp, in agreement with what can be expected on the basis of Figure 1. Data on NGC 3657, a galaxy with similar properties, again do not show a warp (van der Hulst and Bosma, in preparation). Note that both these galaxies could be described as brighter but smaller "cousins" of Malin-1 (Impey and Bothun 1989).

For early type spiral galaxies, two data sources could be used. Data for the Virgo cluster spirals, which are in general of early type, have been collected by Warmels (1988) and Cayatte et al. (1990). Unfortunately, most galaxies in these samples have not been observed with sufficient spatial resolution to determine unambiguously whether they are warped, although for several galaxies there is tentative evidence for it. The other source of data is van Driel's thesis (1987) on S0 and early Sa galaxies, published as a series of papers with various collaborators. If we restrict ourselves only to galaxies in which the HI gas is located in the equatorial plane of the optical image, we notice that 7 galaxies are not warped, and 1 is slightly warped (NGC 5102). For most of these galaxies no surface photometry is available to construct detailed composite disk/halo models, but we note that Athanassoula et al. (1987) found that the halo is more concentrated in earlier type galaxies. For NGC 5102 surface photometry is available, and a decomposition shows that in the case of a maximum disk, the ratio of halo core radius to optical radius is of the order of 2 (van Driel 1987).

Thus we conclude that data on early type disk galaxies seem to confirm the trend we found in Figure 1, i.e., that concentrated haloes inhibit warps. For completeness, we plot in Figure 2 the frequency of warps as function of Hubble type. Due to variations in halo parameters within one Hubble type, the trend is not as clean as in Figure 1, but warps seem to be less frequent in earlier type disk galaxies.

Comments on warp shapes

Briggs (1990) discusses sixteen galaxies for which sufficient data are in the literature to determine the warp shape with respect to the main plane of the galaxy. For twelve of these galaxies he notes the following trends:

i) The HI is planar within R_{25} (radius of the 25th mag arcsec^{-2} isophote), but the warp becomes detectable within R_{Ho} (Holmberg radius), and appears there consistent with a straight line of nodes measured in the plane defined by the inner regions of the galaxy.

ii) Warps change character at a transition radius near R_{Ho}.

iii) For radii larger than R_{Ho}, the line of nodes measured in the plane of the inner galaxy advances in the direction of galaxy rotation for successively larger radii.

iv) A new reference frame can be found, in which there is a common line of nodes for the inner regions within the transition radius, and a differently oriented straight line of nodes at larger radii. This new frame of reference is typically inclined by less than 10 degrees to the plane of the inner galaxy.

Published (NGC 300, Rogstad et al. 1979) and unpublished data on several other galaxies confirm Briggs's rules as general rules-of-thumb. Even so, while not wanting to diminish this admirable effort at systematization, I would like to make the following remarks:

1) Exceptions can be found to rules 1 and 2, i.e. there are galaxies where the warp does start already within the optical radius. A good example is NGC 7331, even though its warp is of low amplitude (see Bosma 1978, 1981a; Begeman 1987). This galaxy is by far the brightest member in its group (cf. Tully 1988), and the influence of the nearest physical companion, NGC 7320, can be estimated to be negligible.

2) IC 342 (Newton 1980a) contradicts rule 3: the line of nodes seems to wind counter the direction of rotation.
3) The "revised" rule 4 may not be physically significant, but in fact poses an interesting question, namely how much non-circular, and in particular slightly elliptical, motions can go undetected in the current observations while giving rise to the situation envisaged by this rule. Since, as argued above, planar elliptical orbits do have a different signature in the velocity field than tilted circular rings, it is of interest to determine an upper limit of detectability here. In this connection, I would like to call attention to the peculiar case of NGC 7793, for which the new VLA data of Carignan and Puche (1990) clearly indicate a warped HI disk, but where the inner regions seem to have a kinematical position angle 11 degrees different from the photometric one. This was already apparent in the Hα work by Davoust and de Vaucouleurs (1981). Yet no outstanding optical feature, like an outer ring or a pseudo-ring, leave alone a clear oval distortion, can be seen on optical photographs, contrary to cases like NGC 4736 or NGC 2903. This case merits further attention since Carignan and Puche's data indicate a falling rotation curve as well.

Flaring HI disks

The composite disk/halo models in which the HI gas is treated as a separate component allow a calculation of the radial variation of the thickness of the gas layer (cf. Athanassoula and Bosma 1988). This is illustrated in Figure 3 for the galaxy IC 342, for the extreme cases of spherical and flat halos. This figure can be compared to Figure 9 in Wouterloot et al. (1990), in which they derive the increase in thickening of the Galactic gas layer outwards (see also older data discussed by Knapp 1987). Clearly the flaring effect (i.e., the increase in thickening of the disk with increasing radius) is seen in two different regimes: a slow increase in thickening in the inner parts, where the stellar disk dominates the potential, and a strong flaring in the outer parts where the gas mainly feels the halo potential. This effect ought to be visible in external galaxies as well, as Figure 3 demonstrates.

Our survey of 10 edge-on galaxies discussed above was in fact designed to find good candidates for this study. Our recent VLA observations at high resolution (about 4.5 arcsec) of two galaxies, NGC 100 and UGC 9242, do indeed show the flaring, and a detailed

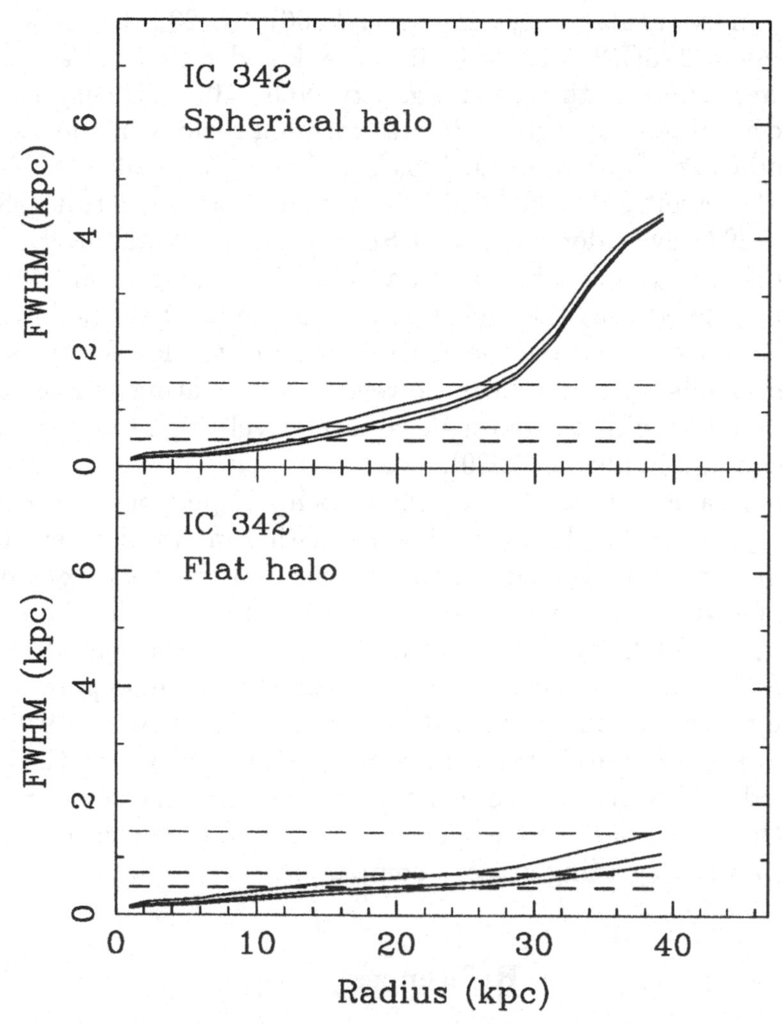

Fig. 3. Predicted flaring of the gas layer of IC 342. In the top panel the dark halo is assumed to be spherical, and in the bottom panel the dark matter scales as the disk matter. The dashed lines indicate values adopted for the scale height of the disk, and the thin lines the predicted gas scale heights. The adopted velocity dispersion of the gas is 10 km s^{-1}.

study of this problem is under way, which hopefully will shed more light on the possible shape of the dark halo.

A potential problem with this analysis is the value to adopt for the velocity dispersion of the gas. Various values are quoted in the literature for this. Low values, of order of 4 km s^{-1} or less, are found from HI absorption studies, either of Galactic HI clouds against extragalactic continuum sources (e.g. Colgan et al. 1988) or, in a few cases, in extragalactic systems (NGC 3067/3C 232, Rubin et al. 1982; NGC 4651/3C 275.1, Corbelli and Schneider 1990). Values of 6 km s^{-1} are found in the face-on galaxy NGC 1058 (Dickey et al. 1990) and from red supergiants in star forming regions in the Large Magellanic Cloud (Prevot et al. 1989). Values of 7 to 10 km s^{-1} are found for the giant Sc's NGC 628 (Shostak and van der Kruit 1984) and NGC 3938 (van der Kruit and Shostak 1982), and the dwarfs IC 1613 (Skillman and Lake 1989) and GR8 (Carignan et al. 1990). It could be argued that the higher values (around 8–10 km s^{-1}) pertain to the warmer phase of the neutral gas, and the lower values to the colder clouds. It is not clear, however, what heating sources are responsible for keeping the outer gas at a relatively high temperature (cf. Corbelli and Schneider 1990).

Whatever value is taken for the gas velocity dispersion, the corresponding gas scale height varies linearly with it at radii where the stellar disk can be neglected. In this connection the very extended HI disks shown at this workshop become of interest: the data for NGC 628 and NGC 5701 show clear structure in the gas at very large radii, where one expects the gas to feel only the halo potential. The same is true for the published data of IC 342 (Newton 1980b), where there is still spiral structure in the gas at radii where the gas disk is predicted to flare. So either the gas there is very cold, or the halos of these galaxies are flattened enough, or a preferred plane is set up by a disintegrating gaseous companion.

References

Athanassoula, E., Bosma, A. & Papaioannou, S. (1987). Astron. Astrophys. **197**, 23.

Athanassoula, E. & Bosma, A. (1988). In *Large Scale Structure of the Universe*, I.A.U. Symp. 130 (eds. J. Audouze, M.-C. Peletan & A. Szalay), p. 391. Dordrecht: Kluwer.

Begeman, K. (1987). PhD Thesis, University of Groningen.

Bosma, A. (1978). PhD Thesis, University of Groningen.

Bosma, A. (1981a). Astron. J. **86**, 1791.

Bosma, A. (1981b). Astron. J. **86**, 1825.

Bosma, A., van der Hulst, J.M. & Athanassoula, E. (1988). Astron. Astrophys. **198**, 100.

Briggs, F.H. (1990). Astrophys. J. **352**, 15.

Burke, B.F. (1957). Astron. J. **62**, 90.

Carignan, C., Beaulieu, S. & Freeman, K.C. (1990). Astron. J. **99**, 178.

Carignan, C. & Puche, D. (1990). Astron. J. **100**, 394.

Cayatte, V., van Gorkom, J.H., Balkowski, C. & Kotanyi, C.G. (1990). Preprint.

Colgan, S.W.J., Salpeter, E.E. & Terzian, Y. (1988). Astrophys. J. **328**, 275.

Corbelli, E. & Schneider, S.E. (1990). Astrophys. J. **356**, 89.
Davoust, E. & de Vaucouleurs, G. (1980). Astrophys. J. **242**, 30.
Dickey, J.M., Hanson, M.M. & Helou, G. (1990). Astrophys. J. **352**, 522.
Gum, C.S., Kerr, F.J. & Westerhout, G. (1960). Mon. Not. R. Astron. Soc. **121**, 132.
Impey, C. & Bothun, G. (1989). Astrophys. J. **341**, 89.
Kerr, F.J. (1957). Astron. J. **62**, 93.
Knapp, G.R. (1987). Publ. Astron. Soc. Pac. **99**, 1134.
Newton, K. (1980a). Mon. Not. R. Astron. Soc. **191**, 169.
Newton, K. (1980b). Mon. Not. R. Astron. Soc. **191**, 615.
Prevot, L., Rousseau, J. & Martin, N. (1989). Astron. Astrophys. **225**, 303.
Rogstad, D.H., Lockhart, I.A. & Wright, M.C.H. (1974). Astrophys. J. **193**, 309.
Rogstad, D.H., Wright, M.C.H. & Lockhart, I.A. (1976). Astrophys. J. **204**, 703.
Rogstad, D.H., McCrutcher, R.M. & Chu, K. (1979). Astrophys. J. **229**, 509.
Rubin, V.C., Thonnard, N. & Ford, W.K. (1982). Astron. J. **87**, 477.
Rupen, M. (1990). Preprint.
Sancisi, R. (1976). Astron. Astrophys. **53**, 159.
Sancisi, R. & Allen, R.J. (1979). Astron. Astrophys. **74**, 73.
Shostak, G.S. & van der Kruit, P.C. (1984). Astron. Astrophys. **132**, 20.
Skillman, E. & Lake, G. (1989). Astron. J. **98**, 1274.
Tully, R.B. (1988). *Nearby Galaxy Catalog.* Cambridge: Cambridge University Press.
van der Kruit, P.C. & Shostak, G.S. (1982). Astron. Astrophys. **105**, 351.
van Driel, W. (1987). PhD Thesis, University of Groningen.
Warmels, R.H. (1986). PhD Thesis, University of Groningen.
Wevers, B.M.H.R. (1984). PhD Thesis, University of Groningen.
Wouterloot, J.G.A., Brand, J., Burton, W.B. & Kwee, K.K. (1990). Astron. Astrophys. **230**, 21.

Behavior of warps in extended disks

FRANK BRIGGS
University of Pittsburgh

JURJEN KAMPHUIS
Kapteyn Astronomical Institute, Groningen

Abstract

The new HI observations of the warped, face-on ScI galaxy NGC 628 reported by Kamphuis and Briggs (this volume) serve to illustrate the kinematical behavior of large spiral galaxies whose warped disks can be detected far outside their optical dimensions. The outlying gas follows inclined orbits that appear to be differentially regressing.

The model derived for NGC 628 from the HI observations can be re-oriented so that the inner, optical disk is viewed exactly edge-on. In this way, a model for an observed "kinematical" warp can be used to predict a range of appearances of "morphological" warps.

Introduction

In a VLA survey of edge-on galaxies, Bosma (this workshop) finds that warps are common and are easily identified in more than half of his sample. This result again raises the big problem with warps that was pointed out when the warp in our own Galaxy was first recognized (Kahn and Woltjer 1959). Theorists have long argued that differential regression of the inclined orbits in the flattened potential of the stellar disk will cause warps to *corrugate* and damp to a single plane in an uncomfortably short time. The problem is how to make warps last long enough that they can appear to be common.

Theoretical models have been constructed in which warps are treated as normal bending modes in self-gravitating disks that are subjected to inclined, flattened halo potentials (cf. Sparke and Casertano 1988). In this picture, the inner disk precesses in the potential of the halo, but the warp remains coherent, which in turn means that the line of the nodes of the inclined rings remains straight when measured in either the plane of symmetry of the halo or in the plane of the inner disk.

A large amount of high quality data, which describes the kinematics of warped spiral galaxies, has now accumulated and allows theories for warps to be tested. Systematic analysis of the body of data has been undertaken by Christodoulou, Tohline, and Steiman-Cameron (1988, 1990) and by Briggs (1990). The findings reported by Briggs confirm and extend the picture developed by Rogstad and colleagues (1974, 1976, 1979) in the earliest attempts to fit kinematical models to the warped disks seen in HI observations of galaxies other than the Milky Way: the warps do not exhibit straight lines of the nodes.

Binney (1990) has shown that a steady infall of material can bend a disk and that this can appear as a warp. NGC 628 may fit into Binney's scheme, since the galaxy has a warped disk that can be observed in HI far outside the optical radius, and Kamphuis and Briggs (KB, this workshop) have shown that it may be assimilating fresh, extragalactic material even at the present epoch.

Ring model for NGC 628

The observational results of Kamphuis and Briggs (KB) provide a good example of the velocity field of a *kinematical warp*, and the models extracted from this data are typical of those for galaxies in which high sensitivity HI observations detect gas far outside the optical extent. The model consists of concentric circular rings, specified by inclination with respect to the plane of the sky, position angle of the line of the nodes (LON) measured in the plane of the sky, and rotation velocity. The rings are fitted to the observed velocity field using the `ROTCUR` routine in the `GIPSY` package as described by Teuben (this workshop); the details of the fit to NGC 628 are given by Kamphuis and Briggs (1991).

Figures 1 and 2 present orientation parameters for the rings in three different reference frames. The *sky frame* is the original frame in which `ROTCUR` fits the rings to the velocity field. Thus, the abrupt rotation of $\sim 90^\circ$ plotted in Figure 1 (top panel) for the LON just outside $R_{Ho} = 6'$ represents the violent reorientation of the velocity contours that occurs near the optical radius in Figure 2 of KB. The inclination (Fig. 2) of the rings in the region of rapid LON rotation decreases to nearly zero, indicating that these rings are viewed very close to face-on where small changes in ring orientation can make

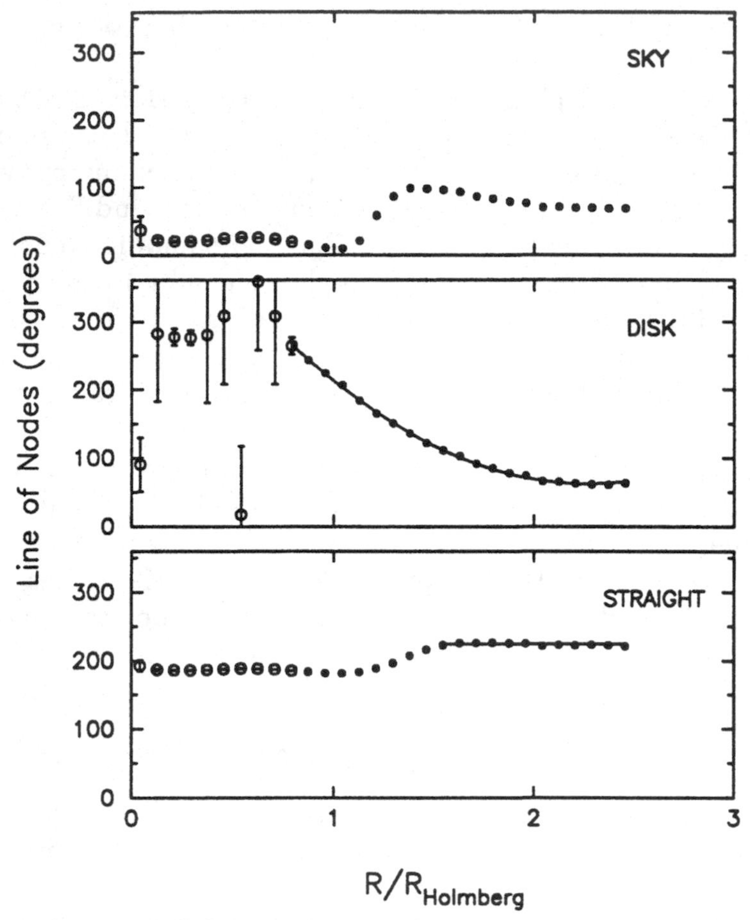

Fig. 1. Radial dependence of the orientation of the lines of the nodes (*LON*) for NGC 628 in three reference frames: the observed frame based on the plane of the *sky*, the frame defined by the inner *disk*, and a frame that produces *straight LON* for both the innermost and outermost rings. Large circles indicate rings with radii less than R_{25}.

large variations in the position of the *LON* seen in the plane of the sky. This description is close to that derived by Shostak and van der Kruit (1984). The apparently abrupt transition observed in the velocity field is a chance occurrence stemming from our particular viewing aspect.

The *disk frame* uses the planar nature of the inner, optical disk to provide a natural reference plane to which ring orientations are referred. Figure 2 (center panel) shows that the *tip* of the rings rises smoothly with respect to the plane of the inner disk. Furthermore, in this reference frame, the orientation of the *LON* rotates by ~180°

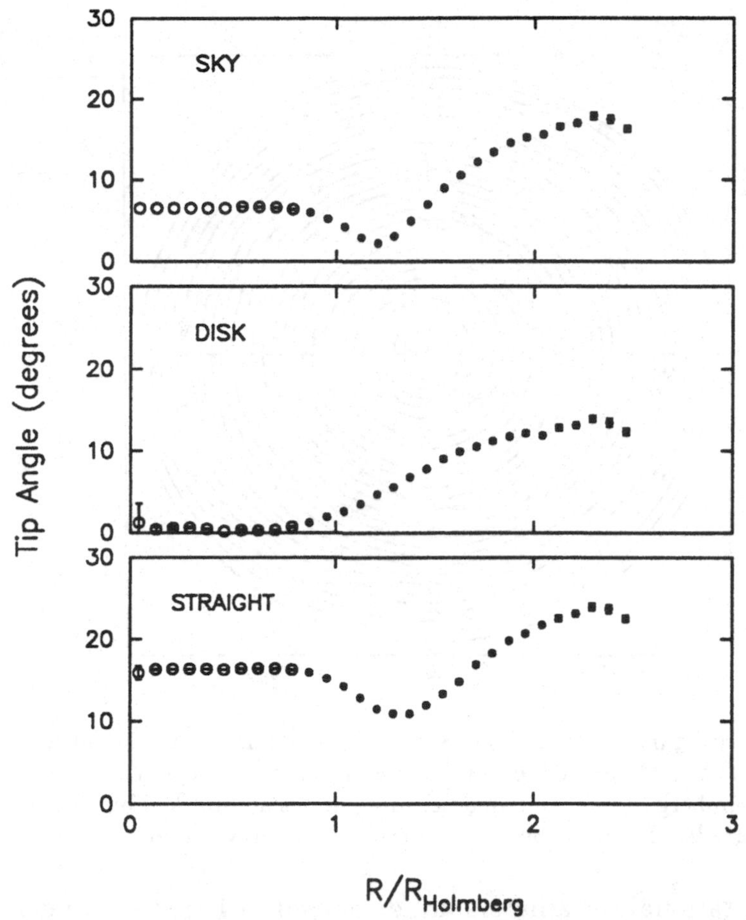

Fig. 2. Radial dependence of the *tip angle* of model rings for NGC 628 measured relative to three different reference planes: the plane of the *sky*, the plane defined by the inner *disk*, and a plane that produces *straight LON* for both the innermost and outermost rings.

between R_{Ho} and the largest radii of the ring model (center panel of Fig. 1).

Figure 3 presents the ring model in a frame where the inner disk is viewed exactly face-on. The node line formed by joining the *LON* for successive rings forms a spiral that advances in the direction of galaxy rotation implied by the sense of winding of the spiral arms. This pattern is consistent with the hypothesis that the outlying gas lies on inclined orbits that are differentially regressing in the flattened potential provided by the stellar disk or by a flattened halo whose symmetry plane is coincident with the disk plane.

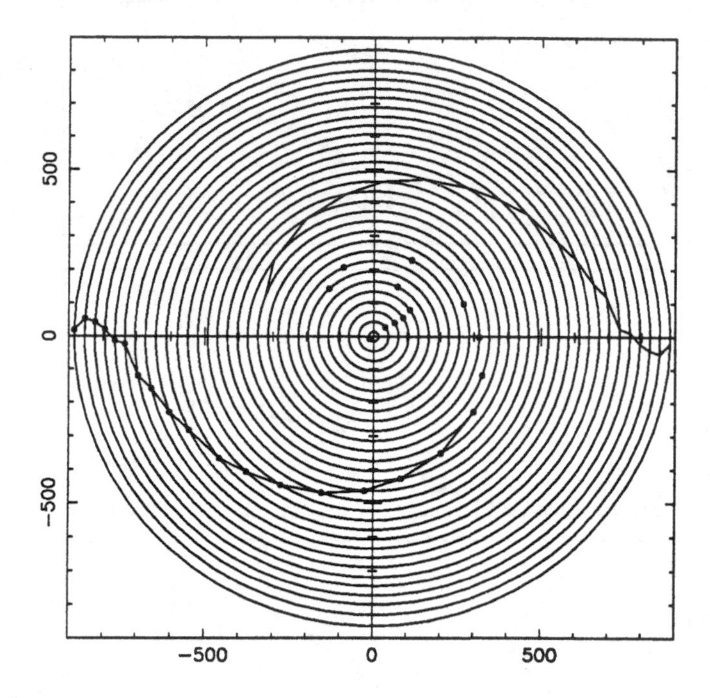

Fig. 3. Ring model for NGC 628 viewed perpendicular to the inner disk. The *LON* are joined by line segments in the region outside the Holmberg radius, and all descending nodes are marked by dots. Axes are labeled in seconds of arc from the galaxy center.

The orientational parameters are displayed in Figures 1 and 2 for a third reference frame labeled *straight*, which has been selected to make the node line straight both for the inner disk (along one *LON*) and also for the outermost rings (along a different *LON*). In each of galaxies reviewed by Briggs (1990) such a frame could be found; many cases were striking in the straightness of the node lines in the two regions and in the abruptness of the transition between the two zones of radius. Figure 4 illustrates how this reference frame could be identified by adopting the *coordinate sphere* representation proposed by Briggs (1990). Here the technique is applied to the sensitive new data with the goal of clarifying Briggs' (1990) *Rule 4*; the outcome will suggest that Rule 4 probably does not have physical significance.

The diagram in Figure 4 is most easily interpreted by visualizing each plotted point as the point at which the normal to the plane of each ring (i.e., a vector aligned along the angular momentum vector) pierces a coordinate sphere centered on NGC 628. The origin of the diagram represents a vector that points toward the observer, and a

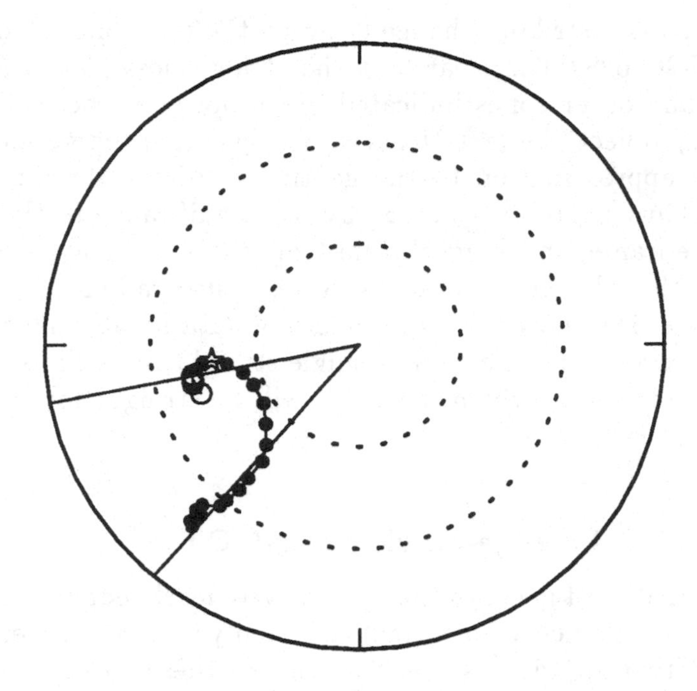

Fig. 4. Coordinate sphere analysis that is used to define a reference frame in which *LON* are straight both for a regime of radius of the inner rings and also for a regime of outer rings where the nodes fall on a differently oriented *LON*. Radial distance in the diagram corresponds to *tip* with respect to the reference plane. The polar angle indicates relative orientation of the *LON* measured in this reference plane. The dashed circles indicate increments of 10° in tip angle. The data points represent the ring orientations for the NGC 628 model. Open circles represent the inner disk, which is here tipped ~17° away from face-on.

point falling there would indicate a ring viewed exactly face-on. Circles of constant tip angle are drawn as dashed curves. A straight line running radially outward from the origin would represent a straight *LON* as the *tip* is increased.

Figure 4 is drawn with the pattern of data points for NGC 628 oriented so that two radial lines come close to many of the points. Each of the two lines represents an orientation of *LON*. In this case, only three points fall more than a few degrees away from one of the two lines; they are the points seen in Figure 1 in the transition zone between the two straight sections of *LON*. Although several of the galaxies in the full sample examined by Briggs (1990) show a more abrupt and striking change between the two zones than is the case for NGC 628, it is clear that the straightness of the *LON* in the inner and outer zones indicated in the lower frame of Figure 1 may simply reflect the fact that two straight line segments form a reasonable approximation to the gentle arc of the data points in Figure 4. Thus, there may be no physical significance to this choice of reference frame, and Briggs' Rule 4 may simply be a restatement of the fact that the *LON* rotate with increasing radius.

In any case, it is clear that the pattern of data points that trace an arc in Figure 4 can not be fit by a single straight line. Thus, there is no reference frame in which the *LON* will be straight for the entire range of ring radii.

An edge-on view of NGC 628

The kinematics of galaxies that are viewed nearly edge-on are difficult to decipher, since lines of sight commonly intercept several rings and ring-fitting algorithms break down. In these cases, extraction of ring orientation parameters may not be possible. To study how galaxies with straight and regressing *LON* would appear if viewed edge-on, Figures 5 and 6 present two edge-on views of NGC 628: one using the regressing *LON* model deduced from our kinematical ring-fitting, the other using a "toy" model in which the *LON* are held strictly straight throughout the entire disk. The run of ring inclination with radius is identical in both models. The top panels in Figures 5 and 6 show the entire ring models. The lower panels break the rings into regimes of line-of-sight velocity ("channel maps"). Due to the symmetry of the models, only the systemic velocity and the velocities higher than systemic need to be shown.

In Figure 5, the galaxy is oriented so that the maximum "morphological warp" will be observed for the rings of largest radius. The inner disk is exactly edge-on. The synthetic model is oriented so that the node line is aligned along the line of sight. The most striking effect of the *LON* rotation is to thicken the appearance of the disk; however, this thickening would be difficult to detect since the outer rings have column density that is much lower than the column density of the gas at small radii, and the effect of the thickening is to place the dim and bright extremes in HI brightness close together in projection on the sky. The rotation of the *LON* can also be seen in the channel maps as a dip below the plane of the inner disk as the gas is traced outward along the major axis; this is a sign of disk cor-

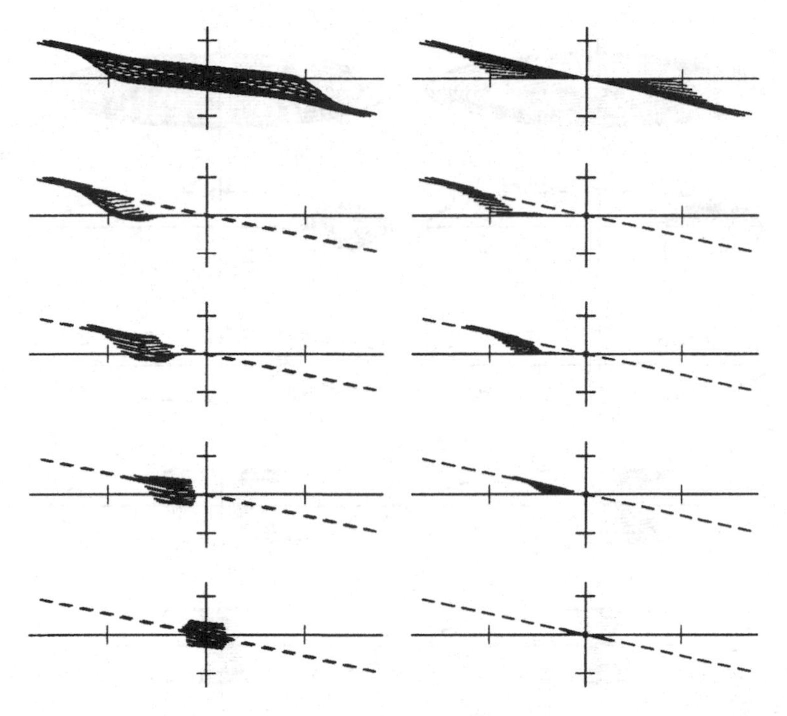

Fig. 5. Ring model for NGC 628 oriented to view the maximum "morphological warp" while holding the inner disk exactly edge-on. The model fitted to NGC 628 is diagramed on the left; the "toy" straight-*LON* model is to the right. The top panel shows the entire extend of each model. The lower four panels show "velocity channel maps" in order of decreasing line-of-sight velocity with respect to systemic; the bottom panel illustrates the ring segments with velocity close to systemic.

rugation as the *LON* of the inner warped rings regress more rapidly than the outer rings.

Figure 6 presents the view in which the edge-on galaxy models of Figure 5 are rotated by 90° about the rotational axis of the inner disk. Thus, the observer sees the rings of largest radii (and maximum warp) tipped above and below the line of sight to the galaxy center. Along the major axis at the extremes of the projected radii, the rings show no signs of the morphological warp. The "toy" model now has its node line oriented perpendicular to the line of sight, and no indication of a morphological warp will be seen at any radius. The model resulting from reorientation of NGC 628 does show deviations at intermediate radii. Thus, observations of edge-on galaxies with antisymmetric wiggles in an otherwise unwarped disk may be

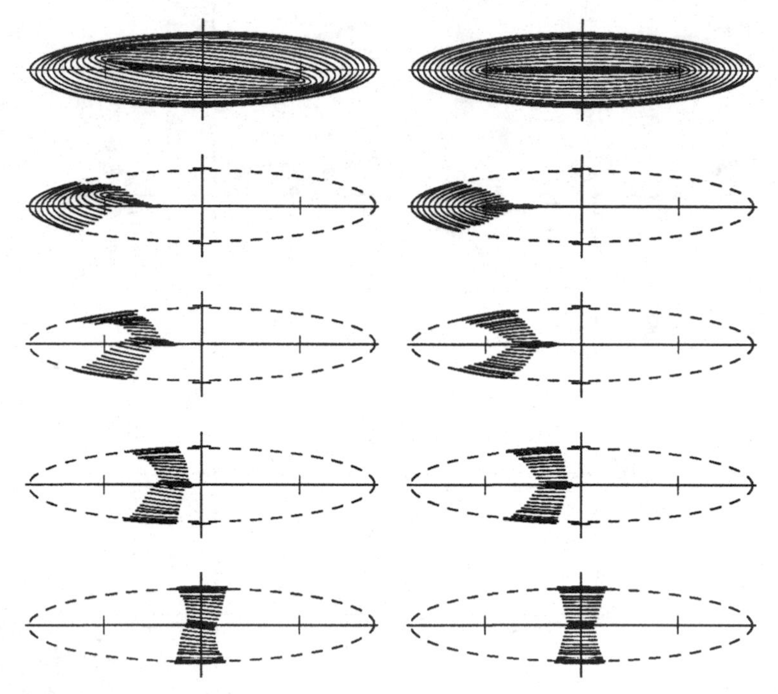

Fig. 6. Fitted kinematic (left) and "toy" straight *LON* models (right) for NGC 628 shown oriented such that the inner disk is edge-on but rotated 90° with respect to the views shown in Fig. 5. "Channel maps" are shown in the lower panels (see Fig. 5).

indicating the *corrugation* that is expected to result from differential regression of the *LON*. The wiggles may be subtle; this was also true in the view presented in Figure 5. The *tip* of the outer orbits would also be difficult to measure since these orbits have low column density gas along the galaxy minor axis that is viewed in close proximity to the bright gas of the inner disk.

Conclusion

Analysis of the velocity field measured for NGC 628 indicates that the *LON* are not straight in any reference system. The effects of the rotation of the *LON* with radius would be difficult to detect if the galaxy were viewed edge-on.

KB find that NGC 628 is in the process of accreting a substantial HI mass in its outer parts, and the galaxy may be warped as a result of the interaction. The system may be an example of Binney's notion that a continuing acquisition of external mass and angular momentum may play an ongoing role in bending extended disks. In such a case, the warp is long-lived because it is continually excited, not because it represents a normal mode of the underlying potential. Bosma's report of a correlation between absence of warping and a concentrated halo raises the question of how to assign cause and effect. Perhaps both characteristics result from an earlier history of accretion or from the ability of the present environment to provide fresh material to the systems where warps are observed.

A further result discussed more fully by Kamphuis and Briggs (1991) concerns the residuals left after the model velocity field has been subtracted from the observed one. Significant residuals remain with power at harmonics higher than the $m = 1$ harmonic fit by inclined circular rings in uniform rotation. The diagnostic methods that Teuben hopes to apply to oval orbits (this workshop) may prove to be powerful tools in the analysis of warped disks such as NGC 628.

Acknowledgements

We wish to thank Penny Sackett, Renzo Sancisi, Linda Sparke, and Thijs van der Hulst for stimulating discussions.

References

Binney, J. (1990). In *Dynamics and Interactions of Galaxies* (ed. R. Wielen), p. 328. Berlin: Springer.

Briggs, F.H. (1990). Astrophys. J. **352**, 15.

Christodoulou, D.M., Tohline, J.E. & Steiman-Cameron, T.Y. (1988). Astron. J. **96**, 1307.

Christodoulou, D.M., Tohline, J.E. & Steiman-Cameron, T.Y. (1990). Submitted to Astrophys. J., Suppl. Ser.

Kahn, F.D. & Woltjer, L. (1959) Astrophys. J. **130**, 705.

Kamphuis, J. & Briggs, F. (1991). This volume.

Rogstad, D.H., Lockhart, I.A. & Wright, M.C.H. (1974). Astrophys. J. **193**, 309.

Rogstad, D.H., Wright, M.C.H. & Lockhart, I.A. (1976). Astrophys. J. **204**, 703.

Rogstad, D.H., McCrutcher, R.M. & Chu, K. (1979). Astrophys. J. **229**, 509.

Shostak, G.S. & van der Kruit, P.C. (1984). Astron. Astrophys. **132**, 20.

Sparke, L.S. & Casertano, S. (1988). Mon. Not. R. Astron. Soc. **234**, 873.

Observational constraints for the explanation of warps

E. BATTANER, E. FLORIDO, M-L. SANCHEZ-SAAVEDRA
Universidad de Granada

M. PRIETO
Instituto de Astrofisica de Canarias

Abstract

Three observational constraints are discussed: a) colour gradients in the warp region, from CCD observations made at La Palma with the 2.5 m Isaac Newton Telescope, b) frequency of the phenomenon and c) possible non-random spatial distribution of warps. A theoretical model in which extragalactic magnetic fields are responsible for warps is also proposed.

Colour dependence of the warp shape

Most theoretical models about warps assume that they are produced by gravitational forces; see for instance the review in Sparke and Casertano (1988). There is in principle a simple test to verify the validity of these models. Gravity produces the same acceleration on any object, i.e., does not make any distinction either between red and blue stars or between gas and stars. Therefore the same warp curve should be expected for red and blue filters as for the neutral hydrogen. Are the red, blue and gas distributions warped by the same amount?

To answer this question, published photometric and radio data can be used. We also report on our own observations made at La Palma, Spain, by means of the 2.5 m Isaac Newton Telescope (Florido et al. 1990a).

Observations made in our own galaxy do not show evidence of a warp in the population of old stars (Ichikawa and Sasaki 1984, Guibert et al. 1978), while it is very clearly depicted by the spatial distribution of young objects (Fick and Blitz 1982, Mayer-Hasselwonder et al. 1982, Reed and Fitzgerald 1984, Kerr 1982, and many others). However, Sasaki (1987) carried out a detailed photometrical study of NGC 5907, and interpreted the difference observed between the optical and the HI warps as an effect of the truncation of the optical disk and the oblique direction of the warp with regard to the line of sight.

Florido et al. (1990a) studied the galaxies NGC 4013, NGC 4565 and NGC 6504. Several morphological properties of warps were found:

a) There is a clear separation between the blue and the red warp curves. The gas warp is larger than the blue warp, which in turn is larger than the red warp. It would seem that this separation is real and not an effect of projection.
   Let us remember that Sasaki (1987) was able to explain the difference in the location of the gas and the optical warps as a projection effect, resulting from the combined effect of the sudden truncation of the optical disk and the extinction. Sasaki adopted the so called "tilted ring model", introduced by Rogstad et al. (1974) and adapted for NGC 5907 by Bosma (1981). The adoption of this geometrical model with its assumed parameters, are partly supported by Kerr's map (1982) of curves of equal altitude over the mean plane for the Milky Way warp. The similarity between the theoretical z-map and the Kerr's map is particularly relevant for the northern warp of our galaxy. In the "tilted ring model", the Sasaki interpretation could be summarized as follows: the external rings (with only gas and absorbing dust) could produce extinction in the internal rings (containing also stars). This effect could reduce the apparent amplitude of the optical warp.
   If this was then the case, as the blue light is more affected by extinction, the blue warp is reduced by a greater amount, so that the order of amplitude of the warps would be "gas-red-blue", and not "gas-blue-red", resulting in the observations by Florido et al. (1990a). Unfortunately, Sasaki did not show in his paper the warp profiles for the two different plates used. Therefore, if for the same galactocentric radius the order of warp amplitude is "gas-blue-red" then it should indicate a real dependence of the warp angle on the age of the disturbed population, and not on a projection effect. Knowing that stars are born where gas is found, it would seem that the force producing the distortion

has an initial influence on the gas, and that stellar warps are produced as a consequence.

b) If this was the actual order (gas-blue-red), the effect of extinction would be to bring these three warps closer. Therefore, if a separation of the different warps is observed, it cannot be attributed to extinction.

c) The galaxy NGC 4013 shows a well-defined warped dust lane. The amplitude of the warp of this lane is much larger than the one defined by the stellar continuum. This is clearly illustrated in the CCD images. The dust lane cuts the stellar disk with a large inclination in the most external region. Dust is a young component of the galaxy, confirming the age dependence of the different warps.

d) The galaxy NGC 4013 may also be a clear example of a frequent property of warps. The warp curve deviates first in the opposite direction of the HI warp, then returns, crosses the mean galactic plane and then finally reaches the hemisphere in which the HI warp lies. As usually optical images extend out to smaller radii than radio maps, one could have the impression that the optical warps are opposite to the HI warps. Bosma (this volume) also reports such a counter deviation.

There is another morphological feature which could be related to warps. Some disks are "corrugated" showing wave-like distortions with respect to the mean plane. Theoretical arguments have been given (Nelson 1976a, b; Nelson and Matsuda 1980) suggesting a connection between corrugations and warps. Corrugations are a well known phenomenon in our galaxy (Quiroga 1977, Lockman 1977, Spicker and Feitzinger 1986, and others). Florido et al. (1990b) have studied corrugations in NGC 4244 and NGC 5023. Both are warped, but this does not mean a physical connection between the two phenomena as these galaxies were observed during a campaign devoted to warps.

The effect discussed above (counter deviation of the innermost part of the warp) would suggest the existence of a common mechanism. However, the disturbance seems to have an origin within the galaxy, propagating in and out with increasing wavelength. Through the blue filters, corrugations are sharper and larger than when observed through the red filters. However the red filters show larger typical wavelengths. This indicates again that the corrugations of the stellar disk are a result of forces initially disturbing the gas.

The impressive corrugations of NGC 4244 make it an ideal prototype galaxy for studies of this phenomenon (see also van der Kruit and Searle 1981, and Sancisi 1976). It remains an open question as to whether or not corrugations and warps are physically related effects.

Frequency of warped spiral galaxies

Sanchez-Saavedra et al. (1990) have compiled a list of all warped NGC spiral galaxies with large inclinations ($\log_{10} R_{25} > 0.57$) following an inspection of the Palomar Observatory Sky Survey. The observed frequency of warps and the derived actual frequency is a very important parameter restricting possible explanations of the warping phenomenon. This kind of statistical study complements detailed observations of individual warps. A similar list of HI warps should generate much interest (see Bosma, this volume).

The identification of a warp in a plate is a rather subjective task. In order to avoid this problem as much as possible, each galaxy was inspected by three independent observers.

The observed frequency is very high. Out of 86 galaxies in the sample, 42 were observed to exhibit a warp. A high frequency was observed in the study correlating with the results of previous works using smaller samples. Making reasonable assumptions it is possible to estimate the real frequency of warped galaxies in space. In this work it was proposed that the real frequency be greater than 83%. This value seems to be in disagreement with the tidal interaction hypothesis but not with the model of Sparke (1984) and Sparke and Casertano (1988).

Warps are more easily observed in the O (blue) plates than in the E (red) plates. This is due to either an enhanced warp in the young stellar population, or that the blue images could be larger. The frequency of "S-warps" (clockwise) is the same as "N-warps" (counterclockwise) as expected, but it cannot be rejected that their spatial distribution is not uniform. An important fact in the spatial distribution of warps, which is appreciated in this catalogue, is that non-warped galaxies are concentrated around RA $= 12^h$. Due to the aforementioned subjectiveness involved in the identifying of a warp in a plate, it would be of interest to contrast this list with other similar works.

Extragalactic magnetic fields and warps

Recently two direct measurements of the magnetic field strength in intercluster regions became available. Kim et al. (1989) observed the field strength ranging from 0.3 to 0.6 μG in the Coma supercluster, in an interesting region between the Coma cluster and A1367. The measurements were made using synchrotron continuum observations under the adoption of the equipartition of energy. With a different technique based on Faraday rotation of distant radio sources, Vallee (1990) obtained 1.5μG in an intercluster region in the Virgo supercluster. It is clearly obvious that these large extragalactic magnetic fields must have an influence in the peripheral morphology of galaxies with an ionized gaseous component.

Battaner et al. (1990) analyzed the way in which extragalactic magnetic fields can influence the shape of a spiral galaxy. It was shown that one of the expected effects is the formation and maintenance of warps, when the vector **B** has the following value:

a) The modulus must be of the order of about 10^{-8} G or higher.

b) If **B** is either parallel to the rotation axis or perpendicular to it (i.e., contained in the galactic plane) then it is unable to produce warps. For a given value of the modulus of **B**, its direction has maximum efficiency in producing this distortion at 45° from the rotation axis. Therefore, when a warped edge-on galaxy is observed, the direction of **B** must be near the plane of the sky (i.e., almost perpendicular to the line of sight), and forming an angle of approximately 45° with the rotation axis. There are however four directions 45° away from the rotation axis. Just two of them are possible: those which are closer to the rising warps. The warps rise with a tendency to alignment of matter along the magnetic field lines.

If this model is confirmed by observations or by further theoretical work, there would be an interesting possibility of deriving additional information about the extragalactic magnetic field.

Is the spatial distribution of warps random?

The spatial distribution of the orientation of warps could also provide a basic test to the different hypotheses of warps. Most theories would predict a random distribution of warps orientation. However, Battaner et al. (1990) claim that a trend exists in regular orientation. The hypothetical individual extragalactic magnetic field strength around each warped edge-on spiral, would correspond to an arrow forming 45° with the rotation axis, closer to the rising warp. It is likely that these arrows are not randomly oriented.

This conclusion was reached with a small sample of edge-on warped spirals for which the warp was studied and reported in the literature. An extension of this work using a larger sample is in progress. Of course, the arrows could be aligned if the position angle distribution of all galaxies in the sample were not random. Coherent orientation effects of galaxies and clusters have indeed been studied (Hawley and Peebles 1975, Djorgovski 1986, and many others). But these effects were undetectable in the sample of 19 galaxies used by Battaner et al. (1990), where a random distribution of position angles was found.

These galaxies are nearby, with a distance less than about 20 Mpc. Under the hypothesis of "magnetically induced warps" and if, in the unit volume of diameter about 20 Mpc, the extragalactic magnetic field is uniform (or at least, if its random component is of the order of or less than the energy due to the uniform field), a certain alignment of the arrows can be expected. This alignment can also be used in deducing the direction of the extragalactic magnetic field in the 20 Mpc Milky Way neighborhood. We believe that the early theory of Kahn and Woltjer (1959) would also predict a regular orientation, but many other hypotheses should be questioned. Further observations and analyses are necessary.

Conclusions

The detailed study of differential warping for different populations and ages of galactic components could be an important clue in deriving an explanation of this morphological feature. On the other hand, statistical studies in frequency and alignment could provide additional observational restrictions.

References

Battaner, E., Florido, E. & Sanchez-Saavedra, M.L. (1990). Astron. Astrophys., in press.

Bosma, A. (1981). Astron. J. **86**, 1791.

Djorgovski, S. (1986). In *Nearly Normal Galaxies* (ed. S. M. Faber), p. 227. Berlin: Springer-Verlag.

Fick, M. & Blitz, L. (1982). In *Kinematics, Dynamics and Structure of the Milky Way* (ed. W.L.H. Shuter). Dordrecht: Reidel.

Florido, E., Prieto, M., Battaner, E., Mediavilla, E. & Sanchez-Saavedra, M.L. (1990a). Astron. Astrophys., in press.
Florido, E., Battaner, E., Prieto, M., Mediavilla, E. & Sanchez-Saavedra, M.L. (1990b). Mon. Not. R. Astron. Soc., submitted.
Guibert, J., Lequeux, J. & Viallefond, F. (1978). Astron. Astrophys. **68**, 1.
Hawley, D. & Peebles, J. (1975). Astron. J. **80**, 477.
Ichikawa, T. & Sasaki, T. (1984). In *Proceedings of the Second Asian-Pacific Regional Meeting on Astronomy* (ed. B. Hidayat & M.W. Feast), p. 182. Jakarta: Tira Pustaka Pub. House.
Kahn, F.D. & Woltjer, L. (1959). Astrophys. J. **130**, 705.
Kerr, F.J. (1982). In *Kinematics, Dynamics and Structure of the Milky Way* (ed. W. L. H. Shuter). Dordrecht: Reidel.
Kim, K.T., Kronberg, P.P., Giovannini, G. & Venturi, T. (1989). Nature **341**, 720.
Lockman, F.J. (1977). Astron. J. **82**, 408.
Mayer-Hasselwander, H.A. et al. (1982). Astron. Astrophys. **105**, 164.
Nelson, A.H. (1976a). Mon. Not. R. Astron. Soc. **174**, 661.
Nelson, A.H. (1976b). Mon. Not. R. Astron. Soc. **177**, 265.
Nelson, A.H. & Matsuda, T. (1980). Mon. Not. R. Astron. Soc. **191**, 221.
Quiroga, R.J. (1977). Astrophys. Space Sci. **50**, 281.
Reed, B.C. & FitzGerald, M.P. (1984). Mon. Not. R. Astron. Soc. **211**, 235.
Rogstad, D.H., Lockhart, I.A. & Wright, M.C.H. (1974). Astrophys. J. **193**, 309.
Sanchez-Saavedra, M.L., Battaner, E. & Florido, E. (1990). Mon. Not. R. Astron. Soc. **246**, 458.
Sancisi, R. (1976). Astron. Astrophys. **53**, 159.
Sasaki, T. (1987). Publ. Astron. Soc. Jpn. **39**, 849.
Sparke, L.S. (1984). Astrophys. J. **280**, 117.
Sparke, L.S. & Casertano, S. (1988). Mon. Not. R. Astron. Soc. **234**, 873.
Spicker, J. & Feitzinger, J.V. (1986). Astron. Astrophys. **163**, 43.
Vallee, J.P. (1990). Astron. J. **99**, 959.
van der Kruit, P.C. & Searle, L. (1981). Astron. Astrophys. **95**, 105.

Warps in S0s: observations versus theories

GIUSEPPE GALLETTA
Università di Padova

Warps and galaxy morphology

Many galaxies possess warped gas planes inside their structure, but the morphological appearance of this peculiarity assumes different aspects depending on the host galaxy. In elliptical galaxies, whose stellar body is very extended outward, dust and gas appear to be concentrated mainly in the central regions, in a tilted series of rings or in warped disks (Bertola and Galletta 1978). This confinement to the inner parts is a reason why many of these features remain often undetected, until the galaxy is observed with a short exposure time and/or by means of CCD frames. The wide and well studied dust-lane in NGC 5128 (Cen A), for instance, owes a large part of its fame to the closeness of the host galaxy (just 3 Mpc, as indicated by the magnitude of the SN1986g). In disk galaxies, on the contrary, rings and warped gas planes appear in the outer regions only: in spiral galaxies, non-negligible warps of the outermost equatorial plane are often visible at 21cm (Sancisi 1976); in S0s, generally poor of neutral gas, a series of rings of luminous matter is sometimes observed to circumscribe the external regions (Schweizer et al.1983), crossing almost perpendicularly the galaxy plane. These *polar rings* are often warped and appear to lack matter in the innermost parts.

Apart from this morphological difference, dust-lane elliptical galaxies and polar-ring S0s are joined by many similar aspects that make the warped structures observed in them substantially different from the warps observed in spirals. As a first common point, the completely different kinematics observed between gas and stars suggests an outer origin for the gas (see Galletta et al. 1990), which could have been accreted from a companion galaxy or from the environment. This is opposite to that believed for the nature of warps in spiral galaxies, where the gas implied is essentially coeval with the galaxy itself. Another important point is that the confinement to the outer regions observed in S0s does not reflect a real difference in size for the gaseous structure. In fact, the extension of the gas and dust distribution observed in NGC 5128 (~10 kpc in diameter) and NGC 5266 (24 kpc), two good cases of ellipticals with dust (Bertola et al. 1985, Varnas et al. 1987), is similar to that measurable for A0136-0801 (17 kpc) and NGC 4650A (20 kpc), nice cases of polar-ring S0s (Schweizer et al. 1983). In these latter it is the stellar component, sharper and less extended than in elliptical galaxies, that generates this morphological difference. In addition, both the cases of polar ring S0s above considered are quite small galaxies.

Warps and rings in accreted gas

If gas accretion is the phenomenon that links polar-ring S0s with dust-lane ellipticals, it is interesting to investigate what behaviors it can assume in different cases.

A first mechanism determining the observed configuration is the tendency to a stability plane. If the potential is triaxial, like in the case of many elliptical galaxies (or for bars in disk galaxies), the theoretical models (Tohline et al. 1982, Steiman-Cameron and Durisen 1982, van Albada et al. 1982, Merritt and de Zeeuw 1983, Lake and Norman 1984) predict that two stability planes could exist: the "classical" equatorial plane (formed by the major and intermediate axis of the galaxy) and the plane perpendicular to the major axis (including the minor and intermediate one). When some gas is attracted by a stellar system, it tends to settle on one of these principal planes, depending on the angle at which it enters the galaxy and on the degree of triaxiality of the latter. The settling time depends on many parameters, ranging from few 10^8 to $\sim 10^{10}$ years (Steiman-Cameron and Durisen 1988). The whole family of ellipticals with dust lanes (Hawarden et al. 1981) presumably represents cases of settling along one of the above planes for gas coming into the system at various inclinations. In the case of an oblate potential, like that generated by a disk, only the equatorial plane is stable, and the gas is always expected to settle on it in a quite short time; so the persistence of a polar ring must be explained with additional hypotheses. They always start from gas accreted in orbits highly inclined with respect to the equatorial plane. As a first possibility, the observed gas configuration may be unstable or transient, and will collapse in a time which is proportionally longer with increasing inclination (*time* $\propto$ $\cos[inclination]^{-1}$, Richstone and Potter 1982). A second possibility is that the quantity of gas present is high enough to induce an effect of self-gravitation (Sparke and Casertano 1988), locally predominating over the gravitational field of the underlying galaxy. A third possibility is that the small stellar disk of the S0 is surrounded by an extended and triaxial halo, which stabilizes the warp (Binney 1978). This case requires gas planes with inner parts already settled on the stable plane and outer regions warped outward, as observed in A0136-0801 and other cases.

From the observational point of view, there are no systematic measurements of the mass included in the warped gas layer. Estimates of the dust mass in dust-lane ellipticals, based on IRAS observations, gives quite low values, between 10^4 and 10^5 $M_\odot$ (Zeilinger et al. 1989). Assuming a ratio 1:100 between dust and gas, as in the case of our Galaxy, the total mass of gas will be not greater than 10^7 $M_\odot$, a quite low value for self-gravitation. But there are indications that a higher mass of molecular clouds could be present inside some of these galaxies (see Zeilinger et al. 1989 and references therein). This point will be discussed later. In any case, more accurate estimates for the mass of gas in polar rings would be very useful in discriminating between models.

Another point to be clarified is why among the many polar-ring S0s known (Whitmore et al. 1990) there are no clear cases of barred galaxies, which represent on the contrary 50% of the population of disk galaxies. The bar is presumably so small a perturbation of the gravitational potential that its incompatibility with polar rings could be accepted only if these latter were in a very unstable situation, so that a slight variation of the potential due to the bar tumbling would be enough to destroy the precarious equilibrium of the polar rings. But this explanation is not completely satisfying, and a more detailed analysis of this aspect would be desirable.

A second mechanism determining the observed morphology is the interaction with the pre-existing environment. If little or no gas exists in the host galaxy, like for ellipticals or S0s, the accreted material interacts with stars only by means of dynamical friction (Chandrasekhar 1943, Tremaine 1981), a process that for small clouds would last more than 10^{10} years. So we could presume that the gas accreted by an early-type galaxy will evolve toward an equilibrium configuration without strong interaction with the stellar component. Studies of S0s where a recent accretion of gas happened agree with this assumption, indicating an unperturbed stellar structure (Galletta et al. 1990). A much different evolution would take place for material accreted by a spiral or by a galaxy with a non-negligible gas content. The arrival of foreign gas can affect very much the morphology and the kinematics of the native gas, by means of collisions between clouds. Two alternative scenarios are expected:

i) The gas could be heated by the shock and evaporates, eventually producing an X-ray halo, after an active phase of star formation. The galaxy may appear for some time like one of the many irregular objects existing in Arp's Atlas (Arp 1966), or may present HI features similar to the clouds with anomalous velocities detected in M101 or in other spirals (van der Hulst & Sancisi 1988). Then, depleted of gas, it may appear as a normal S0. Unfortunately, in a normal galaxy the X-ray emission due to accreted gas (a few 10^8 $M_\odot$) will be superimposed on the intrinsic X-ray flux, and could result being indistinguishable from it, even for galaxies at the distance of Virgo cluster.

ii) As an alternative, the energy of the collision could be completely dissipated by the enhancement of the star formation (and SN rate) and by other cooling processes, able to convert the atomic gas into molecular gas (Young and Knezek 1989). If this process is predominant, the energy input of the gas accretion would paradoxically produce for some time cold molecular clouds. Again, a survey of the molecular gas, with the same limits of detection because of the distance discussed for X-ray emission, would be highly useful to the understanding of this phenomenon.

If some kind of process like one of those described above has taken place in the polar-ring S0s and in some of the dust-lane ellipticals which present rings (e.g., NGC 5266, Varnas et al. 1987), this could explain the lack of gas in the central regions and the absence of these kind of structures in spiral galaxies, at least in the inner regions, where the gas density is still high. In all of these cases, because of a higher content of primordial gas, the accreted gas has been locally or globally converted into a very hot medium (if heated by the shock) or, conversely, into dense molecular clouds (if cooled by other processes).

To complete the picture of warps and rings made by interactions, we need to consider cases where gas is accreted by the galaxy at low inclination with respect to the equatorial plane.

Particular cases of gas accretion

Gas accreted near the galactic plane will quickly settle on it, possibly changing the morphology of the galaxy. So, in the case of gas-free galaxies, a normal elliptical will be observed as a dusty S0 with a big bulge and a gasless S0 will be eventually converted into an early-type (or anaemic) spiral. We don't know, in principle, how many cases of these 'masked' galaxies there are. But for an elliptical with a dust lane along the apparent major axis (Hawarden et al. 1981, Bertola et al. 1988) the difference could be immediately discovered if no trace of stellar disk is detectable, or when the gas appears to rotate in direction opposite to the stars (*gas counterrotation*, Bertola and Bettoni 1988, Bertola et al. 1988). In S0s, the presence of counterrotation (Galletta 1987, Rubin 1988) or of unstable gas structures could reveal the presence of accreted gas.

A typical case is represented by the edge-on SB0 NGC 4546 (Galletta 1987, Bettoni et al. 1990). In this system, a disk of gas extending a few kpc is rotating inside the stellar disc, but in the opposite direction. The relative gas–stars circular velocities are as great as $350\,\mathrm{km\,s^{-1}}$. Observed at 21 cm, the galaxy shows an HI content of $\sim 10^8 M_\odot$, within a stellar component with a mass $\geq 4 \times 10^{10} M_\odot$. While the velocity field of the stellar component appears unperturbed, with orbits slightly elongated on the bar major axis, both the ionized and the neutral gas appear asymmetric on one side of the galaxy. The gas velocity field presents many irregularities, and inner orbits (retrograde with respect to the stars but *prograde* with respect to the outer gas disc) are possibly elongated perpendicularly to the bar. In addition, a bright Hα knot with diameter of $\sim$175 pc is visible near the centre. Among the possible interpretations for such a complicated velocity field, NGC 4546 could be another case of galaxy with a warped gas plane, not yet settled onto the galaxy equator. Within the picture of gas accretion, also the Hα spot near the center could be the remnant of the nucleus of a small galaxy captured by NGC 4546.

A second case of an S0 with a warped gas plane is represented by NGC 2217 (Bettoni et al. 1990). Contrary to NGC 4546, this galaxy is seen almost face-on ($i = 21°$) and the stellar velocities, reaching $150\,\mathrm{km\,s^{-1}}$, indicate high rotational velocities. The galaxy mass is estimated to be $\geq 1.4 \times 10^{11} M_\odot$, in comparison with an HI mass of $\sim 6 \times 10^8 M_\odot$. In NGC 2217, a gas counterrotation appears in the inner regions, while in the outer ones gas and stars show similar velocities and are probably co-planar. The analysis of the gas motions in the space suggests a model composed of a warped and twisted disk, with the innermost part perpendicular to the bar major axis and the outermost one lying on the main galaxy plane. This shape of the gas disk fits fairly well the gas velocity field in the simple hypothesis of circular orbits. On the basis of this model, and contrary to the case of NGC 4546, the gas within the galaxy is not in retrograde motion with respect to the stars, but this kinematic anomaly is generated in the projected velocities by the peculiar shape of the gas disk and by the orientation of the galaxy. Like NGC 4546, the warps could be generated by gas accretion, but with higher inclinations with respect to the galaxy plane. Is this gas the result of a polar ring destroyed by the presence of the bar, and NGC 2217 a lacking case of barred galaxy with polar rings? Following another interpretation, this peculiar configuration could alternatively be explained as the result of orbits excited at the 2:1 vertical resonance of the bar (Binney 1981, Sparke 1984), but in this case there is no need for the gas being an acquisition from the environment.

NGC 5084: a borderline case

Another interesting case of inclined rings is represented by the spiral galaxy NGC 5084 (Zeilinger et al. 1990), which could be an intermediate case between the warped spirals and the gas-accreting systems. This galaxy presents the typical spindles of a stellar disk seen edge-on and an almost straight and luminous lane, probably composed of gas, tilted by $\sim 5°$ with respect to the disk major axis. The lane appears superimposed on an HI ring seen edge-on (Gottesman and Hawarden, 1986), which contains $4.5 \times 10^9 M_\odot$, an order of magnitude greater than the gas detected in dust-lane ellipticals or S0s with counterrotation. Geometrical considerations suggests that the equatorial plane of the galaxy is warped, with the HI and the luminous lane distributed in a ring marking the outer regions.

Another peculiarity of this galaxy is the mass-to-light ratio, increasing from ~ 30 solar units near the nucleus to ~ 45 at distances around $7'$ (= 32 kpc), at the edge of the HI ring (Zeilinger et al. 1990). It suggests the existence of a wide and dark halo, producing a total galaxy mass within $7'$ of $7.3 \times 10^{11} M_\odot$. The presence of such a dark halo, twisted with respect to the galaxy plane, offers a mechanism that could stabilize the warped disc, as discussed by Binney

(1978), and is in agreement with the changes of position angle observed from the central bulge to the outer regions (Zeilinger et al. 1990). It can also explain the presence, at a similar redshift, of a small galaxy, MCG-04-32-007, that appears to be perturbed by the presence of NGC 5084. If it is really close to it, this small object lies at a projected distance of $\sim 50\,\mathrm{kpc}$ in a direction along its minor axis.

The origin of the ring of NGC 5084 could be different from that of the polar rings observed in S0s. In fact, the huge quantity of matter that should be present in the halo of NGC 5084 is similar to that found in other supermassive disk galaxies (Saglia and Sancisi 1988) and can hardly derives from the accretion and disruption of a single dwarf system. In the hypothesis that this gas is coeval with the galaxy, NGC 5084 is just another case of warped galaxy, confirming the fact that polar rings are not observed in massive disk systems. If, on the contrary, the origin of the ring is similar, and derives from gas accretion at low impact angles, we cannot exclude that the massive structure of this and of the other systems has been built by means of sequential accretions of mass from the environment, self-induced by the progressive deepening of the galaxy potential well. Multiple gas accretion, happening on short time-scales in comparison to the life of the galaxy, could destroy the previous rings and allow the galaxy to grow. In this picture, the bigger disk galaxies could hardly present long-living polar-rings, which could persist only in smaller S0s, as observed.

Of course, we must be careful in interpreting all the similar features within a single scenario, and the inclusion of NGC 5084 among the gas accreting galaxies is only speculative, with the present data. New observations of these supermassive galaxies, as well as the discovery of more systems with inclined rings or gas counterrotation could give in the future new clues to complete this puzzle still in composition.

Acknowledgements

The author would like to thank L. Sparke, A. Bosma, K. Freeman and B. Whitmore for the interesting discussions on this subject made during the meeting.

References

Arp, H. (1966). *Atlas of Peculiar Galaxies.* Washington D.C.: Carnegie Institution.
Bettoni, D., Fasano, G. & Galletta, G. (1990). Astron. J. **99**, 1789.
Bettoni, D., Galletta, G. & Oosterloo, T. (1990). Mon. Not. R. Astron. Soc., in press.
Bertola, F. & Bettoni, D. (1988). Astrophys. J. **329**, 102.
Bertola, F., Buson, L. & Zeilinger, W.W. (1988). Nature **335**, 705.
Bertola, F. & Galletta, G. (1978). Astrophys. J. **226**, L115.
Bertola, F., Galletta, G. & Zeilinger, W.W. (1985). Astrophys. J. **292**, L51.
Bertola, F., Galletta, G., Kotanyi, C. & Zeilinger, W.W. (1988). Mon. Not. R. Astron. Soc. **234**, 733.
Binney, J. (1978). Mon. Not. R. Astron. Soc. **183**, 779.
Binney, J. (1981). Mon. Not. R. Astron. Soc. **196**, 455.
Chandrasekhar, S. (1943). Astrophys. J. **97**, 255.
Galletta, G. (1987). Astrophys. J. **318**, 531.
Galletta, G., Bettoni, D., Fasano, G., Oosterloo, T. (1990). In *Paired and Interacting Galaxies*, IAU Coll. 124 (eds. J.W. Sulentic & W.C. Keel). Dordrecht: Kluwer.
Gottesman, S.T. & Hawarden, T.G. (1986). Mon. Not. R. Astron. Soc. **219**, 759.
Hawarden, T.G., Elson, R.A.W., Longmore, A.J., Tritton, S.B. & Corwin, H.G., Jr. (1981). Mon. Not. R. Astron. Soc. **196**, 747.
Lake, G. & Norman, C. (1984). Astrophys. J. **270**, 51.
Merrit, D. & de Zeeuw, T. (1983). Astrophys. J. **267**, L19.
Richstone, D.O. & Potter, M.D. (1982). Nature **298**, 728.
Rubin, V.C. (1988). In *Large scale motions in the Universe: a Vatican study week* (eds. V.C. Rubin and C.V. Cohen), p. 541. Princeton: Princeton University Press.
Saglia, R.P. & Sancisi, R. (1988). Astron. Astrophys. **203**, 28.
Sancisi, R. (1976). Astron. Astrophys. **53**, 159.
Schweizer, F., Whitmore, B.C. & Rubin, V.C. (1983). Astron. J. **88**, 909.
Sparke, L.S. (1984). Mon. Not. R. Astron. Soc. **211**, 911.
Sparke, L.S. & Casertano, S. (1988). Mon. Not. R. Astron. Soc. **234**, 873.
Steiman-Cameron, T.Y. & Durisen, R.H. (1982). Astrophys. J. **263**, L51.
Steiman-Cameron, T.Y. & Durisen, R.H. (1988). Astrophys. J. **325**, 26.
Tohline, J.E., Simonson, G.F. & Caldwell, N. (1982). Astrophys. J. **252**, 92.
Tremaine, S. (1981). In In *The Structure and Evolution of Normal Galaxies* (eds. S.M. Fall and D. Lynden–Bell), p. 67. Cambridge: Cambridge University Press.
van Albada, T.S., Kotanyi, C.G. & Schwarzschild, M. (1982). Mon. Not. R. Astron. Soc. **198**, 303.
van der Hulst, J.M. & Sancisi, R. (1988). Astron. J. **95**, 1354.
Varnas, S.R., Bertola, F., Galletta, G., Freeman, K.C. & Carter, D. (1987). Astrophys. J. **313**, 69.
Whitmore, B.C., Lucas, R.A., McElroy, D.B., Steiman-Cameron, T.Y., Sackett, P.D., Olling, R.B. (1990). Astron. J. **100**, 1489.
Young, Y.S., and Knezek, P.M. (1990). Astrophys. J. **347**, L55.
Zeilinger, W.W., Bertola, F. & Galletta, G. (1989). In *Dusty Objects in the Universe*, in press.
Zeilinger, W.W., Galletta, G. & Madsen, C. (1990). Mon. Not. R. Astron. Soc., in press.

Warps and bulges

JO PITESKY
Department of Astronomy, UCLA

Abstract

A review of the observational literature shows that stellar disk warps occur more frequently in galaxies that have smaller nuclear bulges. We have modeled the effect of a freely-moving bulge potential on the shape of normal warping modes in self-gravitating disks, and have examined the effect of changing the bulge mass, scale length and rotation rate on these modes. In these modes, the bulge is tipped relative to the central disk. The warping mode remains discrete, but flattens out as the bulge rotation rate increases. The warp amplitude is also dependent on the length of the bulge effective radius.

Introduction

Although the most dramatic observations of warped galaxies are seen in extended HI disks, the optical disks of some galaxies also show warping. Published observations of stellar warps suggest that most optically detected warps occur in late-type galaxies (Figure 1). Preliminary statistical studies of the frequency of optical warps in spiral galaxies, though limited in extent (Sánchez-Saavedra et al. 1990), confirm this trend. Without differentiating between barred and unbarred systems, 13 of the 86 galaxies in the Sánchez-Saavedra et al. sample are early-type disks (S, Sa, or Sab). Considering only those galaxies which have a clear optical warp on either a red or blue plate, only 8%, or one of the early-type galaxies shows a warp. By comparison, 27%, or 19 of the 70 late-type (Sb through Sd) spirals are clearly warped.

It is plausible that warping in a disk galaxy is affected by the presence and size of the nuclear bulge, as the bulge mass in early-type

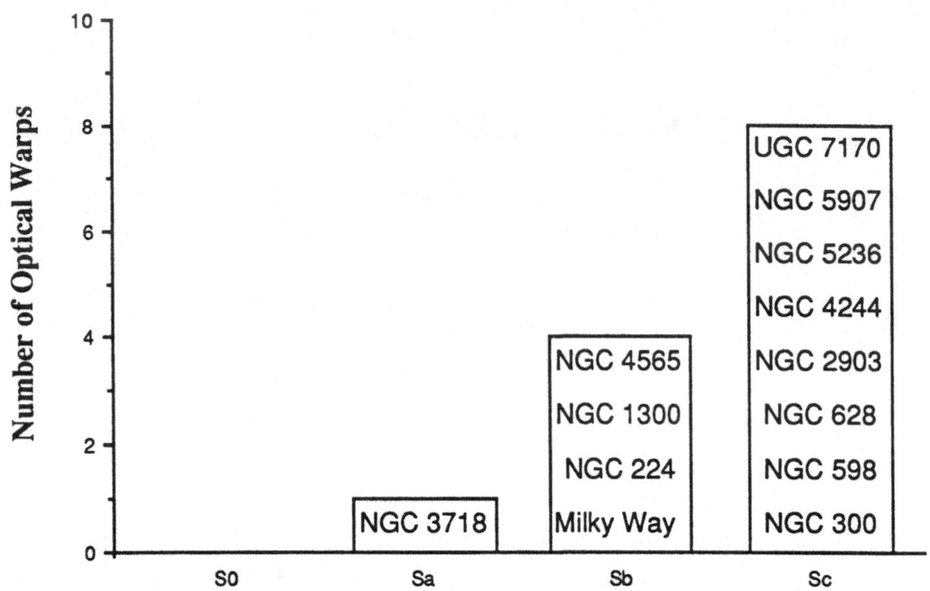

Fig. 1. Number of optically warped galaxies discussed in the literature as a function of Hubble type.

systems can be comparable to the disk mass. The bulge itself may tip relative to the inner disk; some bulges appear to be misaligned with their companion disks, most obviously, in M31. Examining the motion of a freely moving bulge in a realistic galactic potential may therefore give insight into two interesting phenomena.

Warp models including a bulge

Any galaxy may be forcibly bent, but the distortion will not keep its shape unless the bending is supported as a normal mode. In most cases, *isolated* disks will have no discrete bending modes unless the disk's outer edge is sharp (Hunter and Toomre, 1969). Sparke and Casertano (1988; hereafter SC) have shown that a finite disk surrounded by an oblate massive halo may support a time-independent discrete bending mode. Thus a warp, once excited, may persist for many rotation periods. Though the disk's self-gravity keeps the warp from winding up, it is primarily the external mass distribution of the halo that determines the shape of the warp. Here, we discuss a modified form of the SC mode calculations, with the addition of a bulge component.

The galaxy model consists of three separate components: *i)* a thin disk consisting of concentric massive rings which are free to tilt, *ii)* a bulge of fixed oblateness which is also free to tilt, and *iii)* an oblate massive halo of fixed shape and orientation. By treating the disk as a collection of massive, rigid rings, the equations of motion for each ring and the bulge can be simultaneously solved to find a normal mode where both the warped disk and the bulge precesses rigidly about the halo symmetry axis.

The disk is taken to be a standard exponential which is smoothly truncated over a small region near the disk edge. A massive oblate halo stratified on ellipsoids of constant density surrounds the disk. The halo density follows a "pseudo-isothermal" form, with an inner core of nearly constant density, but a potential corresponding to a flat rotation curve in the outer parts. The bulge is given a density distribution (Young, 1976) approximately corresponding to the de Vaucouleurs surface density distribution. The bulge is also oblate; its shape is assumed to be fixed but it is allowed to tilt about its center.

The equations of motion are as follows: a thin, self-gravitating disk of surface density $\sigma(r)$ in cylindrical polar coordinates (r, ϕ, z) is taken to be rotating about the z-axis with rotation speed $\Omega(r)$. Initially, the disk lies in the plane $z = 0$, which is also the halo symmetry plane, but it is then deformed by a small vertical displacement to a height $Z(r, \phi, t)$. Vertical motion of the disk follows the same equation of motion as in SC:

$$\frac{D^2 Z}{Dt^2} = \left(\frac{\partial}{\partial t} + \Omega(r)\frac{\partial}{\partial \phi}\right)^2 Z = F_{ext} + F_{self},$$

where F_{self} is the self-gravity of the bent disk as described in SC. Now F_{ext} includes the external force from the bulge, as well as that of the halo:

$$F_{ext} = F_{halo} + F_{bulge} \approx -\mu_H^2(r)Z - \mu_B^2(r)(Z - Z_B),$$

where Z_B is the height at r, ϕ of the bulge mid-plane above the plane $z = 0$, and $\mu_B(r)$ and $\mu_H(r)$ are the vertical frequency due to the bulge and halo potential, respectively, evaluated in the bulge or halo midplane. "Integral-sign" warps have $m = 1$, so the height Z of any rigid ring precessing rigidly about the z-axis at an angular rate ω can be written as $Z(r, \phi, t) = \mathrm{Re}[h(r)e^{i(\omega t - \phi)}]$. When the bulge precesses along with the disk, the height of its equator above $z = 0$ can be written similarly, as $Z_B(r, \phi, t) = \mathrm{Re}[r\,\alpha(t)e^{-i\phi}] = \mathrm{Re}[h_B(r)e^{i(\omega t - \phi)}]$ where $|\alpha|$ is the "tip" angle of the bulge with respect to the halo symmetry plane. The modified disk equation of motion is then

$$\Big([\omega-\Omega(r)]^2-\mu_H^2(r)-\mu_B^2(r)\Big)h(r)+\mu_B^2(r)h_B(r)$$
$$=-G\int_0^\infty \sigma(r')[H(r,r')-K(r,r')]h(r')r'dr'$$
$$+G\,h(r)\int_0^\infty \sigma(r')H(r,r')r'dr'$$

where H and K have the same definitions as in SC (eq. 5).

The bulge is treated as a rigid, oblate body which is spinning about its symmetry-axis. From Goldstein (1980, § 5.5), the equation of motion in an inertial frame for a rotating body with an applied torque $\mathbf{N}$ is

$$\frac{d\mathbf{L_B}}{dt}=\mathbf{N}.$$

The total angular momentum $\mathbf{L_B}$ of the bulge can be expressed as the sum of $\mathbf{L_{rot}}=I_3\mathbf{\Omega_{SB}}$, the angular momentum due to the bulge rotating at speed $\mathbf{\Omega_{SB}}$ about its instantaneous short axis, and $\mathbf{L_{tip}}=I_1\mathbf{\Omega_{tip}}$, the angular momentum from the bulge tipping, or changing its orientation with respect to the halo midplane. Here I_1 and I_3 are the moments of inertia of the bulge about its long and short axes, respectively. In cartesian coordinates (x,y,z), the bulge short axis points toward $\mathbf{n}=(-\mathrm{Re}[\alpha(t)],\mathrm{Im}[\alpha(t)],1+0(\alpha^2))$, which is also the orientation of $\mathbf{L_{rot}}$. As the bulge tips, $\mathbf{n}$ will change, so that $\mathbf{\Omega_{tip}}$ is given by

$$\frac{d\mathbf{n}}{dt}=\left(-\mathrm{Re}\frac{d\alpha}{dt},-\mathrm{Im}\frac{d\alpha}{dt},0(\alpha^2)\right)=\mathbf{\Omega_{tip}}\times\mathbf{n}$$

Hence, $\mathbf{\Omega_{tip}}=(\mathrm{Im}\frac{d\alpha(t)}{dt},-\mathrm{Re}\frac{d\alpha(t)}{dt},0(\alpha^2))$, which is directed along the line of nodes where the bulge equator crosses $z=0$. Up to first order in tip angle α, the bulge will have total angular momentum

$$\mathbf{L_B}=\mathbf{L_{tip}}+\mathbf{L_{rot}}=\left(I_1\mathrm{Im}\frac{d\alpha}{dt}-L_{rot}\mathrm{Re}\,\alpha,\ -I_1\mathrm{Re}\frac{d\alpha}{dt}-L_{rot}\mathrm{Im}\,\alpha,\ L_{rot}\right),$$

and

$$\frac{d\mathbf{L_B}}{dt}=\left(I_1\mathrm{Im}\frac{d^2\alpha}{dt^2}-L_{rot}\mathrm{Re}\frac{d\alpha}{dt},\ -I_1\mathrm{Re}\frac{d^2\alpha}{dt^2}-L_{rot}\mathrm{Im}\frac{d\alpha}{dt},\ 0\right)=\mathbf{N_D}+\mathbf{N_H}$$

where $\mathbf{N}=\mathbf{N_D}+\mathbf{N_H}$, the sum of the torques due to the warped disk and the halo.

The torque due to the disk is most easily found as the negative of the torque of the bulge on the warped disk. The ith ring in the disk feels a torque from the bulge of

$$-\frac{m_i r_i}{2}[\mu_B^2(r_i) - \Omega_B^2(r_i)]$$

$$\times\Big(\mathrm{Im}\ e^{i\omega t}[h(r_i)-h_B(r_i)],\ -\mathrm{Re}\ e^{i\omega t}[h(r_i)-h_B(r_i)],\ 0[h_b(r_i)h(r_i)/r_i]\Big),$$

where $\Omega_B(r_i)$ is the circular rotation speed due to the bulge potential, alone. The disk torque on the bulge $\mathbf{N_D}$ is the negative of the sum of all such terms:

$$\mathbf{N_D} = \sum_i \frac{m_i r_i}{2}[\mu_B^2(r_i) - \Omega_B^2(r_i)]$$

$$\times\Big(\mathrm{Im}\ e^{i\omega t}[h(r_i)-h_B(r_i)],\ -\mathrm{Re}\ e^{i\omega t}[h(r_i)-h_B(r_i)],\ 0[h_b(r_i)h(r_i)/r_i]\Big).$$

This leaves only the calculation of $\mathbf{N_H}$, the halo torque on the bulge. The bulge and the halo are oblate spheroids with density stratified on surfaces of constant ellipticity. For the halo, $\rho = \rho(m^2)$, $m^2 = x^2 + y^2 + z^2(a/b)^2$, and $\rho(m^2) = \rho_0(1 + m^2/a^2)^{-1}$. Note that a, the halo semi-major axis, is set equal to the halo core radius r_c, and has units of length. Similarly, the bulge density $\rho_B = \rho(m_B^2)$, $m_B^2 = x'^2 + y'^2 + z'^2(a/b)_B^2$, where (x', y', z') are cartesian coordinates with z' aligned along the bulge pole $\mathbf{n}$. The density ρ_B has a more complex form corresponding to the de Vaucouleurs surface density law. In terms of the bulge mass M_B and the bulge effective radius r_e, $\rho(m_B^2) = \gamma M_B/r_e^3(r_e/m_B)^{7/8}\exp[-\beta(m_B/r_e)^{1/4}]$, where γ and β are numerical constants.

At a point (x, y, z), the halo exerts a force with radial and vertical components (Binney and Tremaine, §2.3, 1988)

$$F_r(x,y,z) = -2\pi G(b/a)a^3\sqrt{x^2+y^2}\int_0^\infty \frac{\rho(m^2)d\tau}{(\tau+a^2)^2\sqrt{\tau+b^2}}$$

$$\equiv -K\sqrt{x^2+y^2}F_1(x,y,z),$$

$$F_z(x,y,z) = -2\pi G(b/a)a^3 z\int_0^\infty \frac{\rho(m^2)d\tau}{(\tau+a^2)\sqrt{(\tau+b^2)^3}}$$

$$\equiv -KzF_2(x,y,z),$$

where K is a constant and τ is given implicitly by

$$\frac{m^2}{a^2} = \frac{x^2+y^2}{\tau+a^2} + \frac{z^2}{\tau+b^2}.$$

Instead of calculating the torque due to this force for an arbitrary line of nodes, the y and y' axes are taken to lie along the line of nodes where the bulge symmetry plane crosses $z = 0$; then the total torque lies entirely along the y-axis. The halo then exerts a torque per unit mass at a point (x, y, z) of $Kzx[F_1(x,y,z) - F_2(x,y,z)]$.

The halo torque on the bulge is found by dividing the bulge into oblate shells of mass $4\pi m_B^2 (b/a)_B \rho(m_B) dm_B$, calculating the torque on each shell, and then summing over the entire bulge. If the bulge tips through small angles, the halo torque on the bulge N_H may be approximated as

$$N_H \approx (N_H)_{a=0} + \alpha\left(\frac{\partial N_H}{\partial \alpha}\right)_{a=0} + 0(\alpha^2) = f_{bh}\alpha + 0(\alpha^2).$$

The bulge equation of motion may then be written as

$$\left\{ I_1\omega^2 - L_{rot}\omega - f_{bh} + \sum_i \frac{m_i r_i^2}{2}\left(\mu_B^2(r_i) - \Omega_B^2(r_i)\right)\right\}\frac{h_B(r)}{r} + \sum_i \frac{m_i r_i}{2}\left(\mu_B^2(r_i) - \Omega_B^2(r_i)\right)h(r_i) = 0$$

The bulge and ring equations of motion can now be solved as an eigenvalue problem. In all of the cases explored below, α and $h(r)$ are always found to be real, so there are no growing or decaying modes.

Preliminary Results

The primary interest of this study is in examining how varying the bulge parameters r_e, M_B, and Ω_{SB} affects the bulge orientation and warped disk shape. Hence, the disk parameters (the outer radius and truncation length) are kept fixed, and only two sets of halo parameters are used. The disk scale length provides the unit of length, and its mass is the mass unit; timescales are measured such that $G = 1$. Specifically, the disk has a radius of 5 scale lengths, with the density being smoothly truncated to zero over the last 0.5 scale lengths. A massive halo with ellipticity $(b/a) = 0.8$ surrounds the disk. The halo is characterized by its core radius r_c, which is set equal to the semi-major axis a, and by v_∞, the circular velocity at large radii in the "flat" portion of the rotation curve. A flattened bulge with $(b/a)_B = 0.3$ is used. Since the angular momentum of the bulge is determined by its internal structure, this is parametrized by taking $\Omega_{SB} = f\Omega_{tot}(r_B)$, where $\Omega_{tot}(r_B)$ is the rotation speed at the radius r_B due to the bulge, halo and disk contributions, r_B is defined as $\frac{1}{2}M_B r_B^2 = I_1$, the bulge moment of inertia about the long axis, and f is a number between 0 and 1. For these preliminary results, f_{bh}

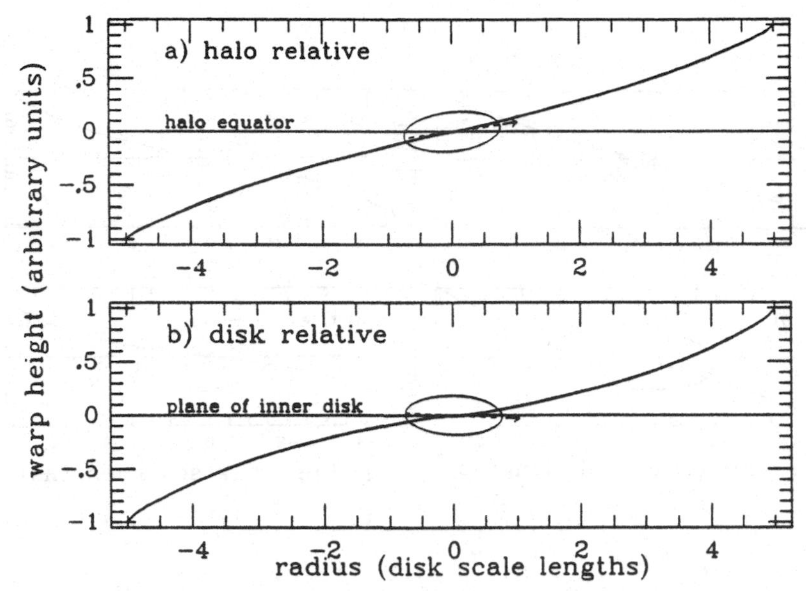

Fig. 2. Normal warping mode with moving bulge component, and $r_c = 2.0$, $v_\infty = 0.71$. The bulge semi-major axis length is equal to the bulge scale length r_e. Dashed arrows indicate the bulge equator.

is calculated from the halo torque on a ring with the same moment of inertia as the bulge (i.e., $f_{bh} = M_B r_B^2 [\mu_B^2(r_B) - \Omega_B^2(r_B)]$), and was used instead of the full integration over the oblate bulge.

Figure 2 shows the normal mode shape of the disk with halo parameters $r_c = 2.0$, $v_\infty = 0.71$, and a moving bulge component. The rotation curve due to the bulge, disk and halo has $v(r) \propto 1/r$ at large radii. Using Milky Way parameters as a starting point, the bulge has $r_e = 0.76$ disk scale lengths and $M_B = 0.45$ disk masses. A solid horizontal line indicates the $z = 0$ halo or inner disk midplane in each figure. The bulge is represented by an ellipse of the appropriate flattening, with major axis length r_e. The bulge solid body rotation rate $\Omega_{SB} = 0.1\Omega(r_B)$.

The most noticeable feature of the mode is the position of the bulge, which tips below the central disk plane — as does the bulge of M31. We can understand this behavior physically by considering what the motion of a ring making up part of the disk would be if it was forced to rotate more slowly than the local circular speed. If that ring (now rotating at speed Ω_{ring}) feels no change in torque, its precession rate $\Omega_{prec}(r) \approx (\Omega_H^2 - \mu_H^2)/2\Omega_{ring}$ will increase in magnitude. As the halo is oblate, the ring will tend to regress faster than the rest of the warp pattern. If the ring tips so that it is less inclined

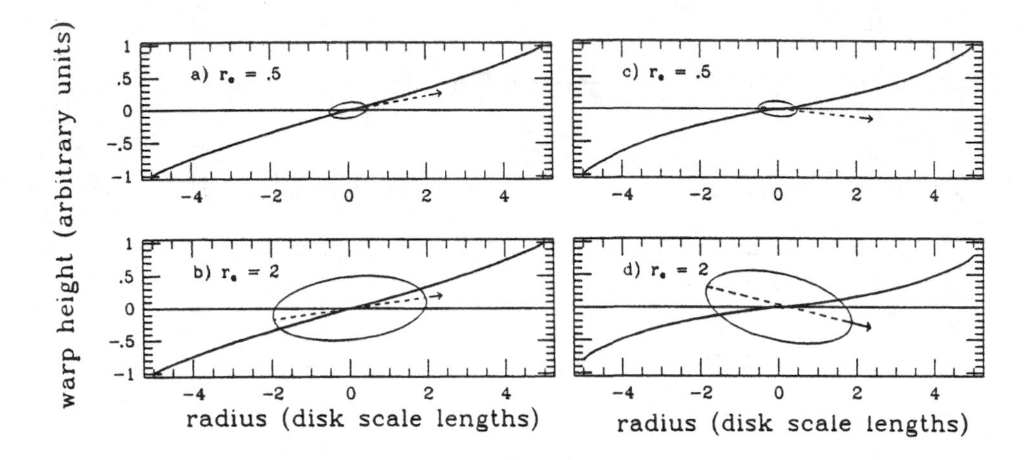

Fig. 3. Normal warping mode when varying the bulge effective radius r_e for halo parameters $r_c = 2.0$, $v_\infty = 0.71$. The left and right handed panels show the disk and halo-relative views, respectively. $M_B = 0.2$ disk masses. Panels a) and c) show a bulge with $r_e = 0.5$; panels b) and d) have $r_e = 2.0$. Dashed arrows trace the bulge equator. $\Omega_{SB} = 0.1\Omega(r_B)$.

than the disk, the disk self-gravity will partially counteract the halo oblateness and slow the regression. Thus, a slowly rotating ring or bulge should tend to lie between the angle of the inner disk and the halo equator.

The amount of tipping also depends on the size of the effective radius of the surrounding halo. Figure 3 shows the effect of varying r_e while keeping the bulge mass constant (here, $M_B = 0.2$ disk masses) in the same halo potential as for Figure 2. The bulge solid body rotation rate Ω_{SB} is kept at $0.1\Omega(r_B)$. Both the halo (left hand side) and inner disk (right hand side) relative views are shown for each system. When r_e is increased from 0.5 to 2.0, the bulge tip grows. Increasing the halo core radius (Fig. 4; $r_c = 6.0$, $v_\infty = 1.05$) changes the amount of bulge tip (as can be seen most clearly in the halo-relative views), but shows the same trend of greater tip for larger r_e. The amount of bulge tip, therefore, is related not only to how fast the bulge is spinning, but also to the surrounding halo potential and to the size of the bulge.

SC have already shown that the magnitude and shape of a disk warp depend on the mass and core radius of the halo. As Figure 5

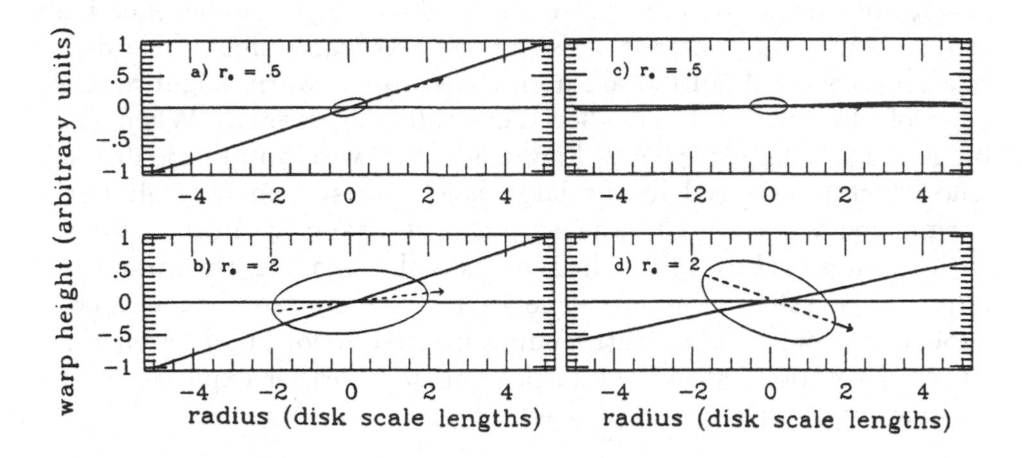

Fig. 4. The bulge-disk system as in Fig. 3, but with halo parameters $r_c = 6.0$, $v_\infty = 1.05$.

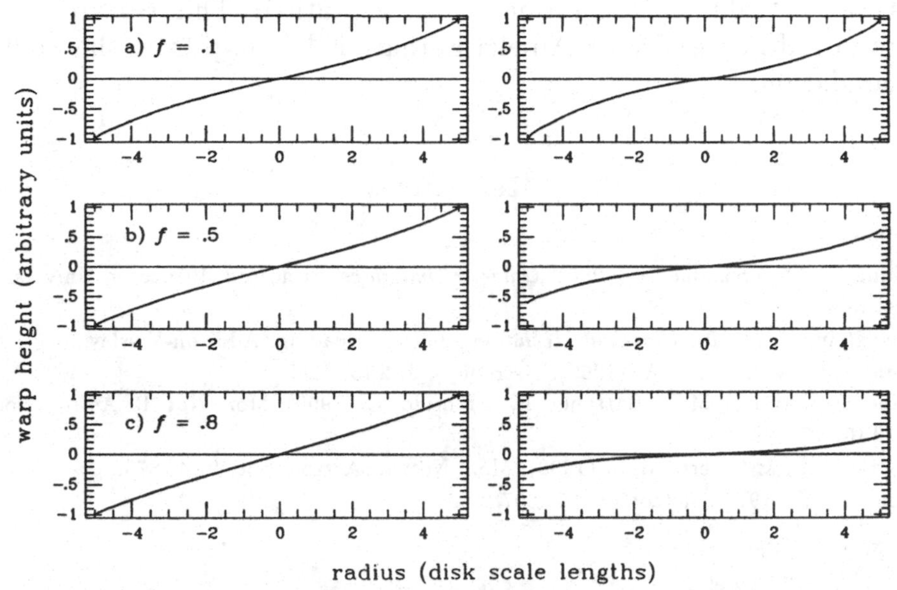

Fig. 5. As Fig. 2, but Ω_{SB} is varied as $\Omega_{SB} = f\Omega(r_B)$ (see text). Halo-relative plots are shown on left, and disk-relative plots on the right.

shows, the warp is affected by the speed at which the bulge rotates (the bulge component is not shown, but has $M_B = 0.45$, $r_e = 0.76$, with halo parameters $r_c = 2.0$, $v_\infty = 0.71$). The disk-relative warp heights are normalized to the warp height at the edge when $f = 0.1$. As Ω_{SB} increases, the warp flattens out. Changing the bulge effective radius (with Ω_{SB} constant) also has a significant effect. The disk relative-views in Figure 3 are normalized to the warp height at the edge of the disk in Figure 3c (the small bulge case). When the bulge is made larger ($r_e = 2$, Figure 3d), the warp amplitude drops. The effect is reversed in the large halo core system (Fig. 4, also normalized as was Fig. 3): the warp almost disappears in the system with smaller r_e (Fig. 4c), while there is still a significant warp in the large bulge system (Fig. 4d). The warp shape and amplitude, then, depend not only on the surrounding massive halo, but also on the size and rotation rate of the nuclear bulge. Further exploration of this parameter space is in progress.

Acknowledgements

I would like to thank Linda Sparke for her advice and encouragement, as well as valuable suggestions on presentation. This research was supported in part by an Amelia Earhart Fellowship from the Zonta Foundation.

References

Binney, J., & Tremaine, S. (1988). *Galactic Dynamics*. Princeton: Princeton University Press.

Goldstein, H. (1980). *Classical Mechanics*, 2nd ed. Reading: Addison-Wesley.

Hunter, C., & Toomre, A. (1969). Astrophys. J. **263**, 116.

Sánchez-Saavedra, M.L., Battaner, E. & Florido, E. (1990). Mon. Not. R. Astron. Soc. **246**, 458.

Sparke, L.S. & Casertano, S. (1988). Mon. Not. R. Astron. Soc. **234**, 873.

Young, P.J. (1976). Astron. J. **81**, 807.

Time evolution of galactic warps

PETER HOFNER & LINDA S. SPARKE
Washburn Observatory, University of Wisconsin-Madison

Abstract

We have modeled the time evolution of a galactic disk initially tilted with respect to an oblate dark halo. The disk is modeled as a set of concentric rings free to tilt about the center, embedded in a fixed halo potential. When a disk that has a discrete mode of bending is initially inclined with a constant angle to the symmetry plane of an oblate dark halo, we find that the inner part of the galactic disk settles towards the mode shape. Even if the disk as a whole has no discrete bending mode, the inner part still settles into a coherent warp with no sense of spirality. We ran models with parameters corresponding to the observed systems NGC 4013, NGC 4565 and NGC 2903. Comparing our models with the observed shape of the warp allows us to place constraints on the core radius and to find a lower limit for the oblateness of the dark halo.

I. Introduction

Many spiral galaxies show considerable deviation from the plane of symmetry in the outer region of the galactic disk. This is most frequently seen in the HI disk, but sometimes also in the visible stellar disk. When a warped galaxy is observed edge-on, the warp shows a characteristic integral sign shape. If a galaxy is seen more face-on, warps can be recognized by patterns in the velocity field; in this case the sense of spirality of the line of nodes can also be determined. Briggs (1990) reviewed the properties of warp structure for a sample of 12 intermediately inclined galaxies, for which high resolution HI 21 cm data are available. He found that the warp usually sets in at about R_{25} and that, if viewed from the inner plane of the galaxy, the warp has a straight line of nodes between R_{25} and the Holmberg radius. Further out, the line of nodes twists in a leading sense.

Galactic warps have been modeled as normal bending modes by Sparke and Casertano (1988). The dynamical system which they considered consists of a disk modeled as a set of concentric circular rings, which are free to tilt about the center. The disk is embedded in a dark, oblate halo whose density contours are flattened with constant ellipticity ϵ. Inside the core radius of the halo r_c, the density is nearly constant, while further out the rotation curve due to the halo potential becomes flat. The forces acting on each ring are the gravitational force of the dark halo and the self-gravity of the galactic disk. Dissipation and radial mass flow are ignored in this model. The calculations are linear in the tip angle, so that amplitudes are arbitrary.

The work of Sparke and Casertano showed that discrete normal modes of bending can exist in realistic galactic potentials and that the observed warps in the edge-on galaxies NGC 4565 and NGC 4013 can be reproduced. The shape of the warp depends mainly on the core radius of the dark halo. A warp corresponding to a normal mode is neutrally stable (neither growing nor decaying) and hence must have a straight line of nodes. Briggs's results indicate therefore that only the inner portion of the disk can be modeled as a time-independent bending mode. Sparke and Casertano did not consider how a warp would evolve in time after being excited and how it might appear while settling to the shape of the normal mode. We address this question here, using the same dynamical system as Sparke and Casertano, but considering its time-evolution.

II. Settling to the normal mode

We first consider two models close to those which Sparke and Casertano found to have normal modes resembling the observed warps in the two galaxies NGC 4013 and NGC 4565. For the density distribution in the disk we take an exponential law with a scale length of 2.3 kpc for NGC 4013 and 5.5 kpc for NGC 4565 which is truncated smoothly at the edge. The halo parameters are chosen to give a good fit to the rotation curve. As initial condition we choose a uniform tilt of the galactic disk relative to the symmetry plane of the halo.

Figure 1 shows the time evolution of this initially inclined disk for a model of NGC 4013. The model has the parameters $r_c = 16.1\,\mathrm{kpc}$,

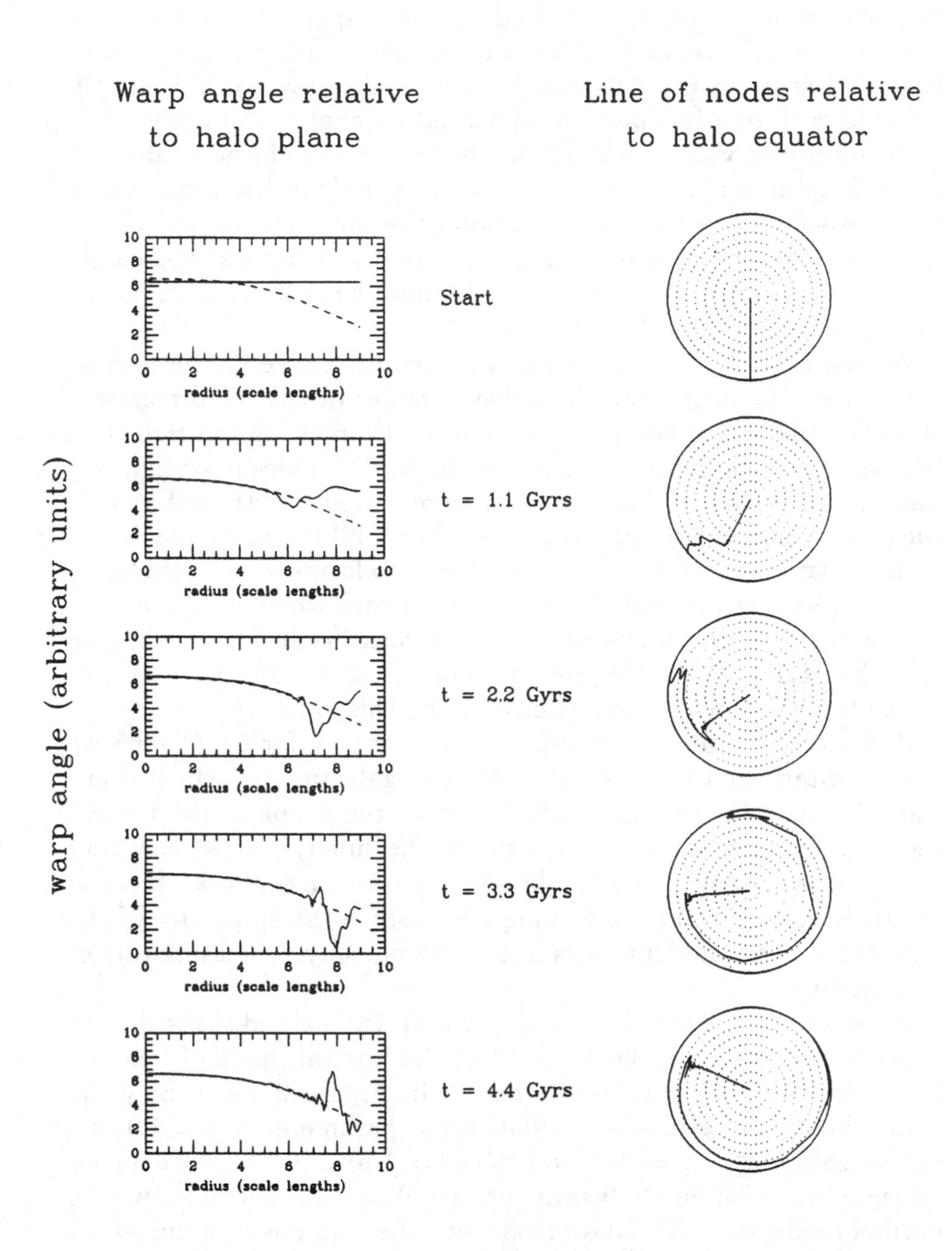

Fig. 1. Time evolution of a model for NGC 4013.

$v_\infty = 263\,\mathrm{km\,s^{-1}}$, $\epsilon = 0.2$ and $M_{\mathrm{disk}} = 4.7 \times 10^{10} M_\odot$. The left panel displays the total tip angle of the disk with respect to the symmetry plane of the halo versus the radius of the disk in scale lengths. Note that this is not a cross-section cut through the warp, but the tip angle of each ring independent of azimuthal angle. The dashed line is the discrete normal mode for this system. The right panel shows a polar diagram for the line of the ascending node of the warped disk with respect to the symmetry plane of the halo; the dotted circles represent radii in units of the disk scale length. The sense of galactic rotation is counter-clockwise. The time interval between consecutive frames corresponds to 1.1×10^9 years.

We see that the disk settles to the smooth shape of the normal mode from the inside out. The line of nodes of the warp regresses slowly in the oblate potential of the halo. Comparing the right and left panel of each frame we find that the line of nodes is straight for radii where the disk has settled to its normal mode. In the outer parts there is initially a trailing spiral, but after 3 billion years the line of nodes is twisted in a leading sense (last two frames). The transient trailing phase is because of the large halo core which causes a non-monotonic rate of precession; see the discussion in § 5.3 of Sparke and Casertano. A trailing line of nodes is also seen in IC 342, in which the halo core is large (Bosma 1991, this volume).

Performing a similar experiment with a model for NGC 4565 we find qualitatively the same behavior: The galactic disk which is initially inclined at constant angle settles to the shape of the normal mode, the line of nodes is straight in the inner, settled part and twists to a leading spiral in the outer part. In this case the line of nodes never develops a trailing sense, since the small size of the halo core in this model results in a monotonically decreasing rate of precession.

When the disk is extended to larger radii (we extended the disk to avoid reflections from the outer edge) the normal mode of bending found for NGC 4565 ceases to exist. In Figure 2 we display the time evolution of a model for NGC 4565 (parameters: $r_c = 4.0\,\mathrm{kpc}$, $v_\infty = 180\,\mathrm{km\,s^{-1}}$, $\epsilon = 0.2$ and $M_{\mathrm{disk}} = 1.5 \times 10^{11} M_\odot$) where we extended the disk out to 9 scale lengths. The dashed line shows the normal mode found for NGC 4565 when the disk has a radius of 5.5 scale lengths only. Although the disk as a whole has no normal mode of bending, the line of nodes is again straight in an inner, settled part and twisted in a leading sense in the outer parts.

The settling process occurs through dispersive bending waves which

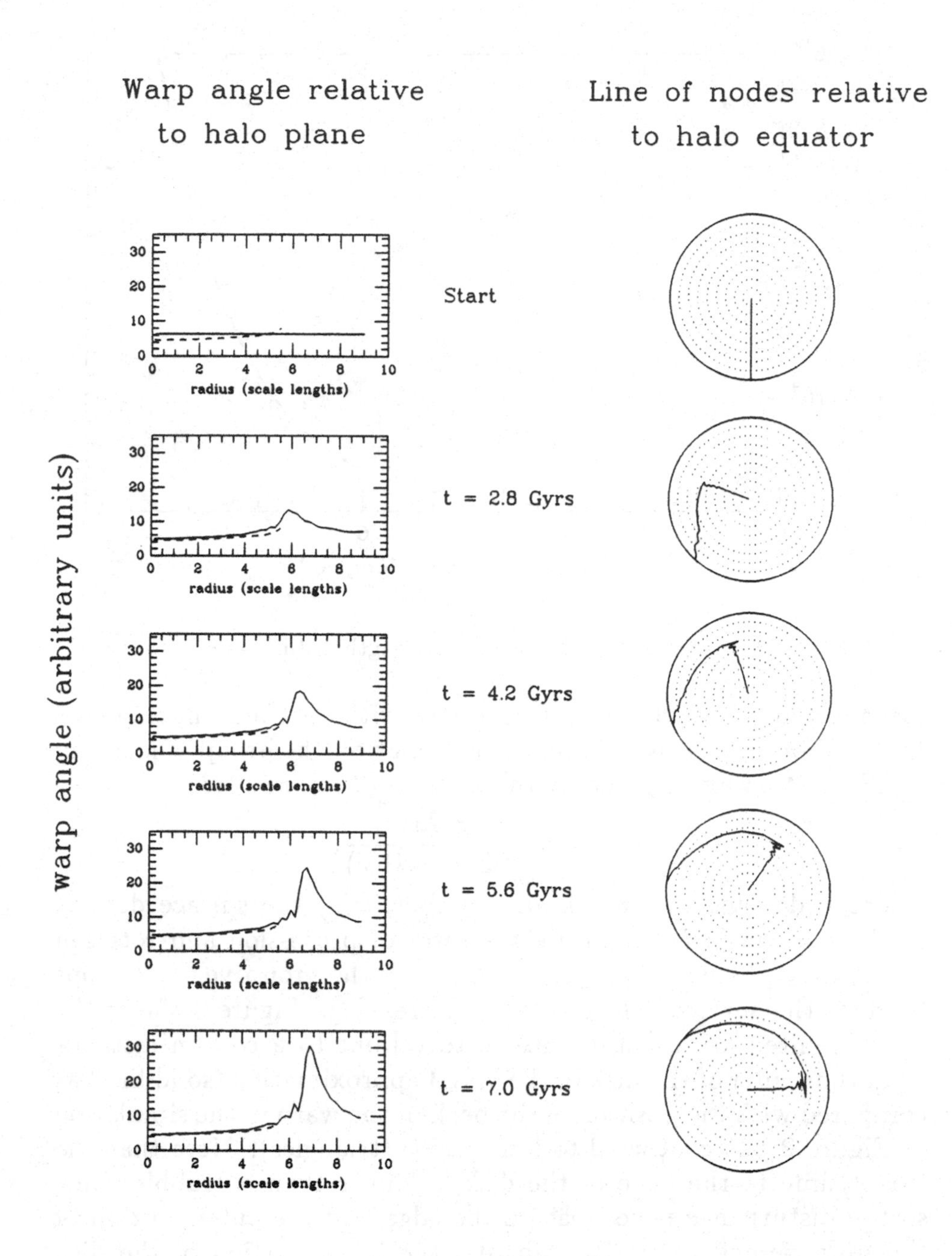

Fig. 2. Time evolution of a model for NGC 4565.

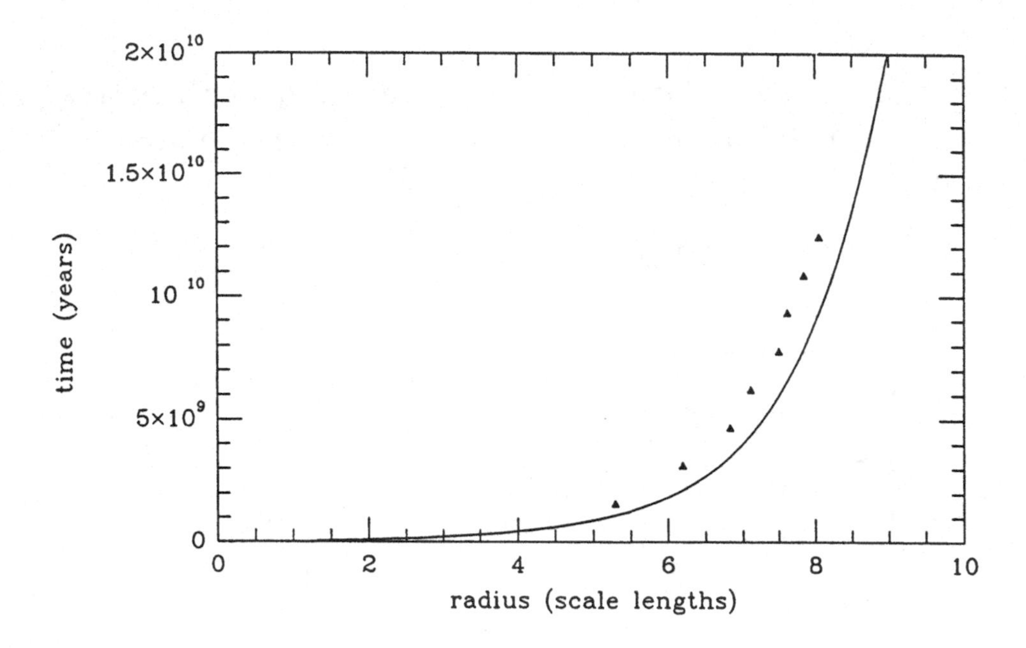

Fig. 3. Propagation time for short wavelength disturbance

transport outwards the energy associated with the transient response. In the short-wavelength WKB approximation the group velocity c_g of a bending wave is given by the formula (Toomre 1983)

$$c_g = \mathrm{sgn}(k)\frac{\pi G\sigma(r)}{\omega(r) - \Omega(r)},$$

where k denotes the radial wavenumber, $\sigma(r)$ the surface density in the disk and $\omega(r), \Omega(r)$ are the rates of precession and rotation of the disk. As waves propagate outward, the group velocity drops because the surface density $\sigma(r)$ is decreasing. Figure 3 shows the time required for a disturbance to travel out to a certain distance from the center of the disk in the WKB approximation (solid line) as compared with the position of the peak of the warp in the simulation of Figure 2 but continued to later times (triangles). Note that the travel time to the edge of the disk is longer than a Hubble time, so the disturbance never reaches the edge and the outer portion of the disk cannot be settled, whereas the inner portion of the disk cannot "know" the extent of the disk. This suggests that *even if no normal mode of bending exists*, the inner portion of the disk can still show a coherent behavior. Furthermore, since the group velocity depends mainly on the galactic rotation speed and the surface density in the disk, both of which are quantities that can be inferred from observations, it is possible to date the warp by observing the settled portion of the warped disk. Typical time scales are a few billion years.

III. Models for NGC 2903

For the warped galaxy NGC 2903 surface densities derived from HI 21cm data and F-band surface photometry were kindly made available to us by Kor Begeman. Assuming a constant value for the mass-to-light ratio M/L in the disk, the photometry data can be combined with the HI measurements to derive a total surface density profile for the disk. This consists of a visible component, which here follows an exponential law with a scale length of 2.0 kpc, and a gasous component as given from the HI data. The disk is truncated smoothly over the last scale length, to avoid the reflection of wave energy from the outer boundary. For given halo parameters core radius and ellipticity we fix the mass-to-light ratio M/L by fitting the rotation curve by eye. We get good fits for core radii in the range 1.2 to 4 disk scale lengths, the same range as found by Begeman (1987) who fitted the rotation curve using a spherical halo.

Again starting with the disk initially tilted at a constant angle, we followed the time evolution for models with different sizes of the halo core. We find that a sizeable warp only appears for the smaller core radii. Figures 4a and 4b show the time evolution of two models for NGC 2903 with small (a) and large (b) core radius. Model (a) has the parameters $r_c = 4\,\mathrm{kpc}$, $M/L = 2.6\,M_\odot/L_{F\odot}$ and $v_\infty = 177\,\mathrm{km\,s^{-1}}$, whereas model b has the parameters $r_c = 8\,\mathrm{kpc}$, $M/L = 3.1\,M_\odot/L_{F\odot}$ and $v_\infty = 200\,\mathrm{km\,s^{-1}}$. In both cases the flattening of the halo was given by $\epsilon = 0.2$. After 9 billion years the warp in model (b) (larger core radius) has a straight line of nodes out to 8 scale lengths and has settled into a shape which is close to a constant tilt and which does not match the observed tip angle. In model (a) (smaller core radius) the line of nodes is straight within a radius of 8 scale lengths and twisted to a leading spiral further out. Here, the warp settles into a shape that matches the observed warp reasonably well. The observed tip angle is reproduced even better if we start the simulation with a larger tilt of the disk at the outer edge. Since we do not know the real initial conditions it is not worthwhile to try to match the observation

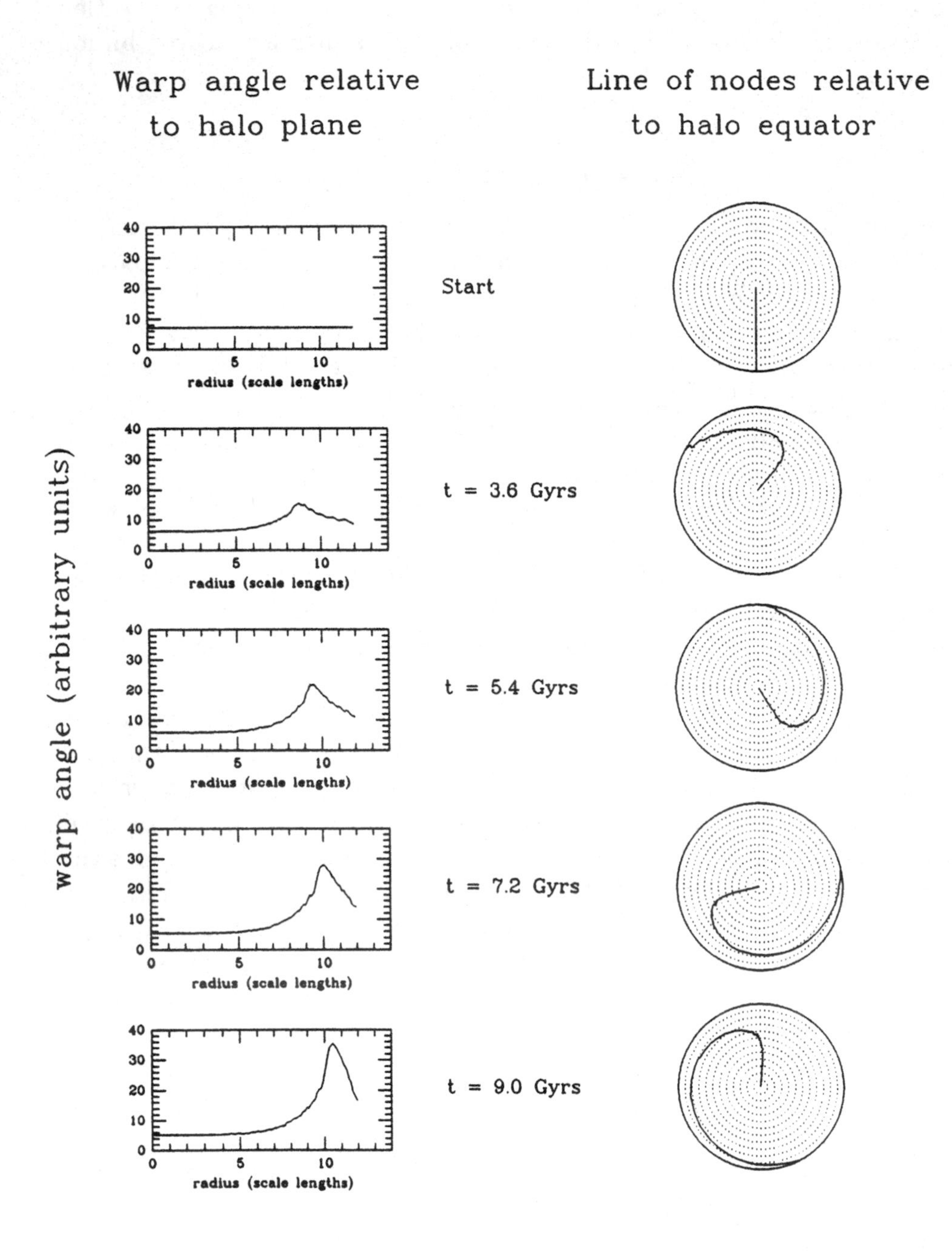

Fig. 4a. Time evolution for model (a) of NGC 2903: $r_c = 2$ scale lengths.

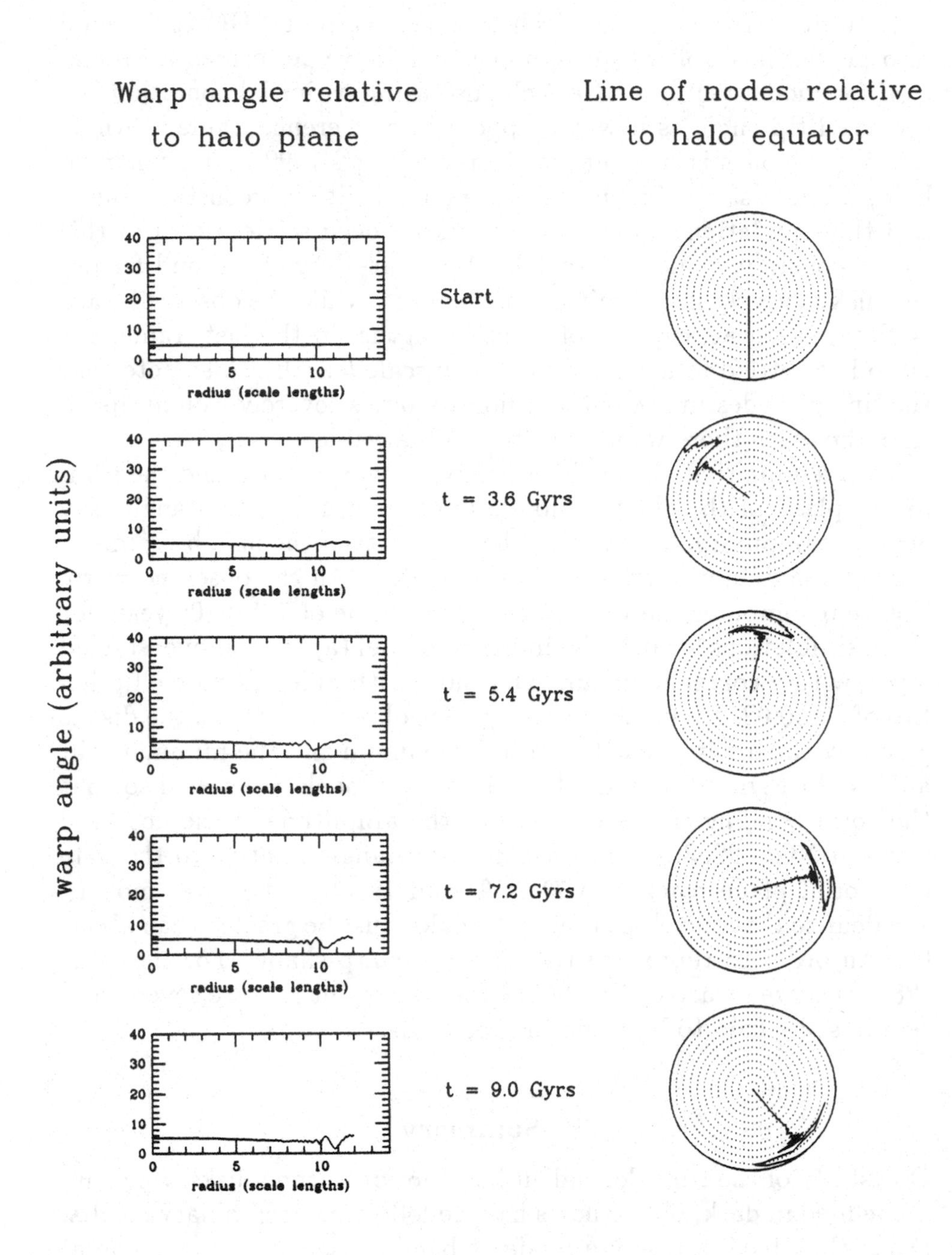

Fig. 4b. Time evolution for model (b) of NGC 2903: $r_c = 4$ scale lengths.

in detail, but the simulations show that we can put limits on the size of the halo core.

To address the question of whether the warp of NGC 2903 could also persist in a spherically symmetric halo we performed a simulation for model (a) in a spherical halo, using as initial condition the observed tip angle as it would appear in a reference frame in which the line of nodes is straight (as given by Briggs 1990). In a spherical halo, the lowest bending mode is the trivial tilt with constant angle, and the simulation shows that the warp indeed settles down to this tilt of constant angle. After 3.6×10^9 years (Fig. 5, second frame) the disk is nearly flat within 7 scale lengths, while the observed warp is clearly seen at a radius of 5 scale lengths; in the last frame, the warp has settled to a rigid tilt out to 9 scale lengths. Also note that the line of nodes in the outer region becomes severely wound up, so that the warp there would hardly be observable.

If we require that the angle between the warped disk and the symmetry plane of the oblate halo does not exceed a certain angle, say 30°, we can put a lower limit on the ellipticity of the halo by comparing the maximum warp angle in the model with the observed warp. Figure 6 compares the warp angle after a time of 7.2×10^9 years for three simulations for models similar to model (a) from above, started again with a constant tip angle but now with different halo ellipticities of $\epsilon = 0.02, 0.1, 0.2$ respectively. The observed warp in the disk is given by the difference between the maximum and the minimum tilt angles. In Figure 6 the amplitudes of the models are scaled so that the total bending in the disk is 15°, the amplitude of the observed warp in NGC 2903. The maximum warp angle relative to the halo equatorial plane reaches $74.7°, 24.9°$ and $18.7°$ in the three models. We deduce that the ellipticity of the halo must be greater than about 0.08 in order to reproduce the observed warp shape. For the large (28°) observed warp of NGC 4013 the constraint is more severe, and requires the halo to be as flat as about E2.

IV. Summary

Our study of the time dependent behavior of warped disks which are embedded in dark, oblate halos has the following preliminary results: Disks that have a discrete mode of bending settle to the shape of the mode from the inside out. The line of nodes is straight in the settled portion of the disk, and eventually twists to a leading spiral in the outer unsettled region. Typical timescales for the settling

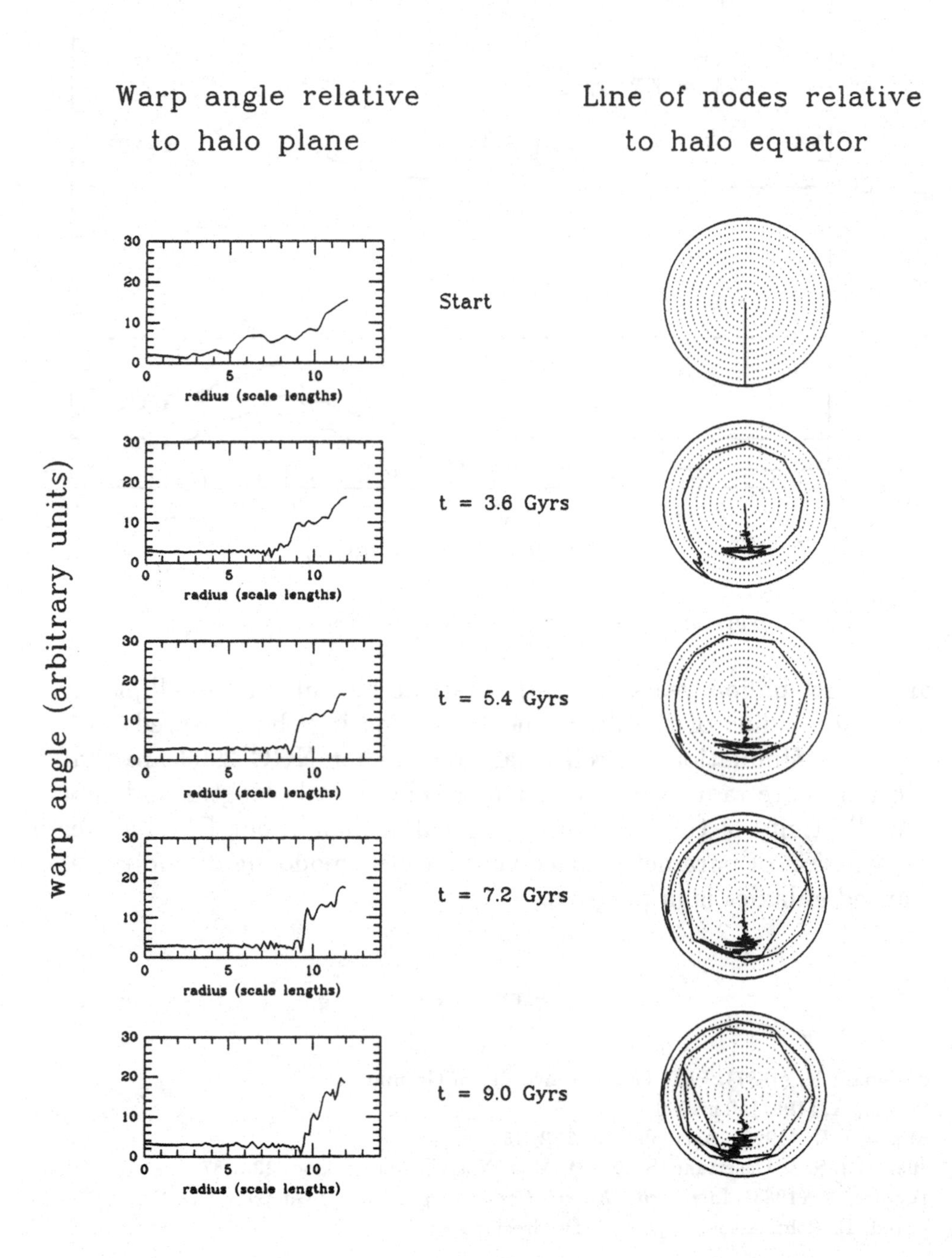

Fig. 5. Time evolution for a model of NGC 2903 in a spherical halo.

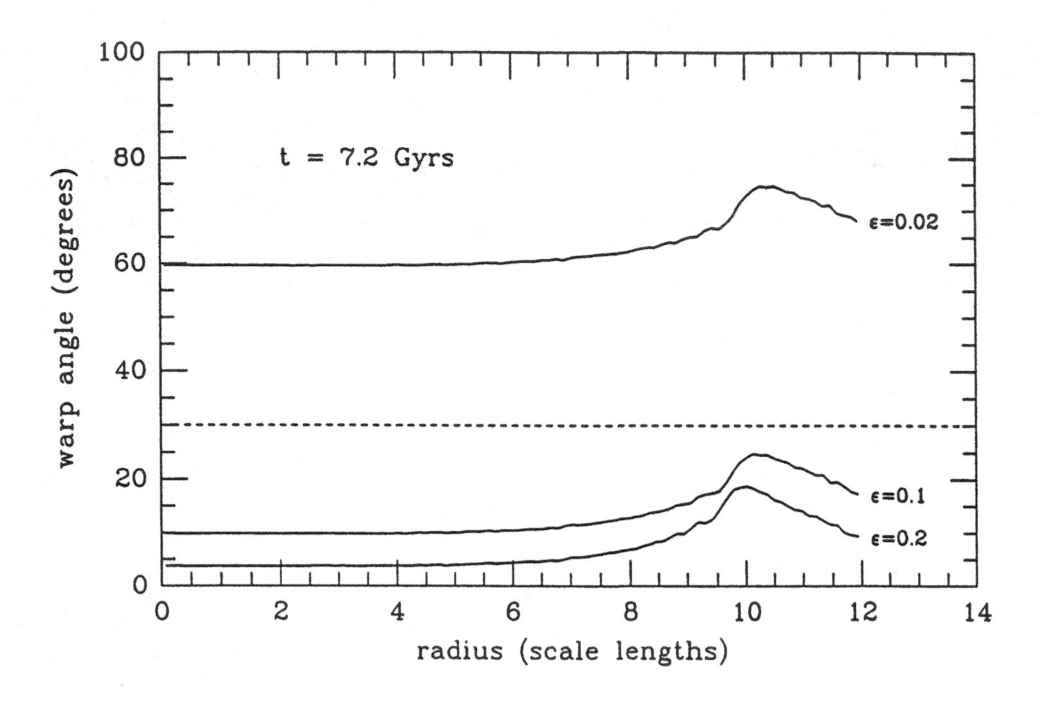

Fig. 6. Warp angle for different halo ellipticities.

are a few billion years. Constraints can be put on the ellipticity and core radius of the dark halo by comparing the observed warp with our simulations. Preliminary results for NGC 2903 are that the halo core radius must be smaller than 4 scale lengths and that the ellipticity of the halo must be greater than about E1. Further exploration of parameter space and detailed modeling of additional warped galaxies is in progress.

References

Begeman, K. (1987). PhD Thesis, University of Groningen.

Bosma, A. (1991). This volume.

Briggs, F.H. (1990). Astrophys. J. **352**, 15.

Sparke, L.S. & Casertano, S. (1988). Mon. Not. R. Astron. Soc. **234**, 873.

Toomre, A. (1983). In *Internal Kinematics and Dynamics of Galaxies*, IAU Symp. 100 (ed. E. Athanassoula), p. 177. Dordrecht: Reidel.

Are warps normal modes?

STEFANO CASERTANO
University of Pittsburgh

Introduction

Ever since their discovery, galactic warps have attracted the attention of theorists. One reason may be the apparent simplicity of their structure, which seems to point to an obvious, large-scale mechanism for their generation; yet the efforts to discover this 'obvious' cause have remained inconclusive for three decades.

Tidal forces have long been a favorite suspect. A weak tidal interaction would naturally produce a distortion with azimuthal wavenumber $m = 1$, like observed regular warps. Besides, strong tidal interactions between galaxies can be observed, and it is natural to assume that weaker interactions must be fairly common; thus, no *ad hoc* mechanism need be introduced. In fact, tidal interactions were suggested as a possible mechanism for the production of warps even as the first warp was reported, by Kerr (1957) for our own Galaxy. There, the Large Magellanic Cloud lies alluringly in the right direction to distort the plane of the Galaxy in the sense observed.

Yet *present-time* tidal forces fail to explain observed warps quantitatively. Warps are very common (Bosma 1991),and tidal interactions appear statistically insufficient to produce as many distortions as observed. Some galaxies with major warps lack nearby companions which could have caused the observed distortion (Sancisi 1976, Krumm and Shane 1982). Even when companions are available, as for the Galaxy, the strength of the interaction is insufficient to account for the observed distortion at the present time (Hunter and Toomre 1969).

If tidal interactions had been more common in the past, they might have produced a large number of warps which could now be detected. However, massless warps are expected to precess at a rate which varies with radius, and to lose their orderly appearance over a time that is short compared to the Hubble time, especially in regions close to the stellar disk. Self-gravity can change this picture, because the disk can organize itself so that every part of it will precess at the same rate. Such a configuration is called a *normal mode.*

Normal modes of bending for a galactic disk have been discussed since Lynden-Bell (1965), who first considered self-gravity in the context of warps. Hunter and Toomre (1969) showed that the difficulty is that the typical galaxy has *too many* modes—in fact, an infinite number, infinitely close to each other, and all singular at the edge. Any realistic perturbation would excite many modes, each precessing with a different frequency, and the organized shape will be lost in a very short time. However, Sparke and Casertano (1988), following earlier suggestions by Dekel and Shlosman (1983) and Toomre (1983), found that, if the disk is embedded in a slightlyflattened halo, it *can* have a mode separated from all others: the "modified tilt mode".

The "modified tilt mode"

Inside a spherically symmetric halo, the disk can be tilted rigidly by an arbitrary angle and retain its configuration. This is the trivial "tilt mode", which has zero frequency (the figure does not precess) and no bending. If the halo is slightly flattened, the angular frequency of vertical oscillations μ_h about its symmetry plane is larger than the angular rotation velocity in the plane Ω_h. Therefore the plane of the disk precesses about the symmetry axis of the halo with the free precession rate $\Omega_{\rm prec}(r) \approx [\Omega_h^2(r) - \mu_h^2(r)]/2\Omega(r)$, where $\Omega(r)$ is the angular velocity of rotation of disk material. Note that $\Omega_{\rm prec}$ is negative for an oblate halo, which means that the orbital plane of test particles would actually *regress.*

The free precession rate $\Omega_{\rm prec}$ varies with radius, and a disk made of test particles would be subject to differential precession. If the disk is bent so that self-gravity compensates for the variation of $\Omega_{\rm prec}$, the disk will precess rigidly as a whole with frequency $\omega_{\rm prec}$ independent of radius: this is the "modified tilt mode". A key characteristic of this mode is that the disk plane is not aligned with the halo plane. The precession frequency of the mode is average free precession frequency weighted with the angular momentum density:

$$\omega_{\rm prec} = \frac{\int_0^\infty \Omega_{\rm prec}(r)\Omega(r)\sigma(r)r^3 dr}{\int_0^\infty \Omega(r)\sigma(r)r^3 dr}, \tag{1}$$

where $\sigma(r)$ is the surface density of the disk. Sparke and Casertano (1988) have studied in detail the properties of this mode; for more information, and a more rigorous derivation of equation (1), see the original paper.

The shape of the warp depends on the properties of the halo. If the halo has a high-density core and its flattening continues into the core, the free precession frequency $\Omega_{\rm prec}(r)$ decreases outward, and the disk must bend sharply away from the plane of the halo to support a relatively fast precession in the outer parts. This type of warp, named Type I by Sparke and Casertano (1988), cannot be supported for a very extended disk, because the disk itself lacks the mass to keep the precession fast in the outer parts. If instead the halo has a large, low density core, the free precession frequency $\Omega_{\rm prec}$ will initially increase with radius, and the disk will bend *towards* the halo plane. Because the precession frequency is lower and less support is needed from the disk, this type of warp (Type II in Sparke and Casertano) can be sustained even for extended disks. Eventually, at larger and larger galactocentric distances, $\Omega_{\rm prec}$ will decrease again, and the disk will bend back away from the plane of the halo.

The most specific prediction made by Sparke and Casertano (1988) is that galaxies with concentrated halos and large disks should not have warps. This prediction sems to be borne out very well by the statistics of Bosma (1991), which therefore greatly strengthen the viability of our model.

However, two problems must be faced if the modified tilt mode is to explain galactic warps. The first is that the modified tilt mode, like all neutrally stable modes of bending, requires the warp to have no winding. The line of nodes, where material at each radius encounters the reference galactic plane, must be straight. Briggs (1990) finds that this is approximately true only for the inner region of the HI disk, out to approximately the Holmberg radius. Further out, where the warp amplitude is largest, the line of nodes *is* wound, normally in the leading sense. We will see below that this winding may be explained by the settling of the warp towards the mode.

The other, related problem, is that a warp will not start in a normal mode. Whatever the mechanism that produces a warp, (off-axis infall, tidal interaction, merging), it is unlikely to produce a warp with the right shape for a modified tilt mode. How does the galaxy evolve towards the shape predicted and required by the modified tilt mode?

Settling towards the mode

In the linear regime, the shape of the disk can be expressed as the superposition of bending waves which propagate inwards or outwards in the disk. In the large-wavenumber approximation, the equivalent of the semi-classical WKB approximation, each wave obeys the dispersion relation

$$(\omega - m\Omega(r))^2 = \mu_{\rm tot}^2 + 2\pi G\sigma(r)|k| \,. \tag{2}$$

Here ω is the angular frequency of the wave, $m \geq 0$ its azimuthal wavenumber ($m = 1$ for bending waves), $\Omega(r)$ the angular frequency of rotation of the disk, $\mu_{\rm tot}$ the vertical frequency of oscillation due to the halo and to the non-local disk contribution, and k the local wavenumber. The phase of the wave is $(\omega t - m\phi - \int k\, dr)$, and the rotation is in the positive sense ($\Omega > 0$). With this sign convention, a wave with positive frequency ω rotates in the same direction as the disk (prograde). The sense of winding is leading if $k < 0$, trailing if $k > 0$.

A packet of bending waves propagates with the group velocity $c_g = \partial\omega/\partial k$. From the dispersion relation (2),

$$c_g(\omega, k) = \frac{2\pi G\sigma \,\mathrm{sgn}\,(k)}{\omega - m\Omega} \,. \tag{3}$$

Thus, wave packets that precess faster than the disk rotates ($\omega > m\Omega$) propagate outwards ($c_g > 0$) if trailing, inwards if leading; retrograde waves ($\omega < m\Omega$) propagate outwards if trailing, inwards if leading. However, the wave amplitude does not remain constant as the waves propagate. Bertin and Mark (1980) show that, if the *action* associated with a bending wave is to be conserved as it propagates, its *amplitude* H must scale with

$$H(r) \propto 1/\sigma(r) r^{1/2} \,. \tag{4}$$

The amplitude of each wave will increase as it propagates outwards, towards regions of lower disk density, and decrease as it moves inwards.

In this description, a *mode* is a standing wave, or the superposition of two waves of equal amplitude and frequency, one leading and the other trailing.* The resulting perturbation has a straight line of nodes. If the disk has a discrete mode, such as the modified tilt mode, its initial configuration can be expressed as the sum of the mode plus a combination of waves. If the disk has initially a straight line of nodes, for each frequency the leading and the trailing wave will be initially present with equal amplitude. The inward-propagating waves will decrease in amplitude and eventually be either reflected (and transformed in their outward-propagating counterpart) or absorbed at vertical resonances; the outward-propagating waves will move outward and grow in amplitude. Therefore, as time progresses a larger fraction of the disk will be cleared from the bending that is not associated with the mode, and the disk itself will settle in the mode shape in that region. This is very clearly illustrated in the time-dependent calculations of Hofner and Sparke (1991).

The time scale for settling at radius r can be estimated as the total time for waves propagating from the center of the galaxy to reach the radius r. Because the group velocity is proportional to the local density, this time becomes very long at large radii; the time also depends on the spectral distribution of the initial perturbation, since the group velocity varies with the frequency of the wave. Hofner and Sparke (1991) estimate that, for a typical massive galaxy, the settling time approaches a Hubble time for radii around 6-8 disk wavelengths. Galaxies smaller than that will have almost entirely settled into their mode over a Hubble time; for galaxies with larger disks, the outer regions have not had time to settle into their mode.

One should point out that what is described here is *dispersive* settling , which is due to the dispersive nature of the medium (the self-gravitating disk). In this process, energy is carried outward, not eliminated. The *dissipative* settling described by several participants to this meeting (see Christodoulou and Tohline 1991, Quinn 1991, Rix and Katz 1991, Steiman-Cameron 1991, and references therein) is physically different and is due to some kind of viscosity. Dissipative processes will become important in a warped disk when the wavelength of the bending is comparable to either the disk thickness or the epicyclic radius of a typical particle; they could affect the late stages of the settling in the outer regions, where the disk can become sharply bent (Hofner and Sparke 1991). Unfortunately, to the best

* This is a necessary but not sufficient condition. A mode also requires proper boundary conditions to replenish the waves that reach the disk boundaries and to close the cycle in phase.

of our knowledge the gas-dynamical problem of settling to the mode configuration has not been addressed yet.

Spectral distribution of the initial perturbation

The shape of the disk in the unsettled region depends on the spectral distribution of the initial perturbation, and specifically on the part of it that is orthogonal to the disk mode. At each radius, the distortion will be the sum of the mode (with straight line of nodes) plus the wave packets that have propagated out to that radius from the inside and outside. Because the amplitude increases in the outward propagation and decreases in the inward propagation, outgoing wave packets will contribute the most.

Precession is usually slow compared to rotation, and it could be expected that most of the initial perturbation will consist of waves with $|\omega| < \Omega$, for which the relevant frequency $\omega - m\Omega$ would then be negative. The outward-propagating component of these waves winds up in the leading sense, giving a predominantly leading perturbation later on. This agrees with the observational picture (Briggs 1990), and is indeed what Hofner and Sparke (1991) find in most cases from their numerical calculations of the evolution of the warp. On the other hand, only a small number of initial conditions has been explored so far. Other plausible initial conditions should be considered in order to determine the robustness of the prediction that warps will be wound in the leading sense outside the settled region.

And if there is no mode?

Any warp, and especially a Type I warp, can only be sustained out to a certain limiting radius, depending on the mass distribution in the disk and on the halo properties. Outside that radius, the local mass density is insufficient to boost the local precession rate to the value required by the inner regions.

The fate of the bending perturbation in such a galaxy has not been studied in detail. The calculations of Hofner and Sparke (1991) suggest that, inside the region where the warp can be maintained, the disk will try to settle towards the mode. Sparke (1986) studied the similar problem of the evolution of a polar ring, and found that a ring with insufficient self-gravity will eventually break up in smaller pieces, each of which is held together by self-gravity. By analogy, one can imagine that, in the case of a warp, the disk might settle in the warp mode out to some critical radius, depending on the density of both disk and halo, and the gas outside this radius could form a separate annulus, precessing at a rate slower than that of the warped disk. However, this process may well be too slow for the break-up to have occurred in a Hubble time. It is also unclear whether dissipative processes will help or hinder the breaking of the disk in separate pieces.

An additional word of caution is necessary here. The dispersion relation (2), and indeed the whole thin disk approximation that has been used so far, are no longer valid when the wavenumber k is very large ($|kZ| \gg 1$, where Z is some measure of the disk thickness). This regime appears to be approached in some of the time-dependent warp calculations by Hofner and Sparke. This problem must be addressed in order to obtain reliable results on galaxies without modes.

Alternatives

Reviews of the plethora of models proposed through the years to explain galactic warps can be found in Hunter and Toomre (1969), Abramenko (1978), Toomre (1983), Sparke and Casertano (1988) and Binney (1990). Here I would like to mention briefly the ideas presented at this conference by Battaner et al. (1991) and by Ostriker (1991).

Battaner and collaborators (Battaner et al. 1990, 1991, Florido et al. 1990a, b, Sanchez-Saavedra et al. 1990) suggest that intergalactic magnetic fields of order 10^{-8} G could cause a warping in the outer regions of the gas disk. The warp observed in the stellar component would be a result of star formation in the warped gas disk. There are two observable predictions of this model:

1) any warping in the stellar disk should be more apparent in the blue light than in the red light, because it would be due to recent star formation;
2) frequency and orientation of warps should be correlated over linear scales of order of the correlation length of the intergalactic magnetic field.

The first prediction seems to be borne out by the observations. Florido et al. (1990a) find that, in three galaxies with well-defined warps, NGC 4013, NGC 4565 and NGC 6504, blue light and dust lanes bend more than red light. This contrasts with the result of

Casertano et al. (1987), who find that the warp in the gas component of NGC 4565 is perfectly compatible with that of the stars. Sasaki (1987) comes to a similar conclusion for NGC 5907.

For the second prediction, there is very little reliable information on either warp orientation or large scale correlations of the intergalactic magnetic field. Furthermore, there is not much evidence to indicate the typical magnitude of the magnetic field near an isolated field galaxy; especially those with strong warps at relatively small radii, such as NGC 4565 and NGC 5907, might require a field significantly larger than the 10^{-8} G quoted by Battaner et al.

Ostriker and Binney (Ostriker and Binney 1989, Binney 1990, Ostriker 1991) have suggested that infall of gas with angular momentum not aligned with the disk will force a warp in the disk. They point out that, as long as the infall takes place over time scales long with respect to the dynamical time of the disk, the bending is transient: adiabatic invariance prevents disk material from changing its angular momentum, and disk and accreted gas will eventually settle on the same plane. In this picture, a warp requires *ongoing* accretion of gas with a skew angular momentum.

Qualitatively, the skew angular momentum accretion can account for observed warps. The accretion rate of neutral hydrogen must be fairly large, comparable to the total mass of the galaxy divided by a Hubble time; due to the frequency of warps and the fact that accretion was probably faster in the past, the required accretion rate may be uncomfortably large. The numbers are sufficiently uncertain, however, that the model remains viable in this respect.

Another potential problem is the nature of the accreting material. The scant direct and circumstantial evidence available seems to indicate that gas is acquired in discrete lumps via mergers, rather than in slow infall from a smooth gas distribution. Discrete lumps are quite capable of exciting modes of oscillation but, depending on their size and mass, may be less effective at producing a slewing of the potential. Possibly both infall of material with skew angular momentum and the shape of the normal mode play a role in the dynamics of warps, the infall providing the necessary initial conditions and perhaps affecting the outer parts of the disk (where the waves from the center take longer than a Hubble time to arrive) with a transient distortion.

Conclusions

So far, the idea that warps are closely related to normal bending modes of the disk has fared well in the comparison with observations. Besides the specific models for individual galaxies presented by Sparke and Casertano (1988) and by Hofner and Sparke (1991), the general trends found so far in the statistical properties of warps agree with the predictions of Sparke and Casertano (1988) in two major respects: the line of nodes is straight in the inner parts of the disk (Briggs 1990), and concentrated halos inhibit warps (Bosma 1991).

The outer parts of the disk probably have not had time to settle in the warp; this may well explain another trend found by Briggs (1990), namely the leading twist of the line of nodes in the outer parts of warped galaxies. However, the sense of winding predicted for a settling warp could depend on the initial conditions; more work is needed before coming to a definite conclusion in that respect.

Several important issues remain open. The reaction of the halo to the changing potential of the disk can be important because 1) the halo core can become rounder or follow the disk to some extent, thus changing the characteristics of the mode (Toomre 1983), and 2) halo particles can exchange angular momentum with the disk and help perturbations grow (Bertin and Mark 1980, Bertin and Casertano 1982). A more realistic treatment of the disk, including the softening from finite thickness and the radial transport of momentum from radial velocity dispersion, is necessary. Finally, viscous settling to a preferred plane has been studied in great detail (see Steiman-Cameron 1991 and references therein), but not in connection with a bending mode; dissipative processes could possibly hasten the evolution of the gas disk towards the mode.

References

Abramenko, B. (1978). Astrophys. Space Sci. **54**, 323.

Battaner, E., Florido, E. & Sanchez-Saavedra, M.L. (1990). Astron. Astrophys., in press.

Battaner, E., Florido, E., Sanchez-Saavedra, M.L. & Prieto, M. (1991). This volume.

Bertin, G. & Casertano, S. (1982). Astron. Astrophys. **106**, 274.

Bertin, G. & Mark, J.W.-K. (1980). Astron. Astrophys. **88**, 289.

Binney, J.J. (1990). In *Dynamics and Interactions of Galaxies* (ed. R. Wielen), p. 328. Berlin: Springer.

Bosma, A. (1991). This volume.
Briggs, F.H. (1990). Astrophys. J. **352**, 15.
Casertano, S., Sancisi, R. & van Albada, T.S. (1987). In *Dark Matter in the Universe*, IAU Symp. 117 (eds. J. Kormendy & G.R. Knapp), p. 82. Dordrecht: Reidel.
Christodoulou, D.M. & Tohline, J.E. (1991). This volume.
Dekel, A. & Shlosman, I. (1983). In *Internal Kinematics and Dynamics of Galaxies*, IAU Symp. 100 (ed. E. Athanassoula), p. 187. Dordrecht: Reidel.
Florido, E., Prieto, M., Battaner, E., Mediavilla, E. & Sanchez-Saavedra, M.L. (1990a). Astron. Astrophys., in press.
Florido, E., Battaner, E., Prieto, M., Mediavilla, E. & Sanchez-Saavedra, M.L. (1990b). Mon. Not. R. Astron. Soc., submitted.
Hofner, P. & Sparke, L.S. (1991). This volume.
Hunter, C. & Toomre, A. (1969). Astrophys. J. **155**, 747.
Kerr, F.J. (1957). Astron. J. **62**, 93.
Krumm, N. & Shane, W.W. (1982). 116, 237.
Lynden-Bell, D. (1965). Mon. Not. R. Astron. Soc. **129**, 299.
Ostriker, E.C. (1991). This volume.
Ostriker, E.C. & Binney, J.J. (1989). Mon. Not. R. Astron. Soc. **237**, 785.
Quinn, T. (1991). This volume.
Rix, H.-W. & Katz, N. (1991). This volume.
Sanchez-Saavedra, M.L., Battaner, E. & Florido, E. (1990). Mon. Not. R. Astron. Soc. **246**, 458.
Sancisi, R. (1976). Astron. Astrophys. **53**, 159.
Sasaki, T. (1987). Publ. Astron. Soc. Jpn. **39**, 849.
Sparke, L.S. (1986). Mon. Not. R. Astron. Soc. **219**, 657.
Sparke, L.S. & Casertano, S. (1988). Mon. Not. R. Astron. Soc. **234**, 873.
Steiman-Cameron, T. (1991). This volume.
Toomre, A. (1983). In *Internal Kinematics and Dynamics of Galaxies*, IAU Symp. 100 (ed. E. Athanassoula), p. 177. Dordrecht: Reidel.

Disk warping in a slewing potential

EVE C. OSTRIKER
University of California at Berkeley

Abstract

Due to gradual infall of cosmic matter with skew angular momentum, a flattened galactic halo is expected to reorient. As it follows this slewing external potential, the embedded galactic disk warps such that torques from local self-gravity balance those from the external potential, in the slewing frame. Such a warp does not precess, and persists only while the potential slews.

1. Introduction

Ubiquitous galactic warps have long defied theorists to formulate a complete theory. The observed feature that causes all the difficulty and interest is warping at radii where a disk of test particles would wind up in a few Gyr. While incorporation of accreted gas could produce transient (but long-lived) warps in the HI components of extended disk galaxies, those with smaller disks could not long maintain warped edges without self-gravity (Binney 1990). Recent findings suggesting that IRAS stars, aged several Gyr, participate in our own Galaxy's warp (Djorgovsky and Sosin 1989) reconfirm the need for a theory that produces global warping.

The most promising method for creating global warps applies disk self-gravity to prevent wind-up while appealing to misaligned angular momenta to engender the warp (Binney 1990). Sparke and Casertano (1988) showed that warped, uniformly-precessing normal modes exist when galactic disk and flattened halo symmetry axes are misaligned. Hofner (this meeting) numerically investigated the time evolution of a flat disk which was inserted skew into a fixed halo potential. He found that the disk differentially precesses until reaching a warped, normal-mode configuration. The question remains: how could disk and halo angular momenta have become permanently misaligned (as a normal mode requires), particularly if galaxies grow gradually? Binney and May (1986) argued that if a disk slowly forms in a pre-existing spheroidal potential, the two would end up with minor axes aligned. Their argument, based on conservation of actions during adiabatic changes in gravitational potential, applies in the converse case as well.

In this paper, I discuss the possibility that warps are indeed evidence of misaligned angular momenta between galactic components, but that they are only transient phenomena, coincident with significant infall. The scenario presented here was first enunciated by Ostriker and Binney (1989). It argues that a temporary warp must appear in the disk when infalling cosmic material slowly reorients the surrounding halo potential, as skew angular momentum is absorbed. In section 2, I discuss observational evidence of skew-$\vec{L}$ cosmic infall. Section 3 describes the response of a galactic disk to slow off-axis accretion. A summary and prospectus appear in section 4.

2. Evidence of skew-$\vec{L}$ infall

2.1. Magnitude of infall

Cannibalism – accretion of compact matter. As we know from the examples of our own galaxy and Andromeda, typical large spiral galaxies have several much smaller satellites. These are expected to be consumed by their big brothers, with orbits decaying due to dynamical friction with the halo until the satellites reach the disk and sink/break up. The tidal tail of the LMC indicates that it has already been partially disrupted, and polar rings in external galaxies have been interpreted as the remains of small, gas-rich satellites (Sparke 1986; see also Rix; Quinn, this meeting). Tremaine (1980) estimated that the typical large spiral will see an increase of its luminous mass by $\sim 10\%$ in a Hubble time, due to cannibalism.

Taxation – accretion of diffuse matter. Although it is not directly observed, a significant amount of currently infalling diffuse material may be inferred from current low-mass star-formation rates in spiral galaxies. The argument is that the rate is high enough so that the typical spiral would use up its supply of gas in a Hubble time; if gas were not replenished from the outside we would not expect so many healthy spirals. The imputed gas accretion rate is $\sim 0.1\text{–}1 M_\odot\,\mathrm{yr}^{-1}$ (Sandage 1986; Larson, Tinsley, and Caldwell 1980). As a cautionary note, intergalactic HI observations (Briggs, this meeting) suggest that there is not much extant in massive clouds ($10^9\text{–}10^{11} M_\odot$), although the low end of the mass function is the least constrained.

The above rates reflect the expected late-time contributions to a galaxy of luminous matter; the dark-matter infall rate may be taken as a suitable multiple. Here, I have mentioned only observational indications of infall rates. Cosmological theories make their own predictions of $\dot{M}$, merger rates, etc. The CDM cosmogony in particular predicts significant merging of previously-condensed lumps at local density maxima, although the rate declines after $z = 1$ (see e.g., Frenk et al.1988, Ryden 1988, Ryden and Gunn 1987). The galaxy merger rate is expected to be far lower than the (dissipationless dark-matter) halo merger rate, however (Carlberg 1990).

2.2. Direction of infall

The prevalence of polar rings in other galaxies (Whitmore, this meeting), as well as the polar orbit of the LMC around the Milky Way, witness that material joining galaxies at late times may have angular momentum vectors with a component perpendicular to that of the relaxed portion of the galaxy. Furthermore, machine-measured orientations of principal axes of galaxies in clusters indicate some correlation, but not a strong one: Fong, Stevenson, and Shanks (1990) found that 8% of clusters in their sample of 621 have internal correlation at a level expected in only 5% with random galaxy orientations. Thus the length scale for decorrelation of $\hat{L}$ among cosmic condensations is on the order of the intergalactic separation.

New theoretical work on angular momentum correlation around collapsing cosmic overdensities is being performed by Binney and Quinn (1990).

3. Galactic response to slow accretion of $\vec{L}$

3.1. General considerations

In n-body studies of angular momentum distribution within halos in CDM cosmogony, Quinn and Zurek (1988) found that dynamical mixing redistributes angular momentum among shells of different binding energies during collapse. They found that while components of $\vec{L}$ are not strongly aligned before virialization, afterwards the halo is well-characterized by a single rotational axis. In addition, Hernquist's simulation of satellite capture by a disk galaxy (this meeting) vividly shows the reorientation of the principal disk axis as the new angular momentum is incorporated. Both of these studies support the hypothesis that at late times, a disk galaxy has a single dominant rotation axis which is realigned when infalling material so requires. Instantaneously, this reorientation is seen as a slewing of the galactic angular momentum vector about a disk diameter perpendicular to the new component of angular momentum. If accretion is gradual, this axis, and the rate of slew, will change only slowly.

On the subject of disk thickening, note that while gradual infall to a galaxy of low-density material ("cosmic drizzle") is not expected to thicken its disk, impulsive accretion of high-density material (a merger) *is*. This is because compact material sinks through the disk, scattering orbits and disrupting the cold flow, while diffuse material is tidally torn apart at large radii and does not disturb the inner galaxy (Tóth and J. Ostriker 1990; see also Hernquist, this meeting). A flattened halo may help the disk-thickening problem. For the warping scenario discussed here and in Ostriker and Binney (1989), we require gradual accretion so that the potential evolves adiabatically. Impulsive accretion, on the other hand, may be relevant to the excitation of normal-mode warps (Binney 1990).

3.2. Disk test-particle response to slewing Φ_{ext}

In investigating the response of a thin stellar and gaseous disk to cosmic infall, we first consider the stars and gas clouds as test particles in a halo potential that begins to slew. We choose this toy model because a slow rate of change in the direction of infalling $\vec{L}$ compared to $\omega = V_c/R$ results (approximately) in a flattened mass distribution which slews about a diameter with angular velocity $\mathbf{\Omega}$. Here, $V_c(R)$ is the circular speed of test-particle orbits. If the galaxy's spin axis points in the $\hat{z}$ direction, and the infalling angular momentum has

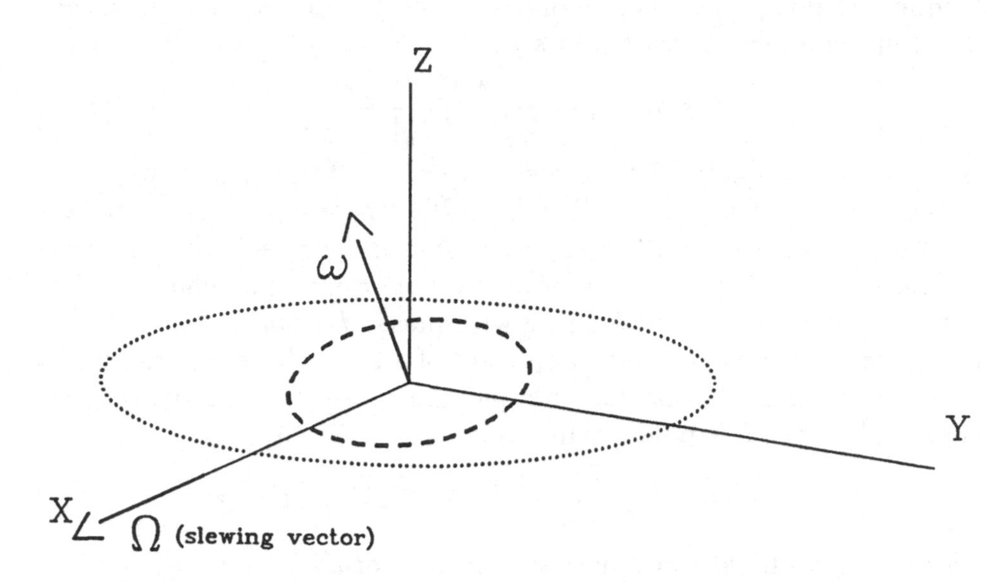

Fig. 1. The orientation of halo symmetry plane (dotted line), test particle orbit (dashed line), and slewing vector ($\hat{\Omega}$) in the slewing-potential model. The torque on the test particle is in the $-\hat{y}$ direction, along its orbital line of nodes. The spin direction of the test particle is shown by $\hat{\omega}$.

skew component $\delta\vec{L}_\perp$ in the $-\hat{y}$ direction, then $\hat{\Omega}$ will be in the $\hat{x}$ direction in the standard right-handed coordinate system (see Figure 1).

We suppose that test-particle orbits lie originally in the halo's symmetry plane, and that the slew rate $\mathbf{\Omega}$ slowly increases from zero to some nonzero constant value. We then apply a theorem of classical mechanics, which states that in a slowly-changing potential, the actions $\oint p\,dq$ of an orbit are conserved (Binney and Tremaine, section 3.6). In this context, the theorem implies that initially circular orbits in the symmetry plane must remain non-precessing and non-nutating. Ostriker and Binney (1989) discuss details related to the solution of the invariant-action problem in a slewing potential. They find that the perturbed orbital plane must be tipped with a constant angle θ_{tip} to the symmetry plane of the slewing potential, so that the orbital plane slews with the potential plane. The line of nodes of the perturbed orbit must be perpendicular to $\hat{\Omega}$, and parallel to $\delta\hat{L}_\perp$ (see Figure 1). The angle θ_{tip} may be found by requiring that the orbit-averaged torque on the test particle equal its angular momentum times the rate, Ω, at which the angular momentum changes. The torque lies along the direction of $\delta\hat{L}_\perp$. Using these requirements, the angle θ_{tip} is easily shown to satisfy

$$\sin(\theta_{tip}) = \frac{2\Omega V_c/R}{\frac{\partial^2\Phi}{\partial z^2} - \frac{1}{R}\frac{\partial\Phi}{\partial R}\big|_{z=0}}. \quad (1)$$

Here, Φ is the external flattened potential. Thus a disk with no self-gravity satisfies $z(R) = R\sin(\theta_{tip}(R))$ where θ_{tip} is given above.

Suppose we have an external potential that decreases with increasing radius (i.e., a spatially external one). For example, consider the potential of a distant massive ring with mass M_{flt} and radius D (this may represent the flattened component of a halo whose ellipticity increases with radius). For this example, the potential at radius $R < D$ is given by the multipole expansion

$$\Phi(R,z) = GM_{flt}\sum_l \frac{(R^2+z^2)^{l/2}}{D^{l+1}} P_l(0)P_l(z/(R^2+z^2)^{1/2}). \quad (2)$$

We use a cylindrical coordinate system with origin at the ring center; the ring lies in the $z = 0$ plane. With this potential,

$$\sin(\theta_{tip}) = \frac{2\Omega V_c/R}{\frac{3}{2}\frac{GM_{flt}}{D^3}\left(1 + \frac{15}{8}\frac{R^2}{D^2} + \ldots\right)} \quad (3)$$

where we include just the two lowest-order multipole terms. If $V_c \propto R$ for small R, then $\sin(\theta_{tip})$ will be approximately constant for these radii. If $V_c \rightarrow const$ for large R, $z(R)$ is near constant and then decreases with increasing R when R/D becomes non-negligible. The orbital plane of a test particle must approach the symmetry plane of the external flattened potential (i.e., $\theta_{tip} \rightarrow 0$) for sufficiently large radii. Thus when a surrounding external flattened potential begins to slew, the embedded disk slews with it while warping into an integral-sign shape. The curvature of the warp is always concave towards the plane of the external potential.

We consider this simple example of a slewing external massive ring with some representative numbers. Suppose that the halo has $M_H \propto R$ out to R_*, with circular velocity V_C. Then in terms of the exponential-disk scale length R_d, we may write

$$\sin(\theta_{tip}) = \left(\frac{R_d}{R}\right)\frac{4}{3}\frac{\Omega R_d(D/R_d)^3}{V_c(R_*/R_d)(M_{flt}/M_H)} \quad (4)$$

for R small compared to D. If $\dot{M}/M \sim (0.1-1)t_H^{-1}$ ($t_H = H_0^{-1}$ is the Hubble time) and infalling $\vec{L}$ is uncorrelated with that of the relaxed galaxy, then we may take $\Omega \sim H_0$. Taking $V_c = 250$ km s^{-1},

$R_d = 3.5\,\mathrm{kpc}$ as reasonable for our galaxy, and supposing (uncertain) $R_*/R_d = 40$ after Caldwell and J. Ostriker (1981); setting $M_{flt}/M_H = 0.1$ and taking (uncertain) $D/R_d = 10$; using $H_0 = 100$ km s^{-1}Mpc^{-1}, we find $\sin(\theta_{tip}(R = 3R_d)) \sim 9°$. This order-of-magnitude calculation suggests that we might expect a warp of several degrees, such as those observed, to arise from the mechanism described herein.

Disk self-gravity

We must now consider what happens when we add disk self-gravity to the scenario described above. In general, the local disk self-gravity tends to keep the disk flat, while the global force from the slewing external potential tends to warp it. The disk must arrange itself so that the total torque from local and global sources on a given ring equals $\Omega \times L$ for that ring. Where the disk surface density is high, in the inner disk, we can expect that the local forces will win and the disk will remain quite flat. But where the disk surface density is low, in the outer disk, a warp should appear. For a disk with exponential surface density, the onset of the warp would be at a radius of a few disk scale lengths. The addition of self-gravity to the picture does not change the general character of the warp, however. Because the torque must be transmitted from the outside to the inside such that the net torque on all rings is in the same direction, θ_{tip} is still a decreasing function of increasing radius. This direction is along the common line of nodes of the disk, and points in the direction of the component of skew $\vec{L}$ that is joining the galaxy, $\delta\hat{L}_\perp$.

4. Conclusions

Evidence from estimated cannibalism rates and inferred gas accretion rates suggests that the typical large spiral galaxy is still acquiring significant amounts of material. This infalling matter adds angular momentum to the galaxy, including a component skew to the current spin axis. Simulations have shown that the galactic axis reorients in time as the infalling skew $\vec{L}$ is shared throughout the galaxy. If accretion is gradual, the reorientation is seen as a slewing about a disk diameter. An initially flat disk responds to an external potential that begins to slew by warping. The sense of the warp is concave towards the plane of the exterior potential. The line of nodes of the warped disk lies along the direction of infalling skew $\vec{L}$.

The slewing-potential model described here was initially devised to explain what was then thought to be a tilt in the nuclear disk of our Galaxy, relative to the outer disk plane (Ostriker and Binney 1989). Modeling the inner Galaxy as a series of concentric rings, and the outer galaxy as a rigid slewing annulus, they found that the inner disk of the galaxy does indeed remain flat. They also found that a gap between inner and outer galaxy corresponding to a posited depression in surface density at $R \sim 2 - 3\,\mathrm{kpc}$ leads to a tilt of a few degrees between the inner and outer galaxy, for $\Omega = 1/10^{10}\,\mathrm{yr}^{-1}$. Although more recent data (Bally et al.1988) have called into question the structure of the inner galaxy upon which those numerical calculations were based, the general formalism developed there and described here is still expected to apply in the outer disk.

Unfortunately, the situation in the outer disk is inherently more complicated than that of the inner disk. The dependence of θ_{tip} on R is quite sensitive to the distorted shape of the outer halo; since this depends on how matter is accreted it is still quite ill-determined. Much more work is needed on setting up the cosmological background conditions, and understanding how infalling matter is incorporated into a local density maximum, before this warp model can be addressed in a truly quantitative fashion. However, the possible difficulties of exciting normal modes through merging (see Binney 1990) without simultaneously thickening the disk (Tóth and J. Ostriker 1990) indicate that a model such as this one deserves more attention. Relying as it does on adiabatic angular momentum accretion, rather than impulsive accretion, it may sidestep some of those difficulties.

I am grateful to J.P. Ostriker for helpful comments on a draft of this paper. This work was supported by an NSF Graduate Student Fellowship.

References

Bally, J., Stark, A.A., Wilson, R.W. & Henkel, C. (1988). Astrophys. J. **324**, 223.

Binney, J.J. (1990). In *Dynamics and Interactions of Galaxies* (ed. R. Wielen), p. 328. Berlin: Springer.

Binney, J.J. & May, A. (1986). Mon. Not. R. Astron. Soc. **218**, 743.

Binney, J.J. & Quinn, T. (1990). Private communication.

Binney, J.J. & Tremaine, S.D. (1987). *Galactic Dynamics*. Princeton: Princeton University Press.

Caldwell, J.A.R. & Ostriker, J.P. (1981). Astrophys. J. **251**, 61.

Carlberg, R. (1990). Astrophys. J. **359**, L1.
Djorgovsky, S. & Sosin, C. (1989). Astrophys. J. **341**, L13.
Fong, R., Stevenson, P.R.F. & Shanks, T. (1990). Mon. Not. R. Astron. Soc. **242**, 146.
Frenk, C.S., White, S.D.M. Davis, M., & Efstathiou, G. (1987). Astrophys. J. **327**, 507.
Larson, R.B., Tinsley, B.M. & Caldwell, C.M. (1980). Astrophys. J. **237**, 692.
Ostriker, E.C. & Binney, J.J. (1989). Mon. Not. R. Astron. Soc. **237**, 785.
Quinn, P.J. & Zurek, W.H. (1988). Astrophys. J. **331**, 1.
Ryden, B.S. (1988). Astrophys. J. **329**, 589.
Ryden, B.S. & Gunn, J.E. (1987). Astrophys. J. **318**, 15.
Sandage, A. (1986). Astron. Astrophys. **161**, 89.
Sparke, L.S. (1986). Mon. Not. R. Astron. Soc. **219**, 657.
Sparke, L.S. & Casertano, S. (1988). Mon. Not. R. Astron. Soc. **234**, 873.
Tóth, G. & Ostriker, J.P. (1990). Preprint.
Tremaine, S.D. (1980). In *The Structure and Evolution of Normal Galaxies* (eds. S.M. Fall and D. Lynden-Bell), p. 67. Cambridge: Cambridge University Press.

Concluding discussion

MODERATOR: KEN FREEMAN
Mt. Stromlo and Siding Springs Observatories and Space Telescope Science Institute

Editors' Comment: *The following is a transcription from audio tape of the concluding workshop discussion led by Ken Freeman. Each of the other five session chairpersons was asked to comment on the most notable aspects of his or her respective session; the discussion was then opened to the floor. Where necessary, we have made the usual editorial changes required to turn loose, oral dialogue into a readable written transcript. Minor editorial insertions, deletions, interpolations and clarifications appear in square brackets. Unfortunately, the poor quality of the audio tape has prevented verbatim transcription of the discussion in its entirety; the transcript is therefore incomplete and we apologize to those speakers (generally those whose voices operated at the same frequencies as the air-conditioning unit) whose comments we were unable to reproduce fully. In an attempt to retain the balance of the discussion, and (we hope) its flavor and content, some editorial summaries are provided for the passages that were most difficult to transcribe. These summaries are enclosed in square brackets; if they are in any way incomplete or misleading, the fault is entirely our own.*

Freeman: There are a few things that have sort of niggled at me during this meeting, positive and negative niggles. One of those positive niggles, I think, was raised by Linda in her talk where she raised for the only time I've heard it raised here the question of using metallicity as an indicator of the genesis and origin of the polar-ring material. It's a very important point because you could gain some quite positive indications that this stuff came from outside; you may, if you were unlucky, get indicators that are ambiguous. For example, if you are able to convince yourself that the metallicity of the parent body or metallicity of the polar-ring material were *greater*, that would be quite strong evidence that the material came from outside, and that's going to lead me onto something else that's sort of worrying me a bit. I don't think that technically it would be a very difficult thing to do for observers: you would do something like J-K photometry on the stellar body which is fairly very well-calibrated for all the populations; you would do direct abundance work on the polar-ring material. In some polar rings, I don't think it would be so incredibly difficult. Certainly in Cen A I don't think it would be at all difficult to get quite realistic abundances. Maybe someone has done this already and I'm not aware of it.

Sparke: Not for Cen A. The only work I'm aware of on the subject is [by] Rick Pogge, now at Ohio State (he was in Texas) and he had some preliminary stuff on the Spindle Galaxy, but he says he doesn't expect to be able to go get 4 meter time at CTIO until he's got some positive results from the northern hemisphere on bright objects from the south.

Freeman: There have been people who regularly make monumental snap-ups with abundances within HII regions because they try short-cuts. People really should think very seriously about this and come up with sensible answers.

Ok, another comment. Excuse me please, this may be gratuitous, but I'm going to make it anyway. And this concerns major-axis dust-lane ellipticals. And what I want to say here is that I think there may be at least a fairly fundamental structural difference between major-axis and minor-axis dust lanes in the following sense. We've seen plenty of minor-axis dust lanes ellipticals during this meeting, Centaurus A, [NGC] 5266 and other objects. In the recent number of major-axis dust lanes that we have studied—mainly work with Richard Wainscoat—I think that things look rather like hamburgers [in a bun], where you just have an elliptical body and a fairly well-defined tight dust lane. And the question again is: What is this? Is this again an accreted thing settling into the short axis plane, or what? What we did here was map these things in alphabetical photometry all the way up through to K, and what we saw was that all the way from U to H, this thing stayed in absorption, just like we see it here. But when we got to K—and it's remarkable that it happened at the same wavelength in all three systems that we did—suddenly this thing stops being an absorption disk and comes out as an emission disk. This I don't think is due to hot dust or anything like that, I'm pretty sure that what we're seeing here is that we're seeing this coming out in emission. I expect that what we're doing there is just seeing through this dust; we're actually seeing quite substantial stellar [light]. Now if that is the right interpretation, then it suggests that these are not things that are just simply being accreted or if they were, then they have gone through a big star formation phase and gas loss phase, and we really know they're something fairly structurally different from the minor-axis objects we're looking at.

de Zeeuw: But Ken, how can you distinguish minor-axis and major-axis cases, because you just showed us Centaurus A, where I don't know anymore whether its a minor- or major-axis dust lane—your rotation axis is right in between.

Freeman: Good point.

de Zeeuw: So why would this be different from the minor axis case? I don't understand.

Freeman: I think it would be difficult at this point, correct me if I'm wrong, to have the dust lane in Centaurus A along the major axis.

de Zeeuw: I agree with that. What I'm getting at is that there is apparently some sort of interaction between the two. Therefore I don't quite see why you make the distinction. Have you looked at any minor-axis ones?

Freeman: No we haven't. Not as yet. But if it turned out that there were a big stellar contribution for the minor-axis ones [it would argue against] accretion.

de Zeeuw: If there is a difference, it might say something about the origin.

Freeman: I'm not aware of any comparable data for minor axis ones.

Steiman-Cameron: I'm still having trouble understanding why if you have two ellipticals, [and] one has a minor-axis and one has a major-axis [dust lane], why you expect one to be different than the other. Is it because you expect one to be due to accretion and the other one not?

Freeman: I don't expect anything. I'm just trying to measure these things and show what they're like. I'm just passing on the information.

Tohline: But if you look at minor axis objects they may look the same.

Freeman: It's possible. It's possible. But I think that it would be embarrassing if that's the case because it would mean that a lot of the underlying mass of the dust lane material would then be stellar.

Tohline: It's low mass?

Freeman: Fairly low mass, edge-on nominally.

Galletta: I have some comments. If you consider from the point of view of the optical images there is another problem. For instance, you expect that if the infall is almost isotropic, you just obtain an equal number of major axis and minor axis dust-lane galaxies. [Actually, for a minor-axis dust-lane galaxy, an S0 with a small polar ring, for instance, you can easily see the underlying stellar S0, but even if there is an underlying stellar disk in a major-axis dust-lane elliptical you cannot see it simply because it is obscured by the dust lane.] But despite this, when we got to the center of the major and minor dust-lane ellipticals, looking at the plates of the Palomar atlas or the ESO atlas: well, most major-axis dust lane ellipticals will have a stellar disk, [unlike] minor axis dust-lane ellipticals. [That] is also [true] from the optical point of view, not only infrared, just visual inspection of the plates gives this. Many major-axis dust-lane galaxies appear to be S0s, the definition [of an S0 being] that they have a stellar disk, while almost all of the minor-axis dust lane ellipticals appear not to have any stellar disk.

de Zeeuw: Ken, [the galaxies you observe, are they] emission line objects, [or do they have any gas]?

Freeman: One of them has got HI in it.

de Zeeuw: And you find that they [all components] rotate in the same sense?

Freeman: Yes, they do, yes. But about the other two, I just don't know; both the galaxies rotate, I know that, but I can't [say anything more]. I'll just want to make one more comment and then I'll shut up and let someone else have a go. One thing that's worried me also through this meeting is why we don't see polar rings around spirals. There is obviously the explanation that we don't have interactions of the ring material with the disk material, but there are many spirals in which the HI is quite low density by the time you get to the Holmberg radius, and I don't see any obvious reason why we shouldn't see polar-ring objects in at least some of these spirals. I don't have any ideas, I was curious to get other people's advice, [to comment on why that's so.]

Galletta: Regarding the comments by Bosma and by you, I just realized possibly a good idea about polar rings and S0s. Well, Bosma expressed the idea and the observation that polar-ring S0s are in general small galaxies. We don't ever see a big S0 with polar rings.

Well, let me [suggest]the hypothesis that if they are small, maybe they have non-standard halos. They are perfectly oblate like a sphere or a disk, and they are not self-gravitating. In this hypothesis, the polar ring is highly unstable—then a small perturbation can destroy it. So what happens of this? I draw on a comment not made by me, that there are no barred S0s with polar rings. Despite that fact that they are edge-on, if you look at a sample by Whitmore there are no bars inside. The bar is not a big perturbation of the potential, but in the hypothesis that the polar ring is highly unstable, just a spinning or a tumbling bar inside would destroy it. The second point, the point of Ken Freeman: there are no spirals in polar rings, despite the fact that the polar ring lies outside the optical radius. Then, if we accept that these galaxies are small, then probably there is no triaxial halo to stabilize them and they are perfectly oblate. Then polar rings are highly unstable: just spiral arms or bars which are the visual perturbation of the potential [would] destroy them. Why not?

Casertano: What makes you think that small galaxies have no dark halos? It seems to me that ...

Galletta: OK, it is just a hypothesis.

Casertano: It seems to me that the little evidence that we have is that halos are more dominant in dwarf galaxies or low mass galaxies than they are large galaxies.

Galletta: Maybe these halos are oblate. Because if you want to stabilize the polar ring, you have [to have] a self-gravitating triaxial halo and if you [don't] have this—any small perturbation is lethal.

de Zeeuw: Didn't Linda Sparke show that you could get a warped polar ring in a perfectly oblate halo?

Steiman-Cameron: That was with self-gravity.

de Zeeuw: [Why shouldn't we see some spiral galaxies with *big* polar rings?]

Whitmore: [Something like NGC 4650A.]

de Zeeuw: For example.

Whitmore: Isn't there enough gas far out in a spiral to damp out [a polar ring], even at low density? I don't know what the cross section would be, [but] it doesn't take too much gas to destroy a polar ring.

Steiman-Cameron: You're talking about low density gas hitting low density gas at 150 to 200 $\mathrm{km\,s^{-1}}$.

Tohline: Look, there's no reason why the gas has to come in at the radius that you see the ring. In general, it's going to come in along some orbit that may bring it in closer. In an S0 it is clear that that radius [can be smaller] than in other spirals.

Whitmore: In various circumstances, wouldn't there be enough gas, from what we know about spirals, to damp it out?

Sparke: Probably for an individual lump of gas coming through it would be OK, as long as it is massive enough, but as the gas comes through, OK, [it messes up something on one side and then] messes up something else on the other side—[but not necessarily symmetrically since] the galaxy will rotate. Really you've got to get rid of all the gas in the spiral at radii which are greater than the inner radius of the polar ring, and so you may end up sweeping up quite a lot of gas. There's also the environmental question, in the way that [one] would say that ellipticals don't have big polar rings because they're in very dense environments so that something bangs the ring every 10^9 years. S0s have polar rings because they're in a dense enough environment to accrete things. Spirals galaxies just don't have enough stuff coming by to make rings ...

Steiman-Cameron: Another speculation is that if you assume that all polar rings are old and therefore the material has made several plane crossings during its lifetime. [Then] in a spiral galaxy, [even if] there's not a really disastrous disk-plane crossing for the polar ring, it may damp a little bit, so that the ring shrinks down and disappears on a time scale of less than a Hubble time.

Freeman: There's a hell of a lot more spirals than there are S0s; there must be an awful lot of tidal rings...

Schneider: Maybe as kind of a related question, pulling it together: de Vaucouleurs noticed years ago that there are more outer rings around early-type spirals and S0s than late-type, and we've heard today that most of the warps are around late-type galaxies. Couldn't they all be various cases of [the same phenomenon]?

(Silence)

Freeman: OK, well let's move on. I was going to ask a number of people to throw up some questions, then [open it up], sort of like a Quaker meeting ...

(Laughter)

I'd like to ask Albert [Bosma] to start off.

Bosma: I had two things; one is about Cen A. The first thing

I have is: do we really need to fine-tune tilted rings, circular ring models for the dust-lane, or is the approach that Tim [de Zeeuw] described—just doing a coarse elliptic orbit model to the dust and gas and trying to figure something out about it—[more appropriate]? I think that the data that you showed indicate the answer. The other thing I was struck by in the first session was this question of the very extended HI disks. Kamphuis showed the disk in [NGC] 628, with beautiful spiral structure in the HI out to at least two Holmberg radii, and Steve Schneider showed not only his ring around Leo, but he previously showed NGC 5701, where again he showed this structure in the HI at radius bigger than two Holmberg radii. I find this surprising because you expect there that the gas disk is already substantially flared so I would simply wonder how those sharp features can occur. I actually want to draw on Lars to help; where it is just the gas and halo, what do you expect: would the gas form classical spiral arms [in a] round halo?

Hernquist: I think if [the gas] was self-gravitating, you might be able to do it, but probably not otherwise, unless it is a sort of tidal perturbation that produced the structure.

Bosma: Well, the only tidal effect would be from [the] small blob-like companion, gas companion. Would that really be enough?

Hernquist: If you just have a tidal force field you can make it look sort of like a localized spiral arm, like a long tail ...

Casertano: I also have a case of an observation of a galaxy which is isolated and does not have a companion, in which the gas seems to have a spiral structure [that] follows the spiral structure of the stellar component out to one and a half Holmberg radii on two sides.

Bosma: I'm talking about gas at say, two Holmberg radii, where the halo would dominate ...

Kuijken: *(Inaudible)*

van der Hulst: I would just like to say that I can name at least three other examples which show the same thing out to two Holmberg radii or at least beyond the Holmberg radius: [NGC] 6946, which is very extended all over, IC 342, [which shows spiral structure] actually in the region where it becomes warped, and M 101, which is very extended in HI to the southwest [and] shows what looks like spiral structure—actually there you can argue maybe it's tidally induced. Looks like this is a phenomenon that no one has yet worried about. The maps begin to show this more and more.

Bosma: M101 is a special case: you can throw that out.

(Laughter)

Casertano: Every galaxy is special.

Bosma: No, M 101 is really special. [For] IC 342 there also is a companion ...

van der Hulst: There is a small companion.

Bosma: *(Inaudible)*

Freeman: Thijs?

van der Hulst: Yes, the point I wanted to raise, actually was already raised in the discussion you brought up: why are there no polar rings around spirals? I think we should remember that we have only six bona fide polar rings. My impression is [that] especially what Brad showed is a collection of objects in all these different categories. It is a zoo: there's lots of different things, and some things I would not have even expected to be there. I don't have a clear idea anymore what a polar-ring is; NGC 4650A has become a prototype, but I think we should be very careful. We have only six, and perhaps we should all go home and ask ourselves, first of all: What are polar rings; can we use statistics to come up with numbers [to address] how many are there, how rare is this, what kind of galaxies have polar rings? So I'm confused about polar rings. I'm equally confused—and that's a subject for the chairman of the session yesterday—about the modeling: as an observer, I'm not sure which models to believe and which ones not to. Among the theoreticians there is a lot of discussion. So that was one point.

Another point: I think we have to go back to lots of galaxies that have been observed in detail—because now we see that the polar rings are used as an excellent tool to study the properties of extended disks, to study the shape of potential—to go back to the all the velocity fields that are sitting around several places, including the place where I am *(laughter)* and do the kind of analysis that Peter [Teuben] has shown us, which may be very powerful; these are only the sort of remaining 10% effects, but the data are there. As an observer, I'm always trying to get an idea when I go home for projects that are worth doing, and I think this is a potential thing to go out and do.

One last thing, which I think has been addressed very carefully every now and then, but not very much, is where these polar ring come from. What should we do in order to get a better handle or a better understanding of the origin? It's hard to get companions to form a polar ring; on the other hand, we do see them. There, as far as I am concerned, is a very open question. Maybe we should study those things that you suggested are in the making?

Whitmore: Yeh, that's why I put them in there. On your first point though, one of the reasons that there are only six [polar-ring galaxies] is that's the amount of time we got from the allocation committee to look at polar rings. Every time we have looked at one that we thought was going to be an S0 with a polar ring, it turned out to be. And I would guess that the 27 in Category B essentially are all polar rings. So there are more than I thought were probably out there. But along the same line, we have to be a little careful of the "typical" polar-ring, in the way that we obviously tend to study the ones that are the most probable. In the B category, you also come up with a lot [of galaxies] with polar ring extents that, instead of being $3 \times R_{25}$ or something like that, are $1 \times R_{25}$. We have to be a little careful of thinking of what a typical polar ring is.

And this last point I want to make is the one I'm real excited about too: for a lot of years people have modeled polar rings [applying simple hydrodynamic consistency checks rather than constructing fully consistent, dissipative N-body simulations to study their origin and evolution. At this meeting we are seeing, for the first time, attempts in this direction.] I was hoping to see some of the banana warps, I didn't see any banana warps: Does anyone have any hopes of making a banana warp in a polar ring?

(Silence)

No, OK.

Tohline: [We need to know the] initial conditions for the formation of a polar ring. I thought it would fun to take a vote to assign Hernquist or Rix with some ideal initial conditions ... If we were to assign them a homework problem [what would it be?]

de Zeeuw: You would try to do a simulation that forms the prototypical polar-ring galaxy.

Sparke: [How about taking the prototypical polar-ring galaxy and running it backwards?]

(Laughter)

Freeman: [Let's not forget dissipation.]

Steiman-Cameron: There is the one system that was in my video, François [Schweizer]'s object, [NGC 3808,] in which it shows material being tidally pulled out of the passing system. It looks like it has almost formed a polar ring, except that it hasn't cut the umbilical cord yet to the originating galaxy. I mean it looks like a polar-ring galaxy.

Whitmore: The point there is that it should just start with a small companion; the majority of tidally stripped objects have nearby galaxies. So maybe we should do everything.

van der Hulst: Except for the six or seven polar rings—don't they still have fairly messy spirals nearby?

Whitmore: Well, [NGC] 4650A, the prototypical [polar-ring galaxy], is in a cluster actually, and, for example, it's got [NGC] 4650 quite nearby. There is no real strong trend, in that they all have lots of companions, but they do seem to tend to have smaller companions. One thing that came up, for example, is that we did a compact Hickson group, at a very, very low level and it's got four galaxies almost like [those in] Stephen's Quintet, one of them at very low light levels. So, I think you can start with fairly messy surroundings.

Steiman-Cameron: [NGC] 4650A is in the Centaurus Chain, where there's five or six fairly good-sized galaxies, I think all of which are weird in some respect or another.

Whitmore: I don't know if you already mentioned it, but a lot of them are in a line. It's kind of scary. I can't remember the exact geometry, but

(... gesturing toward an imaginary line of galaxies in the air ...)

it's kind of a polar ring there and there, and there's a galaxy here and here and here, and kind of one here or something.

Tohline: The other point here is that if we're going to solve this initial condition with a gaseous companion, the other question is what potential we want to toss in. One you could try is a spherical potential with a ring, or an S0 potential with a spherical halo, or a triaxial [mass distribution].

de Zeeuw: One thing we may actually want to do is to assign the problem with the Hernquist potential.

Hernquist: You might choose to make one of the galaxies disappear. Maybe it was destroyed by the encounter and will never create a polar ring.

Sparke: You'll probably want to try all of them.

Hernquist: Yes.

Sparke: Well, I don't know. Maybe we could figure it out, maybe we could *observe* something ...

(Laughter)

People keep trying things, and they keep not working. Which suggests either that the codes aren't really what the ought to be or [they're] on the wrong track.

Steiman-Cameron: But partly it seems like the problem is not that nothing is working, but too many things are working.

Sparke: Too many things work once you get it into a ring, but it still does not appear that you can get something coming in to form a nice ring.

Steiman-Cameron: Well, in Rix's simulations ...

Rix: Keep in mind how I start it, I [vary the potential so that the companion will be disrupted,] partly [by] turning on and off self-gravity, which isn't really fair.

Sparke: The trouble is it seems [to know when to turn itself] on or off. Which is worrying.

Tohline: And also, to come back to the question on the settling problem and theoretical modeling: there is a slightly different approach. It seems to me that we have four ways now, four numerical ways, to try to do some modeling: [there is] the SPH stuff, the Steiman-Cameron and Durisen approach, [which is] purely viscous, ... *(inaudible)*... and does not destroy angular momentum, and then Quinn's stuff [shown] yesterday, which actually had collisions destroy angular momentum.

We've got also to agree on one particular problem to set up, and all try to work on it with some degree of freedom.

Freeman: What if it's the wrong problem?

(Silence)

Freeman: Is there anything more on specific models? I just want [to mention this]—just as objects that may be polar-ring precursors—there's a set of objects that de Vaucouleurs and I worked on years ago. They are distinguished by having the following property: there is an S0, there is always a very well-defined S0, with a stellar bulge and a stellar disk, and next to it is just this chaotic shambles, which is about the same luminosity, the velocity difference is typically sort of around [100–200] $\mathrm{km\,s^{-1}}$, and that's it. Here are four of them right off the top of my head; they seem to be very common. I'm not aware of any spiral shambles counterparts, although [I haven't seriously looked for them.] These are very strong objects, always just with the properties that I mentioned.

Linda, did you want to make a couple of comments?

Sparke: Yes, I liked the results coming out of the supercomputer, the N-body codes and things. But what I was struck by was the difference between what we saw yesterday afternoon—in which it takes about a year to figure out a set of initial conditions, and then you have to get some Cray time and run the simulations, and then you've got to try another set of initial conditions—the contrast between that and what I watched happening in Wisconsin, where Marc Balcells and Alan Stanford are running this, what is it called, this multiple 3-body algorithm, which is a restricted form of the N-body code, where you follow two and only two galaxies. They've been doing simulations which match velocity cuts across the two galaxies, and that's quick enough that you can do it on a Vax. But they can only follow two systems and they get it probably wrong. What I wondered: was is there any possibility of a hybrid approach which will run overnight on the Vax or in half an hour or ten minutes on the Cray, so that you can explore a range of initial conditions but still do a lot better than the 3-body model of gravity?

Hernquist: I think if you want to include full self-gravity there is no alternative.

Sparke: But are there any ways of approximating—you don't need quite full self-gravity—I mean like running the tree code where you keep [only certain terms] or something like that?

Teuben: Is this the real restricted 3-body ...

Sparke: No.

Teuben: ... or is this a new thing?

Sparke: It's Kirk Borne's thing.

Teuben: That's what I thought.

Whitmore: Kirk and Hal Levison are trying to do some work on Hickson [Compact] Groups, where [they follow the collapse of a loose cluster, with each galaxy as one particle, until they collapse enough to begin interacting, at which point they replace each galaxy with 1000 particles.] So they're trying [to do a hybrid approach] in that case, but I'm not sure if they have convinced themselves [that it can be done].

Steiman-Cameron: *(Inaudible)*

Sparke: The point is that you are trying to model young systems, and you've got two things that you are pushing for: one is accuracy, the other is speed. I mean, doing it all is always nice, but you can't afford to do [only] one a year.

Hernquist: The point is, if [you need] self-gravity, you can run all the 3-body simulations you want to and you'll never find anything...

Sparke: But it tells you what almost certainly *doesn't* work.

Hernquist: For example, [what I showed] yesterday, you'll never get that out of a restricted 3-body calculation.

Sparke: You really have to do it right.

Hernquist: Yeh.

Rix: Actually, the computations are not all that expensive, because I'm sure that we can do exploration runs capable of about 1000 particles in, I think it's about, 25 minutes per run. So it's not prohibitive.

Sparke: What you really need [is] to include [more information about exploring parameter space, what initial conditions may lead to a system that looks like observed objects and what definitely won't.]

Steiman-Cameron: Often its not really the computer, it's the man-hours, the person hours, the ...

Katz: You need more good graduate students.

Sparke: Well, that [is not going to be any better] with the Borne stuff ...

Steiman-Cameron: Well, let me rephrase that. If you have been given access to a Cray and the code, ...

Rix: I mean it's my impression that the runs themselves are not really the barrier, its just hard to get [the right initial conditions].

Sparke: So why does Kirk Borne run his codes [on small machines]?

Katz: He never applied for Cray time?

Freeman: Tim?

de Zeeuw: Sure. There was one thing that struck me most this morning—and this applies a bit to these questions earlier on Thursday—that we seem to be in a transition state where we now are moving from having one or two examples of very nice warped galaxies, for example, to having many more observations, and we can start asking statistical questions, for example about the prevalence of warps as a function of Hubble type. [There may be people now who might want to do their PhD on this topic, how warps relate to the structural properties of galaxies.]

What is not so clear to me, however, is what the actual statistics are. Is it feasible to do work on complete samples of [these objects]? I could imagine that you could go to lower sensitivity for the earlier types to have the same detection fraction. The same kind of thing, of course, for 21 cm [work on warps, also] applies to the polar-ring catalog, where again it was not clear to me what possibilities there are that this is not a complete catalog, and to base physical conclusions [by] some theorist on [this] is all fine, but we must not overlook [the possibility of] more systematic [problems with the] data.

Bosma: *(Inaudible as tape is being flipped over)*
So, if you want to do real statistics, you just can't look at these.

de Zeeuw: No, I understand that. The question would be, can you improve the current situation without too much effort. For example, taking what you have and applying a proper set of selection criteria of some sort and folding it into the statistics. [An associated point is, something that you mentioned also, that you can have an image of a galaxy and say, well, there is no warp, but] if you look deeper, say, yes it's a warp.

Bosma: I didn't look deeper, I just looked [more carefully to get rid of] bad data. What you should do in those cases is study the data yourself. You should look at the data, you shouldn't just [take the published conclusions].

Casertano: It sounds like potentially a very wasteful situation. If every one of us should review everyone else's data, nobody would have time to look at their own data.

Bosma: You should make your own assessment of the data, always.

Casertano: Actually, there is something I wanted to mention in this respect. It is, of course, always nice to have a nice, well-defined, statistically complete sample. But, one should always determine where the priorities are, and perhaps we're not yet at the stage where the most important thing is to get a well-defined statistically complete sample. I think at this level it is true that it is a transition stage, there is more coming on, but still some understanding [is] needed. And if I was going to emphasize trying to understand what's going on or making a nice well-defined sample, I would think that we're not yet at the stage where the completeness, in terms of statistics, is the most important item. But that's a personal assessment, if you like.

Bosma: I think the most important thing is to try and figure out whether there's some basic point [that should be looked at.] At a certain point I decided to look at whether [there was some correlation of] halo core radius with warps or not. I had read the paper by Sparke and Casertano, I had some suggestion about properties of halos in spirals, [and I decided to look at them.] I guess the thing Frank [Briggs] did on this leading stuff, that was sort of hanging around, for quite some time, with very small number statistics. [Then there was the] discussion of our Galaxy [for which] people said, well, the line-of-nodes is straight. Then Frank decided to look at at sample of spirals, to see if that actually was the case.

Whitmore: For the polar-ring catalog, it's a kind of hybrid really. We took anything we could possibly find and then we went back and searched [four more plates more systematically]. Like you said: Is there any easy thing you can do? We did the easy thing. But as an advertisement of the future, [in the] next year or two, one of the questions we want to address is interaction rates in groups in a more systematic way, doing the Guide Star Selection plates, and using Tyson's survey plates, to go very, very deep and try to see what the frequencies are for dust-lanes, polar rings, [etc.] So, hopefully within a couple of years we'll have a little better handle on those kind of problems.

Galletta: Another problem about dust-lane ellipticals... Dust-lane ellipticals have the dust well inside the object, and then [we have to consider that] actually we don't have a uniform sky atlas. If you just make this test: take the same region of the sky where there's intersection between ESO plates and Palomar plates, you can immediately realize that in the Palomar case you see very deep, while in the other you see [the dust lane] quite plainly. Then most dust-lane ellipticals will be missed, for instance, on the Palomar plates where the nuclei are big, and will be easily seen in the ESO survey. Then I think [there may be a problem in making] a uniform search, maybe not for polar-rings, because polar rings are outside the object, but for dust-lane galaxies.

Whitmore: That's going to be true of almost all of our deep polar rings. I mean, several of the polar rings were discovered serendipitously by CCD imaging. If you look at them at Palomar, you see nothing at all. So to some degree then, after you've got the [frequencies of occurrence], you have to figure out how to scale them up or whatever, based on better plate material. So it's not an easy thing.

Tohline: I also want to say that if we are going to get anywhere [on modeling] polar-rings, more than two perpendicular axes [are needed].

Whitmore: *(Inaudible)*

van der Hulst: Maybe this is just very naive, but just on the statistics on the warps: Maybe there still are selection effects, maybe we don't understand it at all. If I take Frank's nice [cardboard model] and tilt at different angles edge-on, in one angle, or a range of angles, you can see a warp very clearly; if you turn it 90 degrees you don't see a warp, you see a sort of fattish disk. The question is: have we seen any of those things? I'm not aware of it. On the other hand, more than half of the 20 galaxies that [Albert studied] show warps; [that's] a lot. On the other hand, maybe some of the warps, especially the ones that are one-sided, are not really warps perhaps, but outlying spiral arms or tidal features. This is just a very naive thought, but something perhaps we need to think about.

Rix: There is a real difference, I think, between real disks and Frank's model in the sense that the disk [thickness] increases outward. Since warps are on the outside, this would probably make a big difference.

van der Hulst: Then the question is whether you would expect to find fewer unwarped galaxies than we do; we find quite a lot. I mean if even if they are all warped, then it is potentially embarrassing when you see warps [this often].

Pitesky: We should be able to derive some statistics about how often you expect to see along the line of nodes [so that we can see the warps, since we] obviously expect that the line of nodes is randomly oriented among galaxies.

Freeman: OK, we have time to go into Quaker mode, in which anybody who has something they would like to [raise, should feel free to bring it up.]

Tohline: *(Inaudible)*

Casertano: Rather than raising a question, quite, I would like to have some answers to a question that arose before on differences between the various codes and various numerical methods used to treat the evolution of gas sheets or gas rings or gas configurations. I would like to ask at least Steiman-Cameron if he wants to comment on these different results and different methods. How well [do] the numerical methods reproduce the physics, and if so, which of the two is more appropriate? I really would like to understand this better.

Steiman-Cameron: I could make general comments first. We're using slightly different physics in the different codes, and I'm not sure that the codes are really compatible with getting the same answers. I don't think that we disagree that significantly—even though we have different physics, it's not greatly different.

Christodoulou: But you don't do SPH.

Steiman-Cameron: No, I don't do SPH, but the people with SPH aren't here, so they can't defend themselves.

Casertano: What do you mean?

Katz: Well, we have to do the same calculations, probably.

Steiman-Cameron: I mean for doing the same problem.

Katz: We will have an answer soon, though. Once you form a polar ring, of course, it would either settle into a plane...

Steiman-Cameron: But even with SPH though, Habe-Ikeuchi got their dissipation by turbulent viscosity in shocks; they still got settling on the time scale of one to three precessional periods. And [with] our analytical stuff: one to three precessional periods...

Tohline: [For] low inclination.

Steiman-Cameron: Low inclination, right.

Tohline: I'd say that at low inclination, the SPH and your stuff, and Christodoulou and mine, are all qualitatively different, and maybe close to quantitatively agree, in terms of time scales, except what Quinn showed yesterday, which was a slightly different set-up. But I suspect that the reason why he didn't get much settling was because the collisions flattened the system and widened the system. He also ended up with a very thin ring which really didn't have much differential precession.

Casertano: Absolutely, I think that the fundamental difference is not so much the fact that it settled, but the fact that it ended up as a narrow, thin object, and then as a consequence had a hard time settling. In fact that is what struck me as the most obvious difference: one side approaches some sort of fluid equations, with particles and without particles, and Tom Quinn's approach—unfortunately he's not here. And yes, I realize that there is somewhat different physics, and perhaps the different answers are to be expected, but which ones are more realistic?

Katz: Thomas Quinn's model has totally different physics; he doesn't assume it's a fluid, he assumes that it is a bunch of clouds that rarely collide. You have to ask: is the polar ring really made up of little clouds that rarely collide? I don't think that's very feasible if you look at the observations of polar rings. [With] the densities of the gas in the polar ring, I can't imagine that they have high enough densities to be Jeans unstable to form into little lumps. It seems to me that it would be more likely to be in a smoother kind of gas. Especially if it came from the merger of a some sort of dwarf kind of system, where the gas would be disrupted as it went around the ring. I can't believe that it would really break into little knots that rarely collide.

Steiman-Cameron: Well, in fact, the cloud-fluid [approach] implicitly assumes that you have clouds. I mean, if you have just a pure sheet of atomic or molecular gas that's extremely smooth, you probably won't have enough viscosity or enough, well, at least from my approach, you probably would not ever have a large enough coefficient of viscosity to get any settling. To be honest, in everybody's code in the universe [there is] a coefficient of viscosity; they don't know the number to stick in there. In a big sense, I don't fully understand Tom Quinn's code, so I don't want to say totally why we get such radically different answers. But it struck me that he's still colliding particles with particles.

Katz: But they rarely collide, that's the whole point.

Steiman-Cameron: But, they interact.

Katz: No, they rarely interact.

Steiman-Cameron: They rarely interact?

Katz: Yes.

Steiman-Cameron: Why do they not differentially precess?

Christodoulou: Maybe there is *(inaudible)*.

Steiman-Cameron: But those were particles that were still basically on circular orbits, so they are subject to [differential precession].

Tohline: I would think that the principal difference between, for example, your approach to the problem and Quinn's, is that you don't destroy angular momentum, you don't dump angular momentum in the ring, where as he does. Your disk has to smear out in radius, has to redistribute its own momentum. But in the collision model, the easiest way to look at how it evolves is to take two rings and let one of them precess at high rate and have the two smash into one another, completely destroying the part of the angular momentum that is perpendicular to the symmetry axis of the potential. When you do that you tend to not necessarily smear out in radius, but ...

Steiman-Cameron: But that's instantaneous.

Casertano: Perhaps a last comment of mine is that, well, it looks to me from the little I've seen that polar-rings are not all that broad and they do have some substructure. OK, that could be just formation of stars. But I don't see [broad, smooth sheets]—except perhaps like your folded sheet galaxy—in the regular polar-ring galaxies, it seems more like a narrow structure. I'm not sure.

Steiman-Cameron: That's only true [for] some of them. They're optical rings—when you look at the gas, it's a much bigger sheet. And some of them, even in the optical they're pretty broad.

Whitmore: In many cases you can see lots of blue star formation regions just right at the level of resolution; they're pretty far away. Some pictures you showed yesterday of the new one you did, again it kind of looked like there were lots of small dots. So, it's kind of up in the air, but I'd be surprised if they don't break up into little cloudlets.

(Final comments by Tohline, Steiman-Cameron and Christodoulou were inaudible.)

At this point, Ken Freeman declared the session and the workshop closed, and Linda Sparke, in the name of all participants, presented the organizers with a beautiful color picture of everybody's favorite polar-ring galaxy, NGC 4650A, autographed by all.

NAME INDEX

Names have been indexed only the *first* time they appear in the body of a contribution. Page numbers of an author's contribution appear in boldface.

OBJECT INDEX

SUBJECT INDEX